Mathematics

1

R. Piessens E. de Doncker-Kapenga
C.W. Überhuber D.K. Kahaner

QUADPACK

A Subroutine Package for Automatic Integration

With 26 Figures

Springer-Verlag
Berlin Heidelberg New York Tokyo 1983

Robert Piessens

Computer Science Department, University of Leuven,
Celestijnenlaan 200 A, 3030 Heverlee, Belgium

Elise de Doncker-Kapenga

Computer Science Department, Western Michigan University,
Kalamazoo, Mi 49008, USA

Christoph W. Überhuber

Department of Applied and Numerical Mathematics,
Technical University Vienna, Gusshausstr. 27–29,
1040 Wien, Austria

David K. Kahaner

National Bureau of Standards, Washington, DC 20234, USA

AMS Subject Classifications (1980): 65 D 30

ISBN 3-540-12553-1 Springer-Verlag Berlin Heidelberg New York Tokyo
ISBN 0-387-12553-1 Springer-Verlag New York Heidelberg Berlin Tokyo

Library of Congress Cataloging in Publication Data. Main entry under title:
QUADPACK : a subroutine package for automatic integration. (Springer series in
computational mathematics ; 1) Includes bibliographical references. 1. QUADPACK
(Computer programs) 2. Numerical integration–Computer programs. I. Piessens,
R. II. Series. QA299.3.Q36. 1983. 515.4'028'5425. 83-10359
ISBN 0-387-12553-1 (U.S.)

Printing and bookbinding: Beltz, Offsetdruck, Hemsbach
2141/3140-543210

Acknowledgement

We owe the financial support to the "Nationaal Fonds voor Wetenschappelijk Onderzoek" and the "Onderzoeksfonds van de Katholieke Universiteit van Leuven".
We thank Philip Rabinowitz, James Lyness, Ann Haegemans and May Branders for their comments and suggestions regarding either the manuscript or the routines and their results.
We thank Brian Ford and the NAG staff, and Bob Huddleston, for their faith in our product at an early stage, and their assistance with the implementation of the package as a part of the NAG and of the SLATEC libraries.
We thank everyone involved in testing the routines, in the first place Ian Robinson and Tom Patterson.
We thank our analyst-programmers Marc De Meue, Liliane De Roose and Anita Ceulemans our typists Denise Brams, Lieve Swinnen and Bea Stroobants, and for drawing the figures Rudy de Doncker.

Table of Contents

I. Introduction

1.1. Overview of Numerical Quadrature

The numerical evaluation of integrals is one of the oldest problems in mathematics. One can trace its roots back at least to Archimedes. The task is to compute the value of the definite integral of a given function. This is the area under a curve in one dimension or a volume in several dimensions. In addition to being a problem of great practical interest it has also lead to the development of mathematics of much beauty and insight. Many portions of approximation theory are directly applicable to integration and results from areas as diverse as orthogonal polynomials, Fourier series and number theory have had important implications for the evaluation of integrals. We denote the problem addressed here as numerical integration or numerical quadrature. Over the years analysts and engineers have contributed to a growing body of theorems, algorithms and lately, programs, for the solution of this specific problem.

Much effort has been devoted to techniques for the analytic evaluation of integrals. However, most routine integrals in practical scientific work are incapable of being evaluated in closed form. Even if an expression can be derived for the value of an integral, often this reveals itself only after inordinate amounts of error prone algebraic manipulation. Recently some computer procedures have been developed which can perform analytic integration when it is possible. Unfortunately, these programs are not at all portable and consequently are only available in a very few installations. Furthermore, successful analytic integration of rather simple functions can result in lengthy formulae. Such a formula can involve the evaluation of complicated functions which can only be computed approximately. Thus many expressions for integrals which appear in books or tables as "exact" require approximations before they can be evaluated numerically. One is lead then, naturally, to seek methods which attempt to approximate integrals directly for general integrands rather than by performing problem specific algebraic manipulation.

An approximate integration sum, also known as an approximate quadrature or quadrature evaluation sum, is an expression which is a linear combination of values of the integrand and which serves as an approximation to the definite integral of this integrand. A Riemann sum is a simple example. A quadrature or integration rule or formula is any rule which yields an integral approximation and may thus include nonlinear combinations of integrand or other values. Quadrature rules, or sums, usually occur in families depending upon a parameter such as the spacing between the integrand evaluation points or the number of these points. Historically, some quadrature rules have been considered which utilize not only integrand, but derivative evaluations as well. While these have interesting mathematical properties they do not have much practical utility for solving quadrature problems. Henceforth we will be concerned only with rules which use integrand evaluations, both linearly and nonlinearly.

Quadrature rules are the building blocks upon which quadrature algorithms can be built. We think of an algorithm as a concrete implementation of one or more rules along with a strategy, albeit sometimes primitive, for evaluating the results. A quadrature rule, by itself, is not especially useful. Any approximation to an integral involves errors and unless some assessment of these errors is available no significance can be attached to the approximation. A Riemann sum without further specialization, for example, requires far too many points for practical work. Often during the derivation of the rule a formula will also be found for the error in the approximation in terms of certain characteristics of the integrand. The most common of these expresses the error (of the integral estimate) in terms of a point value of some derivative of the integrand or the integral of a derivative times an appropriate kernel function. Alternatively, errors are often given in terms of a complex path integral, an infinite or asymptotic series, or a norm in an appropriate function space. In a practical sense, though, many of these error formulae are not realistically computable. It is sometimes possible to produce an error bound by taking upper bounds of the various indeterminates that appear in the error expressions. Unfortunately, such bounds can be wildly pessimistic. The determination of practical error estimates is as important as the derivation of new formulae and a subject that lends itself more to engineering and experiment than to traditional mathematical analysis.

As computer simulations have grown in popularity so have the needs for numerical methods for the evaluation of integrals. Every computer center now has some programs which address themselves to these problems. From the point of view of scientists and engineers the diversity of needs is very great. In principle it is possible to approximate any integral by a simple average of integrand values. Such general but primitive methods are too slow by orders of magnitude for any realistic problem except those with very specific kinds of integrands. Hence special methods have been developed which make various assumptions about the integral which is to be approximated. Examples of explicit assumptions are : the integrand has an algebraic singularity at left or right end-point, the interval of integration extends to plus infinity, integrand has a multiplicative factor of $\sin(\omega x)$, etc. An implicit assumption is that practical integrands are composed of piecewise analytic functions. Programs for "general" integrals usually assume that Taylor series or other expansions provide good approximations, at least on small enough intervals. Anyone studying this subject even as a user or consumer of programs should be aware that the number of potential references is staggering. The basic theory is presented in the textbooks of Krylov (1962), Davis and Rabinowitz (1975), and Engels (1980). Both new results and programs are continually being published in journals like Mathematics of Computation, Numerische Mathematik, The Computer Journal, ACM Transactions on Mathematical Software, SIAM Journal on Numerical Analysis, Computing, BIT and many others.

Many of the programs published in the course of the past 20 years are automatic. That is, they are meant to be called with the problem specifications such as the integrand, the interval of integration, the level of accuracy desired, etc., given by the user. Upon return, they provide an estimate of the integral and an error estimate which hopefully satisfy the accuracy requirement determined by the call sequence. Papers which are concerned with automatic integration are listed in the bibliography of de Doncker and Piessens (1976). Important surveys and comparative studies were published by Kahaner (1971), Dixon (1974) and Robinson (1979).

1.2. Scope and Philosophy of QUADPACK

QUADPACK is a collection of FORTRAN programs for the numerical evaluation of integrals. The programs are explicitly designed for univariate (one dimensional) integrals although some multidimensional problems can be treated by suitably combining the basic routines. Many of the programs are automatic and are provided for different classes of problems, and various "driver" programs allow users with diverse requirements access to the package.

The majority of the algorithms are adaptive. That is, the sets of evaluation points for two different integrands (on the same interval) do not generally stand in the relation of the one being a subset of the other. Rather the evaluation points are clustered in the neighborhood of difficult spots of each integrand. For this reason, in all our programs the integrand function is assumed to be available via a FORTRAN FUNCTION. The QUADPACK programs will evaluate the integrand function at various points on the interval of integration. This selection of evaluation points is done by the QUADPACK subroutines rather than by the user. Thus the programs in this package are NOT suitable for the integration of functions which are given in tabular form.

Most algorithmic decisions are generally taken out of the hands of the novice and left to the QUADPACK programs. For example, the detailed mechanisms of where to evaluate the integrand next are internal and not modifiable except in a general way through the call sequences. Similarly, roundoff error can occur in various places during the course of a calculation and the QUADPACK programs attempt to monitor its growth, and adjust the error estimates accordingly.

One important implication of the QUADPACK program organization is often overlooked by novice users. The only possibility for a QUADPACK program to discover the characteristics of the integrand is by examining its values at a finite number of selected points on the interval of integration. The values at these points determine exactly what is to be the further course of the calculation and the estimates that are to be returned. Function behaviour off these points is simply missed unless it is built into the assumptions about the program. Thus a sharp spike in the integrand, unsampled in a general routine, will escape undiscovered. Similarly, a program for integrands with algebraic singularities expects just such behaviour. It is not possible to prevent the misuse of a program which results from giving it inappropriate input information. In some cases the routine simply works harder than necessary, but in others it returns incorrect results.

Almost all the QUADPACK programs behave nonlinearly in that one is not able to deduce the outcome on f + g from the results on f and g separately. This will be true even when the integrand values are used linearly, e.g. in quadrature sums. The most sophisticated routines attempt to assess the course of the calculation and extrapolate the results via a highly nonlinear algorithm. This has been shown to be very effective for the integration of even strongly singular functions.

Chapter II gives some mathematical details about the various approximations and formulae that are used in the QUADPACK programs. In particular, it describes the underlying Gauss-Kronrod quadrature sums which are fundamental to many of the automatic subroutines. These formulae and their accompanying error estimates represent new technology. Known to some specialists since the mid 1970s, they are implemented here for the first time for general widespread use.

1.3. QUADPACK Contents

There are twelve automatic quadrature programs in this version of QUADPACK, each in both single and double precision. These are all Standard FORTRAN subroutines. We use a consistent naming convention which is described in Chapter III. There are six non-automatic user-callable quadrature programs, in single and double precision, which implement Gauss-Kronrod quadrature rules with 15, 21, 31, 41, 51 and 61 evaluation points to obtain an integral and an error estimate. The non-automatic programs are utilized by the automatic ones but they can also be effectively called by users.

Chapter III describes the algorithms used, and goes into detail on tests for roundoff error, lack of convergence, etc. The complete documentation for the use of each integrator is given in its routine listing in Chapter VI.

1.4. Guidelines for Routine Selection

The higher level QUADPACK drivers are designed to be robust and efficient in terms of integrand evaluations. The actual computer time to perform an integration is a combination of the aggregate integrand evaluation time and the overhead time within the subroutine itself. Experience has shown that for simple problems the latter may be a substantial fraction of the total, which is generally low. As the level of difficulty increases the fraction of time devoted to overhead functions invariably decreases. However, the decisions made internally become much more significant to the total time for the computation. Thus the authors recommend that the automatic programs in QUADPACK be considered for moderate to difficult problems. Easier problems can, of course, be integrated using the same programs and most routine calculations should be done this way as well. An important exception occurs in those cases where machine time is at a premium or where a large number of simple problems need to be done repeatedly, say, in an inner loop of a lengthy run. In those cases it is appropriate to consider the lower level QUADPACK programs. These are the non-automatic implementations of some specific Gauss-Kronrod quadrature rules which have already been mentioned in 1.3.

Chapter IV contains extensive guidelines to help users select the correct routine. Various tradeoffs such as computer versus scientist time are examined, methods for increasing the flexibility and efficiency of the programs in the face of difficult integrals are presented and a detailed set of numerical examples and sample user programs are given.

1.5. Machine Dependencies

All the subroutines in QUADPACK are designed to be almost, but not completely, machine independent. In each specific computer center there may be some changes that need to be made to correctly run these programs on different computers.

All the programs in QUADPACK are written in Standard FORTRAN and have been checked with the PFORT verifier of Bell Labs. They ought to compile without difficulties on virtually any machine with an ANSI 66 FORTRAN or FORTRAN 77 compiler. Compilation, however, does not imply that the programs will run correctly on any machine environment. In many of the lower level subroutines various constants needed for the integral approximation are tabulated. In the single precision QUADPACK version, these constants are given to 16 figures. Two double precision versions are supplied with constants given to 16 or to 33 figures. With a few exceptions, the mathematically correct values are irrational and thus the tabulated data are only approximations. The formulae which use this data are thus incapable of providing accurate results to more than, say, 16 or 33 places even for a very simple integrand. This then, becomes a fundamental limitation of virtually all the QUADPACK routines.

There are three explicitly machine dependent numbers which are used extensively throughout the package. They are (1) the machine epsilon, (2) the largest non-overflowing positive real number, and (3) the smallest non-underflowing positive real number. All references to these values are through a single subroutine, QMACO. Thus installing QUADPACK on a new computer involves examining this subroutine and, if necessary, changing the three constants given there.

1.6. Testing and Omissions

All of QUADPACK has undergone extensive testing over the course of its development. The computational field testing was carried out at the University of Leuven, Technical University of Vienna, National Bureau of Standards Washington, and, in part, Los Alamos National Laboratory and Sandia Laboratories. Versions of QUADPACK have been incorporated into the NAG Library and the SLATEC Library. The programs have been run on a variety of computers from a PDP 11/44 to a Cray I. Users have been satisfied with the power and flexibility of the package. Nevertheless it should not be inferred that QUADPACK represents the last word on the subject of numerical evaluation of integrals. Quite the opposite is true. The wide distribution of these programs has spurred research into new and, hopefully better, algorithms. Programming and documentation errors must still exist although we are not at this time aware of any. The authors will welcome information about all such bugs. The organization of the programs quite clearly indicates that many gaps remain to be filled even for the approximation of one dimensional integrals. Furthermore the case of multidimensional integrals is not really treated at all. The theoretical basis for numerical evaluation of integrals in several variables is reasonably well developed. However the organizational, algorithmic and other practical details have only been recently sorted out in the univariate case and it is these which are implemented in the QUADPACK programs. The multivariate situation will have to await further work, which we feel will not be too long in arriving, at least for simple problems in low dimensions.

II. Theoretical Background

2.1. Automatic integration with QUADPACK

It is the objective of all automatic integrators which are part of QUADPACK to calculate a numerical approximation for the solution of a one dimensional integration problem

$$I = \int_a^b f(x)dx \qquad (2.1.1)$$

to within a requested absolute accuracy ε_a and/or a requested relative accuracy ε_r. To this end they all compute one or more sequences

$$\{R_{n_k}, E_{n_k}\}, \qquad k = 1,2,\ldots,N$$

where the first item of each couple is an approximation to I based on n_k function values and the second item denotes a pertinent error estimate. The sequence is truncated after the computation of the N-th couple if it is either assumed by the integrator that the condition

$$|R_{n_N} - I| \leq E_{n_N} \leq \max\{\varepsilon_a, \varepsilon_r |I|\} \qquad (2.1.2)$$

is satisfied, or the integrator presumes the impossibility to attain (2.1.2). The accuracy requirement (2.1.2) was first introduced by de Boor (1971). A strictly absolute accuracy is requested if ε_r is set to zero, and a strictly relative one if ε_a is set to zero. In all other cases one imposes the weaker of the two tolerances.

The procedure for calculating the approximating sequence is called adaptive if the position of the integration points of the n-th iterate depends on information gathered from iterates 1,...,n-1. This is usually achieved by a successive partitioning of the original interval, such that many points are located in the neighbourhood of difficult spots of the integrand, causing there a high density of quadrature points. A non-adaptive integrator works with fixed abscissae, so that the algorithm proceeds in the same way for each problem and only N is chosen dependent on the complexity of the problem.

Both types of algorithms have one or more rule evaluation components (utilizing fixed quadrature rules) for calculating an integral approximation together with a pertinent error estimate over a given (sub)interval, termed local integral approximation and local error estimate respectively. The adaptive algorithm includes strategy components for deciding how to continue the integration procedure and for detecting integrand behaviour which requires special algorithmic actions (singularities for instance are treated by special methods). Furthermore both types of integrators comprise one or more termination criteria. Following Malcolm and Simpson (1975) an adaptive strategy is called locally adaptive if it attempts to achieve (2.1.2) by imposing independent accuracy requirements for all subintervals. As long as the accuracy on a particular subinterval is not reached, it is considered as pending. In a globally adaptive algorithm, all subintervals remain pending until the sum of the error estimates becomes smaller than the error tolerance.

Integrators which do not require the user to provide explicit information about the integrand function will be termed general-purpose integrators. Special-purpose integrators are provided for integrals of the form

$$I = \int_a^b w(x)f(x)dx$$

where w(x) is a weight function like for instance

$$w(x) = \sin(\omega x),\ w(x) = \cos(\omega x)$$
$$w(x) = (x-a)^{\alpha}(b-x)^{\beta},\ldots\ .$$

The type of the weight function and the particular values of the parameters have to be provided explicitly by the user (via parameters of the respective routines).

Section 2.2 outlines the theory underlying the integration methods used by the rule evaluation components of the QUADPACK integrators. The strategy components will be described along with the respective algorithms in Chapter 3.

2.2. Integration methods

2.2.1. Quadrature sums

Throughout this text only Riemann integrals and improper integrals (whose range or integrand is unbounded) will be dealt with.

Given an integral

$$I_w[a,b]f := \int_a^b w(x)f(x)dx \qquad (2.2.1)$$

over the finite or infinite interval $[a,b] \subseteq R$, where w is a weight function, which is integrable over $[a,b]$, a quadrature sum yields an approximation

$$Q_n[a,b]f := \sum_{k=1}^{n} w_k\ f(x_k) \simeq I_w[a,b]f. \qquad (2.2.2)$$

In (2.2.2), the (distinct) numbers $x_1,x_2,\ldots,x_n$ are abscissae or nodes, and $w_1,w_2,\ldots,w_n$ are weights corresponding to these abscissae. Note the difference between the notation $w(x)$ for the weight function and w_k for the quadrature weights, which are not to be confused with $w(x_k)$. We consider only weight functions which are real and for which all moments

$$\mu_i = \int_a^b w(x)\, x^i\, dx, \qquad i \geqslant 0 \tag{2.2.3}$$

exist.

All quadrature sums, which will be dealt with in the following, have real abscissae and weights.

In the case that the subscript w is omitted in the notation it is understood that $w = 1$. The indication of the interval may be omitted if it is obvious or irrelevant which interval is meant. A similar convention will be assumed with respect to n, the number of nodes.

The term _quadrature_ or _integration rule_ or _formula_ will be used alternately to indicate, in general, any formula which yields an integral approximation. Thus, a quadrature sum is always a linear combination of integrand values, whereas a quadrature rule or formula may also be defined by non-linear operations.

The choice of the abscissae or nodes x_k and the weights w_k in (2.2.2) provides 2n degrees of freedom in the construction of the formula, which will always be governed by the idea that, for a given number of points, the _integration error_

$$E[a,b]f := Q_n[a,b]f - I_w[a,b]f \tag{2.2.4}$$

should be _small_. This can, however, be realized in several ways (Krylov (1962), pp. 65-72). Of special interest are quadrature sums of the highest possible _degree of precision_.

A quadrature sum Q is said to be _of_ (_polynomial_, _algebraic_) _degree_ (_of precision_) _d_ if it is exact for all functions f which are polynomials of degree $\leqslant d$ and is not exact for all polynomials of degree $d+1$. We shall always assume $d \geqslant 0$.

The following properties of a quadrature sum (besides having a high degree of precision d) are particularly desirable from a practical point of view :

- all abscissae are inside the integration interval (possibly at the end-points of the integration interval)
- all weights are positive (Krylov (1962, p. 72)).

In fact all quadrature sums which will be dealt with in the following have these properties. If the second property is to be stressed we shall use the term <u>positive quadrature rule or formula</u>, or we shall talk about a <u>quadrature sum with positive weights</u>.

For any given set of distinct, real nodes x_k, $k = 1,2,\ldots,n$, usually chosen as to lie in the interval of integration, weights w_k, $k = 1,2,\ldots,n$ can be determined such that $Q_n[a,b]f$ is of (polynomial) degree of precision $n-1$. A quadrature sum constructed according to this principle is said to be <u>interpolatory</u>. Important examples of interpolatory quadrature formulae are Newton-Cotes formulae, Clenshaw-Curtis formulae and Gauss formulae. Newton-Cotes formulae are based on equidistant nodes. The quadrature formulae of Clenshaw and Curtis (1960), are derived in $[-1,1]$ by using the abscissae $x_k = \cos((k-1)\pi/(n-1))$, $k = 1,2,\ldots,n$. Quadrature sums of the highest possible degree $d = 2n-1$ are uniquely given by the class of <u>Gauss formulae</u> which can be obtained by utilizing all $2n$ degrees of freedom inherent in the choice of n abscissae and n weights.

The theory of integration rules of Gauss type is intimately related to the theory of orthogonal polynomials.

For a given weight function $w(x) \geqslant 0$, integrable on $[a,b]$, and for which all moments μ_i given by (2.2.3) exist, and $\mu_0 > 0$, it is possible to define a sequence of polynomials $\{\omega_n(x)\}$, $n = 0,1,2,\ldots$ where $\omega_n(x)$ is of exact degree n and which satisfies

$$\int_a^b w(x)\,\omega_m(x)\,\omega_n(x)\,dx = 0, \qquad m \neq n. \tag{2.2.5}$$

This sequence of polynomials is called <u>orthogonal</u>. It can be shown that the zeros of $\omega_n(x)$, $n \geqslant 1$ are real, simple and located in the interior of the interval $[a,b]$.

The connection between Gaussian integration rules and orthogonal polynomials is described by the following theorem.

Theorem 2.1

Let $w(x) \geq 0$ *be a weight function over* $[a,b]$ *for which a sequence of orthogonal polynomials* $\{\omega_n(x)\}$ *exists. If the abscissae* $x_1, x_2, \ldots, x_n$ *in* (2.2.2) *are the zeros of* $\omega_n(x)$, *the weights* $w_1, w_2, \ldots, w_n$ *can be determined such that formula* (2.2.2) *is of degree of precision* $2n-1$. *These weights are real and positive. The degree of precision is maximal.*

The orthogonal polynomials and Gaussian quadrature formulae corresponding to the following special selection $[a,b]$ and $w(x)$ are the classical orthogonal polynomials and classical Gaussian quadrature formulae.

	a	b	w(x)	
(i)	$a=-1$	$b=1$	$w(x)=1$	Legendre polynomials $P_n(x)$ Gauss-Legendre quadrature
(ii)	$a=-1$	$b=1$	$w(x)=(1-x^2)^{-1/2}$	Chebyshev polynomials of the first kind $T_n(x)$ Gauss-Chebyshev quadrature of the first kind.
(iii)	$a=-1$	$b=1$	$w(x)=(1-x^2)^{1/2}$	Chebyshev polynomials of the second kind $U_n(x)$ Gauss-Chebyshev quadrature of the second kind.
(iv)	$a=-1$	$b=1$	$w(x)=(1-x)^{\alpha}(1+x)^{\beta}$ $\alpha,\beta>-1$	Jacobi polynomials $P_n^{(\alpha,\beta)}(x)$ Gauss-Jacobi quadrature
(v)	$a=0$	$b=\infty$	$w(x)=\exp(-x)$	Laguerre polynomials $L_n(x)$ Gauss-Laguerre quadrature
(vi)	$a=0$	$b=\infty$	$w(x)=x^{\alpha}\exp(-x)$ $\alpha>-1$	Generalised Laguerre polynomials $L_n^{(\alpha)}(x)$ Generalised Gauss-Laguerre quadrature
(vii)	$a=-\infty$	$b=\infty$	$w(x)=\exp(-x^2)$	Hermite polynomials $H_n(x)$ Gauss-Hermite quadrature.

There are many slightly different definitions of the classical orthogonal polynomials in the literature. We use here the definitions of Abramowitz and Stegun (1964).

Legendre and Chebyshev polynomials are special cases of Jacobi polynomials (to within a multiplicative constant).

An important relation is : $T_n(x) = \cos(n \arccos x)$.

Because of their importance in the field of approximation and series expansion of functions, and of numerical integration, many books and papers on orthogonal polynomials have been published (see Szegö (1959), Tricomi (1955), Fox and Parker (1968)). Tables of Gauss-Legendre, Gauss-Jacobi, Gauss-Laguerre, Gauss-Hermite formulas and of some non-classical Gaussian quadrature rules are given by Stroud and Secrest (1966). Algorithms for the construction of Gaussian quadrature rules for arbitrary w(x) are given by Golub and Welsh (1969) and Gautschi (1968,1970). For constructing classical Gauss quadrature rules, algorithms are available, which take advantage of special properties of the classical orthogonal polynomials. For an interesting survey and bibliography, see Davis and Rabinowitz (1975) and Engels (1980).

2.2.2. Construction of Gauss-type rules with preassigned abscissae

In the sequence $\{Q_n\}$ of Gauss formulae, for different values of n, the respective node sets $\{x_1,\ldots,x_n\}$ do not have points in common (except for the mid-point, which is a node whenever n is odd). This is a serious disadvantage if one wants to estimate the quadrature error of one rule by comparing its result to that of another rule of different degree. In an optimal sequence of rules, each formula would use all the abscissae of its predecessor. A sequence of this type will be called a nested sequence of rules.

In order to construct a nested sequence of rules one could start from the construction of a formula of the form

$$\sum_{k=1}^{n} w_k f(x_k) + \sum_{\ell=1}^{m} v_\ell f(y_\ell) \simeq I_w[a,b]f, \qquad (2.2.6)$$

where the y_ℓ are the abscissae of a given m-point rule.
A degree $d = n+m-1$ of this formula can be achieved trivially by choosing the free parameters according to the interpolatory principle. However, as there are $2n+m$ parameters at our disposal we would expect that under certain conditions a degree $d = 2n+m-1$ can be achieved. General conditions under which all the resulting abscissae are real and inside the interval are presently not known.

Theorem 2.2

Let

$$D_m(x) = \prod_{\ell=1}^{m} (x-y_\ell) \tag{2.2.7}$$

Then for formula (2.2.6) *to be of degree* $d = 2n+m-1$ *it is necessary and sufficient*

- *that this formula be interpolatory and*
- *that* $\omega_n(x) = \prod_{k=1}^{n} (x-x_k)$ *be orthogonal with respect to the weight function*

$$w(x)\ D_m(x), \tag{2.2.8}$$

to every polynomial p_d *of degree* $d < n$:

$$\int_a^b w(x)D_m(x)\omega_n(x)p_d(x)dx = 0, \qquad d = 0,1,\ldots,n-1 \tag{2.2.9}$$

Proofs of Theorems 2.1 and 2.2 are given in textbooks on numerical integration (Davis and Rabinowitz (1975), pp. 77-79, and Krylov (1962), pp. 100-107, 160-166).

Important for our purpose will be the family of Kronrod formulae. They are obtained when $n = m+1$ abscissae are added to the m-point Gauss-Legendre rule (Kronrod (1965)). Indeed, if an m-point Gauss-rule (of degree $2m-1$) is combined with its $(2m+1)$-point optimal extension (of degree of precision $3m+1$, or $3m+2$ through symmetry), a very economical pair of formulae results for the simultaneous calculation of an approximation for an integral and the respective error estimate.

The existence of optimal extensions of the Gauss-rules as well as the property that the abscissae of such pairs of formulae interlace has been proved by Szegö (1934). For the optimal extensions of the Gauss-Legendre and of certain Gauss-Jacobi formulae the positivity of the weights has been proved by Monegato (1976, 1978). The first actual computation of the optimal extensions of m-point Gauss-Legendre rules by adding m+1 points was performed by Kronrod (1965).

Patterson (1968) constructed a sequence of nested formulae, starting with a 3-point Gauss-Legendre rule and adding further rules by the optimal extension of the respective predecessor. He also outlined a procedure for starting from an arbitrary m-point quadrature rule, but in this case the extension procedure has often to be terminated because the zeros of ω_{m+1} may turn out to be either complex or outside the integration interval. Moreover, the positivity of the weights is not guaranteed in general. In the case that the first rule is a Gauss-Legendre rule, a sequence of nested rules with real abscissae, inside the integration interval, and with positive weights, appears to exist.

We shall now discuss some techniques for the construction of Gauss-type rules with preassigned abscissae. To make this discussion more transparent we restrict ourselves to the case of the unit weight function and the interval [-1,1].

(i) First we consider the Gauss-Legendre abscissae y_k, $k = 1,2,\dots,m$ as the preassigned abscissae. Consequently D_m defined by (2.2.7) is the Legendre polynomial P_m of degree m to within a multiplicative constant. Kronrod proceeds by formulating the m+1 orthogonality conditions imposed by Theorem 2.2 as

$$\int_{-1}^{1} P_m(x)\, \omega_{m+1}(x)\, x^d\, dx = 0, \quad d = 0,1,\dots,m. \qquad (2.2.10)$$

By taking the properties of the Legendre polynomials into account, he derives a triangular system of linear equations in the coefficients of ω_{m+1}. Unfortunately this system is seriously ill-conditioned. Patterson (1968) expands ω_{m+1} in terms of Legendre polynomials and outlines a numerically stable method for the computation of the coefficients in this expansion. A computationally simpler algorithm is developed by Piessens and

Branders (1974) who expand ω_{m+1} in terms of Chebyshev polynomials. The m-point Gauss-Legendre rules together with the corresponding (2m+1)-point Kronrod rules, m = 7,10,15,20,25 and 30, are tabulated in the Tables 2.1 to 2.6.

(ii) In order to obtain the optimal extension of an arbitrary rule, Patterson (1968) starts from (2.2.9) in the particular form

$$\int_{-1}^{1} D_m(x)\ \omega_{m+1}(x)\ P_d(x)\ dx = 0, \qquad d = 0,1,\ldots,m, \tag{2.2.11}$$

and expands $D_m\ \omega_{m+1}$ in terms of Legendre polynomials. We shall now describe an alternative technique, making use of a product Chebyshev expansion of $D_m\ \omega_{m+1}$. Using the properties of Chebyshev polynomials (see e.g. Fox and Parker (1968)), we calculate the Chebyshev expansion of the polynomial D_m :

$$D_m = \sum_{k=0}^{m}{}' a_k\ T_k. \tag{2.2.12}$$

The notation $\sum'$ is used to indicate that the first term in the summation must be halved. It will be the objective to calculate the coefficients b_k, k = 0,1,...,m, of the Chebyshev expansion of ω_{m+1} :

$$\omega_{m+1} = \sum_{k=0}^{m+1}{}' b_k\ T_k \tag{2.2.13}$$

with

$$b_{m+1} = 1. \tag{2.2.14}$$

Using (2.2.12-14) we may replace (2.2.11) by

$$\int_{-1}^{1} \left(\sum_{k=0}^{m}{}' a_k T_k(x)\right)\left(\sum_{\ell=0}^{m+1}{}' b_\ell T_\ell(x)\right) T_d(x)\ dx = 0, \tag{2.2.15}$$

where d = 0,1,...,m.

7- POINT GAUSS RULE

ABSCISSAE	WEIGHTS
0.94910 79123 42758 52452 61896 84047 851	0.12948 49661 68869 69327 06114 32679 082
0.74153 11855 99394 43986 38647 73280 788	0.27970 53914 89276 66790 14677 71423 780
0.40584 51513 77397 16690 66064 12076 961	0.38183 00505 05118 94495 03697 75488 975
0.00000 00000 00000 00000 00000 00000 000	0.41795 91836 73469 38775 51020 40816 327

15- POINT KRONROD RULE

ABSCISSAE	WEIGHTS
0.99145 53711 20812 63920 68546 97526 329	0.02293 53220 10529 22496 37320 08058 970
0.94910 79123 42758 52452 61896 84047 851	0.06309 20926 29978 55329 07006 63189 204
0.86486 44233 59769 07278 97127 88640 926	0.10479 00103 22250 18383 98763 22541 518
0.74153 11855 99394 43986 38647 73280 788	0.14065 32597 15525 91874 51895 90510 238
0.58608 72354 67691 13029 41448 38258 730	0.16900 47266 39267 90282 65834 26598 550
0.40584 51513 77397 16690 66064 12076 961	0.19035 05780 64785 40991 32564 02421 014
0.20778 49550 07898 46760 06894 03773 245	0.20443 29400 75298 89241 41619 99234 649
0.00000 00000 00000 00000 00000 00000 000	0.20948 21410 84727 82801 29991 74891 714

Table 2.1

10- POINT GAUSS RULE

ABSCISSAE	WEIGHTS
0.97390 65285 17171 72007 79640 12084 452	0.06667 13443 08688 13759 35688 09893 332
0.86506 33666 88984 51073 20966 88423 493	0.14945 13491 50580 59314 57763 39657 697
0.67940 95682 99024 40623 43273 65114 874	0.21908 63625 15982 04399 55349 34228 163
0.43339 53941 29247 19079 92659 43165 784	0.26926 67193 09996 35509 12269 21569 469
0.14887 43389 81631 21088 48260 01129 720	0.29552 42247 14752 87017 38929 94651 338

21- POINT KRONROD RULE

ABSCISSAE	WEIGHTS
0.99565 71630 25808 08073 55272 80689 003	0.01169 46388 67371 87427 80643 96062 192
0.97390 65285 17171 72007 79640 12084 452	0.03255 81623 07964 72747 88189 72459 390
0.93015 74913 55708 22600 12071 80059 508	0.05475 58965 74351 99603 13813 00244 580
0.86506 33666 88984 51073 20966 88423 493	0.07503 96748 10919 95276 70431 40916 190
0.78081 77265 86416 89706 37175 78345 042	0.09312 54545 83697 60553 50654 65083 366
0.67940 95682 99024 40623 43273 65114 874	0.10938 71588 02297 64189 92105 90325 805
0.56275 71346 68604 68333 90000 99272 694	0.12349 19762 62065 85107 79581 09831 074
0.43339 53941 29247 19079 92659 43165 784	0.13470 92173 11473 32592 80540 01771 707
0.29439 28627 01460 19813 11266 03103 866	0.14277 59385 77060 08079 70942 73138 717
0.14887 43389 81631 21088 48260 01129 720	0.14773 91049 01338 49137 48415 15972 068
0.00000 00000 00000 00000 00000 00000 000	0.14944 55540 02916 90566 49364 68389 821

Table 2.2

15- POINT GAUSS RULE

ABSCISSAE	WEIGHTS
0.98799 25180 20485 42848 95657 18586 613	0.03075 32419 96117 26835 46283 93577 204
0.93727 33924 00705 90430 77589 47710 209	0.07036 60474 88108 12470 92674 16450 667
0.84820 65834 10427 21620 06483 20774 217	0.10715 92204 67171 93501 18695 46685 869
0.72441 77313 60170 04741 61860 54613 938	0.13957 06779 26154 31444 78047 94511 028
0.57097 21726 08538 84753 72267 37253 911	0.16626 92058 16993 93355 32008 60481 209
0.39415 13470 77563 36989 72073 70981 045	0.18616 10000 15562 21102 68005 61866 423
0.20119 40939 97434 52230 06283 03394 596	0.19843 14853 27111 57645 61183 26443 839
0.00000 00000 00000 00000 00000 00000 000	0.20257 82419 25561 27288 06201 99967 519

31- POINT KRONROD RULE

ABSCISSAE	WEIGHTS
0.99800 22986 93397 06028 51728 40152 271	0.00537 74798 72923 34898 77920 51430 128
0.98799 25180 20485 42848 95657 18586 613	0.01500 79473 29316 12253 83747 63075 807
0.96773 90756 79139 13425 73479 78784 337	0.02546 08473 26715 32018 68740 01019 653
0.93727 33924 00705 90430 77589 47710 209	0.03534 63607 91375 84622 20379 48478 360
0.89726 45323 44081 90088 25096 56454 496	0.04458 97513 24764 87660 82272 99373 280
0.84820 65834 10427 21620 06483 20774 217	0.05348 15246 90928 08726 53431 47239 430
0.79041 85014 42465 93296 76492 94817 947	0.06200 95678 00670 64028 51392 30960 803
0.72441 77313 60170 04741 61860 54613 938	0.06985 41213 18728 25870 95200 77099 147
0.65099 67412 97416 97053 37358 95313 275	0.07684 96807 57720 37889 44327 77482 659
0.57097 21726 08538 84753 72267 37253 911	0.08308 05028 23133 02103 82892 47286 104
0.48508 18636 40239 68069 36557 40232 351	0.08856 44430 56211 77064 72754 43693 774
0.39415 13470 77563 36989 72073 70981 045	0.09312 65981 70825 32122 54868 72747 346
0.29918 00071 53168 81216 67800 24266 389	0.09664 27269 83623 67850 51799 07627 589
0.20119 40939 97434 52230 06283 03394 596	0.09917 35987 21791 95933 23931 73484 603
0.10114 20669 18717 49902 70742 31447 392	0.10076 98455 23875 59504 49466 62617 570
0.00000 00000 00000 00000 00000 00000 000	0.10133 00070 14791 54901 73747 92767 493

Table 2.3

20- POINT GAUSS RULE

ABSCISSAE	WEIGHTS
0.99312 85991 85094 92478 61223 88471 320	0.01761 40071 39152 11831 18619 62351 853
0.96397 19272 77913 79126 76661 31197 277	0.04060 14298 00386 94133 10399 52274 932
0.91223 44282 51325 90586 77524 41203 298	0.06267 20483 34109 06356 95065 35187 042
0.83911 69718 22218 82339 45290 61701 521	0.08327 67415 76704 74872 47581 43222 046
0.74633 19064 60150 79261 43050 70355 642	0.10193 01198 17240 43503 67501 35480 350
0.63605 36807 26515 02545 28366 96226 286	0.11819 45319 61518 41731 23773 77711 382
0.51086 70019 50827 09800 43640 50955 251	0.13168 86384 49176 62689 84944 99748 163
0.37370 60887 15419 56067 25481 77024 927	0.14209 61093 18382 05132 92983 25067 165
0.22778 58511 41645 07808 04961 95368 575	0.14917 29864 72603 74678 78287 37001 969
0.07652 65211 33497 33375 46404 09398 838	0.15275 33871 30725 85069 80843 31955 098

41- POINT KRONROD RULE

ABSCISSAE	WEIGHTS
0.99885 90315 88277 66383 83155 76545 863	0.00307 35837 18520 53150 12182 93246 031
0.99312 85991 85094 92478 61223 88471 320	0.00860 02698 55642 94219 86617 87950 102
0.98150 78774 50250 25919 33429 94720 217	0.01462 61692 56971 25298 37879 60308 868
0.96397 19272 77913 79126 76661 31197 277	0.02038 83734 61266 52359 80102 31432 755
0.94082 26338 31754 75351 99827 22212 443	0.02588 21336 04951 15883 45050 67096 153
0.91223 44282 51325 90586 77524 41203 298	0.03128 73067 77032 79895 85431 19323 801
0.87827 68112 52281 97607 74429 95113 078	0.03660 01697 58200 79803 05572 40707 211
0.83911 69718 22218 82339 45290 61701 521	0.04166 88733 27973 68626 37883 05936 895
0.79504 14288 37551 19835 06388 33272 788	0.04643 48218 67497 67472 02318 80926 108
0.74633 19064 60150 79261 43050 70355 642	0.05094 45739 23728 69193 27076 70050 345
0.69323 76563 34751 38480 54907 11845 932	0.05519 51053 48285 99474 48323 72419 777
0.63605 36807 26515 02545 28366 96226 286	0.05911 14008 80639 57237 49672 20648 594
0.57514 04468 19710 31534 29460 36586 425	0.06265 32375 54781 16802 58701 22174 255
0.51086 70019 50827 09800 43640 50955 251	0.06583 45971 33618 42211 15635 56969 398
0.44359 31752 38725 10319 99922 13492 640	0.06864 86729 28521 61934 56234 11885 368
0.37370 60887 15419 56067 25481 77024 927	0.07105 44235 53444 06830 57903 61723 210
0.30162 78681 14913 00432 05553 56858 592	0.07303 06903 32786 66749 51894 17658 913
0.22778 58511 41645 07808 04961 95368 575	0.07458 28754 00499 18898 65814 18362 488
0.15260 54652 40922 67550 52202 41022 678	0.07570 44976 84556 67465 95427 75376 617
0.07652 65211 33497 33375 46404 09398 838	0.07637 78676 72080 73670 55028 35038 061
0.00000 00000 00000 00000 00000 00000 000	0.07660 07119 17999 65644 50499 01530 102

Table 2.4

25- POINT GAUSS RULE

ABSCISSAE	WEIGHTS
0.99555 69697 90498 09790 87849 46893 902	0.01139 37985 01026 28794 79029 64113 235
0.97666 39214 59517 51149 83153 86479 594	0.02635 49866 15032 13726 19018 15295 299
0.94297 45712 28974 33941 40111 69658 471	0.04093 91567 01306 31265 56234 87711 646
0.89499 19978 78275 36885 10420 06782 805	0.05490 46959 75835 19192 59368 91540 473
0.83344 26287 60834 00142 10211 08693 570	0.06803 83338 12356 91720 71871 85656 708
0.75925 92630 37357 63057 72828 65204 361	0.08014 07003 35001 01801 32349 59669 111
0.67356 63684 73468 36448 51206 33247 622	0.09102 82619 82963 64981 14972 20702 892
0.57766 29302 41222 96772 36898 41612 654	0.10053 59490 67050 64420 22068 90392 686
0.47300 27314 45714 96052 21821 15009 192	0.10851 96244 74263 65311 60939 57050 117
0.36117 23058 09387 83773 58217 30127 641	0.11485 82591 45711 64833 93255 45869 556
0.24386 68837 20988 43204 51903 62797 452	0.11945 57635 35784 77222 81781 26512 901
0.12286 46926 10710 39638 73598 18808 037	0.12224 24429 90310 04168 89595 18945 852
0.00000 00000 00000 00000 00000 00000 000	0.12317 60537 26715 45120 39028 73079 050

51- POINT KRONROD RULE

ABSCISSAE	WEIGHTS
0.99926 21049 92609 83419 34574 86540 341	0.00198 73838 92330 31592 65078 51882 843
0.99555 69697 90498 09790 87849 46893 902	0.00556 19321 35356 71375 80402 36901 066
0.98803 57945 34077 24763 73310 14577 406	0.00947 39733 86174 15160 72077 10523 655
0.97666 39214 59517 51149 83153 86479 594	0.01323 62291 95571 67481 36564 05846 976
0.96161 49864 25842 51241 81300 33660 167	0.01684 78177 09128 29823 15166 67536 336
0.94297 45712 28974 33941 40111 69658 471	0.02043 53711 45882 83545 65682 92235 939
0.92074 71152 81701 56174 63460 84546 331	0.02400 99456 06953 21622 00924 89164 881
0.89499 19978 78275 36885 10420 06782 805	0.02747 53175 87851 73780 29484 55517 811
0.86584 70652 93275 59544 89969 69588 340	0.03079 23001 67387 48889 11090 20215 229
0.83344 26287 60834 00142 10211 08693 570	0.03400 21302 74329 33783 67487 95229 551
0.79787 37979 98500 05941 04109 04994 307	0.03711 62714 83415 54356 03306 25367 620
0.75925 92630 37357 63057 72828 65204 361	0.04008 38255 04032 38207 48392 84467 076
0.71776 64068 13084 38818 66540 79773 298	0.04287 28450 20170 04947 68957 92439 495
0.67356 63684 73468 36448 51206 33247 622	0.04550 29130 49921 78890 98705 84752 660
0.62681 00990 10317 41278 81226 81624 518	0.04798 25371 38836 71390 63922 55756 915
0.57766 29302 41222 96772 36898 41612 654	0.05027 76790 80715 67196 33252 59433 440
0.52632 52843 34719 18259 96237 78158 010	0.05236 28858 06407 47586 43667 12137 873
0.47300 27314 45714 96052 21821 15009 192	0.05425 11298 88545 49014 45433 70459 876
0.41788 53821 93037 74885 18143 94594 572	0.05595 08112 20412 31730 82406 86382 747
0.36117 23058 09387 83773 58217 30127 641	0.05743 71163 61567 83285 35826 93939 506
0.30308 95389 31107 83016 74789 09980 339	0.05868 96800 22394 20796 19741 75856 788
0.24386 68837 20988 43204 51903 62797 452	0.05972 03403 24174 05997 90992 91932 562
0.18371 89394 21048 89201 59698 88759 528	0.06053 94553 76045 86294 53602 67517 565
0.12286 46926 10710 39638 73598 18808 037	0.06112 85097 17053 04830 58590 30416 293
0.06154 44830 05685 07888 65463 92366 797	0.06147 11898 71425 31666 15441 31965 264
0.00000 00000 00000 00000 00000 00000 000	0.06158 08180 67832 93507 87598 24240 055

Table 2.5

30- POINT GAUSS RULE

ABSCISSAE	WEIGHTS
0.99689 34840 74649 54027 16300 50918 695	0.00796 81924 96166 60561 54658 83474 674
0.98366 81232 79747 20997 00325 81605 663	0.01846 64683 11090 95914 23021 31912 047
0.96002 18649 68307 51221 68710 25581 798	0.02878 47078 83323 36934 97191 79611 292
0.92620 00474 29274 32587 93242 77080 474	0.03879 91925 69627 04959 68019 36446 348
0.88256 05357 92052 68154 31164 62530 226	0.04840 26728 30594 05290 29381 40422 808
0.82956 57623 82768 39744 28981 19732 502	0.05749 31562 17619 06648 17216 89402 056
0.76777 74321 04826 19491 79773 40974 503	0.06597 42298 82180 49512 81285 15115 962
0.69785 04947 93315 79693 22923 88026 640	0.07375 59747 37705 20626 82438 50022 191
0.62052 61829 89242 86114 04775 56431 189	0.08075 58952 29420 21535 46949 38460 530
0.53662 41481 42019 89926 41697 93311 073	0.08689 97872 01082 97980 23875 30715 126
0.44703 37695 38089 17678 06099 00322 854	0.09212 25222 37786 12871 76327 07087 619
0.35270 47255 30878 11347 10372 07089 374	0.09636 87371 74644 25963 94686 26351 810
0.25463 69261 67889 84643 98051 29817 805	0.09959 34205 86795 26706 27802 82103 569
0.15386 99136 08583 54696 37946 72743 256	0.10176 23897 48405 50459 64289 52168 554
0.05147 18425 55317 69583 30252 13166 723	0.10285 26528 93558 84034 12856 36705 415

61- POINT KRONROD RULE

ABSCISSAE	WEIGHTS
0.99948 44100 50490 63757 13258 95705 811	0.00138 90136 98677 00762 45515 91226 760
0.99689 34840 74649 54027 16300 50918 695	0.00389 04611 27099 88405 12672 01844 516
0.99163 09968 70404 59485 86283 66109 486	0.00663 07039 15931 29217 33198 26369 750
0.98366 81232 79747 20997 00325 81605 663	0.00927 32796 59517 76342 84411 46892 024
0.97311 63225 01126 26837 46938 68423 707	0.01182 30152 53496 34174 22328 98853 251
0.96002 18649 68307 51221 68710 25581 798	0.01436 97295 07045 80481 24514 32443 580
0.94437 44447 48559 97941 58313 24037 439	0.01692 08891 89053 27262 75722 89420 322
0.92620 00474 29274 32587 93242 77080 474	0.01941 41411 93942 38117 34089 51050 128
0.90557 33076 99907 79854 65225 58925 958	0.02182 80358 21609 19229 71674 85738 339
0.88256 05357 92052 68154 31164 62530 226	0.02419 11620 78080 60136 56863 70725 232
0.85720 52335 46061 09895 86585 10658 944	0.02650 99548 82333 10161 06017 09335 075
0.82956 57623 82768 39744 28981 19732 502	0.02875 40487 65041 29284 39787 85354 334
0.79972 78358 21839 08301 36689 42322 683	0.03090 72575 62387 76247 28842 52943 092
0.76777 74321 04826 19491 79773 40974 503	0.03298 14470 57483 72603 18141 91016 854
0.73379 00624 53226 80472 61711 31369 528	0.03497 93380 28060 02413 74996 70731 468
0.69785 04947 93315 79693 22923 88026 640	0.03688 23646 51821 22922 39110 65617 136
0.66006 10641 26626 96137 00536 68149 271	0.03867 89456 24727 59295 03486 51532 281
0.62052 61829 89242 86114 04775 56431 189	0.04037 45389 51535 95911 19952 79752 468
0.57934 52358 26361 69175 60249 32172 540	0.04196 98102 15164 24614 71475 41285 970
0.53662 41481 42019 89926 41697 93311 073	0.04345 25397 01356 06931 68317 28117 073
0.49248 04678 61778 57499 36930 61207 709	0.04481 48001 33162 66319 23555 51616 723
0.44703 37695 38089 17678 06099 00322 854	0.04605 92382 71006 98811 62717 35559 374
0.40040 12548 30394 39253 54762 11542 661	0.04718 55465 69299 15394 52614 78181 099
0.35270 47255 30878 11347 10372 07089 374	0.04818 58617 57087 12914 07794 92298 305
0.30407 32022 73625 07737 26771 07199 257	0.04905 54345 55029 77888 75281 65367 238
0.25463 69261 67889 84643 98051 29817 805	0.04979 56834 27074 20635 78115 69379 942
0.20452 51166 82309 89143 89576 71002 025	0.05040 59214 02782 34684 08930 85653 585
0.15386 99136 08583 54696 37946 72743 256	0.05088 17958 98749 60649 22974 73049 805
0.10280 69379 66737 03014 70967 51318 001	0.05122 15478 49258 77217 06562 82604 944
0.05147 18425 55317 69583 30252 13166 723	0.05142 61285 37459 02593 38628 79215 781
0.00000 00000 00000 00000 00000 00000 000	0.05149 47294 29451 56755 83404 33647 099

Table 2.6

This expression can be simplified by introducing

$$S_d := \sum_{k=0}^{m}{}' \, a_k \, T_k \, T_d. \tag{2.2.16}$$

Taking into account that

$$T_k \, T_d = \frac{1}{2} \left(T_{k+d} + T_{|k-d|} \right) \tag{2.2.17}$$

and after some manipulation with the summation indices, S_d may be expressed as :

$$S_d = \frac{1}{2} \left(\sum_{k=d}^{d+m}{}' \, a_{k-d} \, T_k + \sum_{k=0}^{d}{}^{*} \, a_{d-k} \, T_k + \sum_{k=1}^{m-d} a_{k+d} \, T_k \right). \tag{2.2.18}$$

The notation $\sum^{*}$ indicates a summation with the last term halved. Thus

$$S_d = \frac{1}{2} \sum_{k=0}^{d+m}{}' \, q_k \, T_k \tag{2.2.19}$$

where the coefficients q_k are derived from (2.2.18). Making use of (2.2.16,19) in (2.2.15) yields

$$\sum_{k=0}^{d+m}{}' \sum_{\ell=0}^{m+1}{}' \, q_k \, b_\ell \int_{-1}^{1} T_k(x) \, T_\ell(x) \, dx = 0, \tag{2.2.20}$$

and, taking (2.2.17) into account :

$$\sum_{\ell=0}^{m+1}{}' \, b_\ell \sum_{k=0}^{d+m}{}' \, q_k \left(M_{k+\ell} + M_{|k-\ell|} \right) = 0 \tag{2.2.21}$$

where

$$M_k := \int_{-1}^{1} T_k(x)dx \begin{cases} = 0, \ k \text{ odd}, \\ = 2/(1-k^2), \ k \text{ even}. \end{cases} \qquad (2.2.22)$$

Setting

$$\alpha_{d,k} = \sum_{\ell=0}^{d+m}{}' \, q_\ell \, (M_{k+\ell} + M_{|k-\ell|}) \qquad (2.2.23)$$

we obtain from (2.2.21) :

$$\sum_{k=0}^{m}{}' \, b_k \, \alpha_{d,k} = -\alpha_{d,m+1}, \quad d = 0,1,\ldots,m. \qquad (2.2.24)$$

This system of m+1 equations, in the m+1 Chebyshev coefficients b_k of ω_{m+1}, is well-conditioned.

By means of the technique described above, and starting with the 10-point Gauss rule, a sequence containing the 21-, 43-, and 87-point formulae has been computed. The abscissae and weights of the 43- and 87-point formulae are tabulated in Table 2.7 to 25 digits accuracy. Together with the rules of Table 2.2, these formulae form a nested sequence, used by one of the integrators of QUADPACK.

43- POINT PATTERSON RULE

ABSCISSAE	WEIGHTS
0.99933 33609 01932 08139 40993 23919 911	0.00184 44776 40212 41410 03891 06552 965
0.99565 71630 25808 08073 55272 80689 003	0.00576 85560 59769 79618 41843 27908 655
0.98743 34029 08088 86979 59614 78381 209	0.01079 86895 85891 65174 04654 06741 293
0.97390 65285 17171 72007 79640 12084 452	0.01629 67342 89666 56492 42819 74617 663
0.95480 79348 14266 29925 79192 00290 473	0.02189 53638 67795 42810 25231 23075 149
0.93015 74913 55708 22600 12071 80059 508	0.02737 18905 93248 84208 12760 69289 151
0.90014 86957 48328 29362 50994 94069 092	0.03259 74639 75345 68944 38822 22526 137
0.86506 33666 88984 51073 20966 88423 493	0.03752 28761 20869 50146 16137 95898 115
0.82519 83149 83114 15084 70667 32588 520	0.04216 31379 35191 81184 76279 24327 955
0.78081 77265 86416 89706 37175 78345 042	0.04656 08269 10428 83074 33391 54433 824
0.73214 83889 89304 98261 23548 48755 461	0.05074 19396 00184 57778 01890 20092 084
0.67940 95682 99024 40623 43273 65114 874	0.05469 49020 58255 44214 72126 85465 005
0.62284 79705 37725 23864 11591 20344 323	0.05837 93955 42619 24837 54753 69330 206
0.56275 71346 68604 68333 90000 99272 694	0.06174 49952 01442 56449 62403 36030 883
0.49947 95740 71056 49995 22148 85499 755	0.06474 64049 51445 88554 46892 59517 511
0.43339 53941 29247 19079 92659 43165 784	0.06735 54146 09478 08607 55531 66302 174
0.36490 16613 46580 76804 39895 48502 644	0.06956 61979 12356 48452 86333 15038 405
0.29439 28627 01460 19813 11266 03103 866	0.07138 72672 68693 39776 85591 14425 516
0.22225 49197 76601 29649 82609 28066 212	0.07282 44414 71833 20815 09395 35192 842
0.14887 43389 81631 21088 48260 01129 720	0.07387 01996 32393 95343 21406 95251 367
0.07465 06174 61383 32204 39144 35796 506	0.07450 77510 14175 11827 35718 13842 889
0.00000 00000 00000 00000 00000 00000 000	0.07472 21475 17403 00559 44251 68280 423

Table 2.7

87- POINT PATTERSON RULE

ABSCISSAE	WEIGHTS
0.99990 29772 62729 23449 05298 30591 582	0.00027 41455 63762 07235 00165 27092 881
0.99933 33609 01932 08139 40993 23919 911	0.00091 52833 45202 24136 08433 92549 948
0.99798 98959 86678 74542 74963 22365 960	0.00180 71241 55057 94294 83413 11753 254
0.99565 71630 25808 08073 55272 80689 003	0.00288 48724 30211 53050 13341 56248 695
0.99217 54978 60687 22280 85233 52251 425	0.00409 68692 82759 16486 44580 70683 480
0.98743 34029 08088 86979 59614 78381 209	0.00539 92802 19300 47136 77387 43391 053
0.98135 81635 72712 77357 19169 41623 894	0.00675 82900 51847 37869 98165 77897 424
0.97390 65285 17171 72007 79640 12084 452	0.00814 83773 84149 17290 00028 78448 190
0.96505 76238 58384 61912 82841 10607 926	0.00954 99576 72201 64653 60535 81325 377
0.95480 79348 14266 29925 79192 00290 473	0.01094 76796 01118 93113 43278 26856 808
0.94316 76131 33670 59681 64166 34507 426	0.01232 94476 52244 85369 46266 39963 780
0.93015 74913 55708 22600 12071 80059 508	0.01368 59460 22712 70188 89500 35273 128
0.91580 64146 85507 20959 18264 30720 050	0.01501 04473 46388 95237 66972 86041 943
0.90014 86957 48328 29362 50994 94069 092	0.01629 87316 96787 33526 26657 03223 280
0.88322 16577 71316 50137 21175 48744 163	0.01754 89679 86243 19109 96653 52925 900
0.86506 33666 88984 51073 20966 88423 493	0.01876 14382 01562 82224 39350 59003 794
0.84571 07484 62415 66660 59020 11504 855	0.01993 80377 86440 88820 22781 92730 714
0.82519 83149 83114 15084 70667 32588 520	0.02108 15688 89203 83511 24330 60188 190
0.80355 76580 35230 98278 87394 74980 964	0.02219 49359 61012 28679 63321 02959 499
0.78081 77265 86416 89706 37175 78345 042	0.02328 04135 02888 31112 34092 91030 404
0.75700 57306 85495 55832 89427 93432 020	0.02433 91471 26000 80547 03606 47041 454
0.73214 83889 89304 98261 23548 48755 461	0.02537 09697 69253 82724 34679 99831 710
0.70627 32097 87321 81982 40942 74740 840	0.02637 45054 14839 20724 15037 86552 615
0.67940 95682 99024 40623 43273 65114 874	0.02734 74510 50052 28616 15828 29741 283
0.65158 94665 01177 92253 44222 05016 736	0.02828 69107 88771 20065 99680 02987 960
0.62284 79705 37725 23864 11591 20344 323	0.02918 96977 56475 75250 14461 54084 920
0.59322 33740 57961 08887 52737 70349 144	0.03005 25811 28092 69532 25211 10347 341
0.56275 71346 68604 68333 90000 99272 694	0.03087 24976 11713 35867 54663 94126 442
0.53149 36059 70831 93228 52689 48562 671	0.03164 67513 71439 92940 45860 51078 883
0.49947 95740 71056 49995 22148 85499 755	0.03237 32024 67202 78968 57881 94889 595
0.46676 36230 42022 84487 19667 81659 270	0.03305 04134 19978 50329 07859 44862 689
0.43339 53941 29247 19079 92659 43165 784	0.03367 77073 11637 93004 65810 56957 588
0.39942 48478 59218 80473 21016 65817 923	0.03425 50997 04226 06178 70828 21046 821
0.36490 16613 46580 76804 39895 48502 644	0.03478 30989 50365 14275 07819 97949 596
0.32987 48771 06188 28826 50533 71824 597	0.03526 24126 60156 68103 37827 17998 428
0.29439 28627 01460 19813 11266 03103 866	0.03569 36336 39418 77071 93513 55457 044
0.25850 35592 02161 55180 22809 75429 025	0.03607 69896 22888 70118 55003 18003 895
0.22225 49197 76601 29649 82609 28066 212	0.03641 22207 31351 78756 28011 63687 577
0.18569 53965 68346 65201 59171 41167 606	0.03669 86044 98456 09449 80180 47441 094
0.14887 43389 81631 21088 48260 01129 720	0.03693 50998 20427 90761 45895 86742 499
0.11184 22131 79907 46817 23983 59241 362	0.03712 05492 69832 57611 41199 58413 599
0.07465 06174 61383 32204 39144 35796 506	0.03725 38755 03047 70853 95920 01191 226
0.03735 21233 94619 87081 49981 65437 704	0.03733 42287 51935 04032 12354 49094 698
0.00000 00000 00000 00000 00000 00000 000	0.03736 10737 62679 02341 03212 41766 599

Table 2.7 (cont'd)

2.2.3. Modified Clenshaw-Curtis integration

2.2.3.1. Clenshaw-Curtis integration

Clenshaw-Curtis integration (Clenshaw and Curtis (1960)) handles the numerical computation of the integral of a function f over a finite interval, with weight function w = 1. The idea of Clenshaw-Curtis integration is to approximate f by a truncated Chebyshev expansion, which can be integrated exactly.

If f is continuous and of bounded variation on [-1,1], then it admits there a uniformly convergent expansion

$$f = \sum_{k=0}^{\infty}{}' a_k T_k. \tag{2.2.25}$$

Since

$$T_k(x) = \cos(k\theta), \quad x = \cos(\theta), \tag{2.2.26}$$

and

$$a_k = (2/\pi) \int_{-1}^{1} f(x)\, T_k(x)/(1-x^2)^{1/2}\, dx, \tag{2.2.27}$$

we obtain

$$a_k = (2/\pi) \int_0^{\pi} g(\theta) \cos(k\theta)\, d\theta \tag{2.2.28}$$

with

$$g(\theta) = f(\cos(\theta)). \tag{2.2.29}$$

An efficient method for computing approximations of the coefficients a_k will be described in the following section.

2.2.3.2. Computation of the Chebyshev coefficients

In this section a method for computing an approximation for a given function f in terms of a finite number of Chebyshev polynomials

$$\sum_{k=0}^{d}{}'' \, a_k^{(d)} \, T_k(x) \simeq f(x), \qquad x \in [-1,1] \tag{2.2.30}$$

will be investigated. The notation $\sum''$ indicates a summation with first and last term halved.
The coefficients a_k given by (2.2.28) can be approximated by means of the trapezoidal rule :

$$a_k^{(d)} = (2/d) \sum_{\ell=0}^{d}{}'' \, g(\pi\ell/d) \cos(\pi\ell k/d), \qquad k = 0,1,2,\dots,d. \tag{2.2.31}$$

For selected values of d, expression (2.2.31) can be evaluated efficiently using a variant of the FFT algorithm (Gentleman (1972)). For d = 12 a particularly simple procedure is described by Tolstov (1962). In the following we extend Tolstov's method, in such a way that the coefficients $a_k^{(12)}$ and $a_k^{(24)}$ are obtained simultaneously.

Let

$$x_\ell = \cos(\pi\ell/24), \qquad \ell = 0,1,\dots,11, \tag{2.2.32}$$

and construct the following tables :

	$f(1)/2$	$f(x_1)$	$f(x_2)$	...	$f(x_{11})$	$f(0)$
	$f(-1)/2$	$f(-x_1)$	$f(-x_2)$	...	$f(-x_{11})$	
sum	u_1	u_2	u_3	...	u_{12}	u_{13}
difference	v_1	v_2	v_3	...	v_{12}	

(2.2.33)

	u_1	u_2	u_3	u_4	u_5	u_6	u_7
	u_{13}	u_{12}	u_{11}	u_{10}	u_9	u_8	
sum	s_1	s_2	s_3	s_4	s_5	s_6	s_7
difference	t_1	t_2	t_3	t_4	t_5	t_6	

(2.2.34)

	s_1	s_2	s_3	s_4
	s_7	s_6	s_5	
sum	a_1	a_2	a_3	a_4
difference	b_1	b_2	b_3	

(2.2.35)

	a_1	a_2
	a_4	a_3
sum	c_1	c_2

(2.2.36)

These tables enable us to compute the coefficients $a_k^{(12)}$, $k = 0,1,2,\ldots,12$, and $a_k^{(24)}$, $k = 0,1,\ldots,24$, as follows :

$$
\begin{aligned}
12a_0^{(12)} &= c_1, \\
12a_{12}^{(12)} &= b_1-b_3, \\
6a_{6\pm5}^{(12)} &= (v_1+x_4v_5+x_8v_9) \mp (x_2v_3+x_6v_7+x_{10}v_{11}), \\
6a_{6\pm4}^{(12)} &= (t_1+x_8t_5) \mp (x_4t_3), \\
6a_{6\pm3}^{(12)} &= (v_1-v_9) \mp x_6(v_3-v_7-v_{11}), \\
6a_4^{(12)} &= b_1+x_8b_3, \\
6a_8^{(12)} &= a_1-x_8a_3, \\
6a_{6\pm1}^{(12)} &= (v_1-x_4v_5+x_8v_9) \mp (x_{10}v_3-x_6v_7+x_2v_{11}), \\
6a_6^{(12)} &= t_1-t_5,
\end{aligned}
\qquad (2.2.37)
$$

and

$$
\begin{aligned}
24a^{(24)}_{12\pm12} &= 12a^{(12)}_{0} \mp c_2, \\
12a^{(24)}_{12\pm11} &= 6a^{(12)}_{1} \mp (x_1v_2+x_3v_4+x_5v_6+x_7v_8+x_9v_{10}+x_{11}v_{12}), \\
12a^{(24)}_{12\pm10} &= 6a^{(12)}_{2} \mp (x_2t_2+x_6t_4+x_{10}t_6), \\
12a^{(24)}_{12\pm9} &= 6a^{(12)}_{3} \mp (x_3(v_2-v_8-v_{10}) + x_9(v_4-v_6-v_{12})), \\
12a^{(24)}_{12\pm8} &= 6a^{(12)}_{4} \mp x_4b_2, \\
12a^{(24)}_{12\pm7} &= 6a^{(12)}_{5} \mp (x_5v_2-x_9v_4-x_1v_6-x_{11}v_8+x_3v_{10}+x_7v_{12}), \\
12a^{(24)}_{12\pm6} &= 6a^{(12)}_{6} \mp x_6(t_2-t_4-t_6), \\
12a^{(24)}_{12\pm5} &= 6a^{(12)}_{7} \mp (x_7v_2-x_3v_4-x_{11}v_6+x_1v_8-x_8v_{10}-x_5v_{12}), \\
12a^{(24)}_{12\pm4} &= 6a^{(12)}_{8} \mp (x_8a_2-a_4), \\
12a^{(24)}_{12\pm3} &= 6a^{(12)}_{9} \mp (x_9(v_2-v_8-v_{10}) - x_3(v_4-v_6-v_{12})), \\
12a^{(24)}_{12\pm2} &= 6a^{(12)}_{10} \mp (x_{10}t_2-x_6t_4+x_2t_6), \\
12a^{(24)}_{12\pm1} &= 6a^{(12)}_{11} \mp (x_{11}v_2-x_9v_4+x_7v_6-x_8v_5+x_3v_{10}-x_1v_{12}), \\
12a^{(24)}_{12} &= 6a^{(12)}_{12}.
\end{aligned}
\qquad (2.2.38)
$$

2.2.3.3. Clenshaw-Curtis procedure for integrals with a weight function

For functions with a slowly convergent Chebyshev series expansion, use of the Clenshaw-Curtis method is not advisable. In order to give an idea of how strongly the rate of convergence of the expansion can be influenced for example by a singularity, we quote the following asymptotic estimate (Piessens and Criegers (1974)). For the function

$$f(x) = (1+x)^{\alpha}(1-x)^{\beta} \log((1+x)/2) \log((1-x)/2)h(x) \qquad (2.2.39)$$

with $h(x)$ well-behaved, the following asymptotic behaviour can be shown

$$\pi a_k \sim (-1)^{k+1}\, 2^{\beta-\alpha} \sin(\pi\alpha)\Gamma(2\alpha+3)F_0 k^{-2\alpha-3} - \\ - 2^{\alpha-\beta} \sin(\pi\beta)\Gamma(2\beta+3)G_0 k^{-2\beta-3} \tag{2.2.40}$$

with

$$F_0 := h(-1)(-\log(2k) + \pi \,\mathrm{cotan}(\pi\alpha)/2 + \psi(2\alpha+3)) \tag{2.2.41}$$

and

$$G_0 := h(1)(-\log(2k) + \pi \,\mathrm{cotan}(\pi\beta)/2 + \psi(2\beta+3)). \tag{2.2.42}$$

Here Γ and ψ represent the gamma function and the psi function respectively (Abramowitz and Stegun (1964)).

However, there is a possibility to retain the advantage of the computational ease of the Clenshaw-Curtis method while extending its practical applicability to functions with slowly convergent Chebyshev expansions. This is accomplished by transfering the difficult part of the integrand, which causes the slow rate of convergence of the Chebyshev series (e.g. a rapidly varying factor), into a weight function. We shall see that even certain singular weight functions, becoming infinite somewhere in the integration interval, can be treated successfully by this technique.

We thus write the integrand as a product $w(x)f(x)$, so that the integral $I_w[-1,1]f$ can be approximated as follows :

$$I_w[-1,1]f \simeq \sum_{k=0}^{d}{}'' a_k^{(d)} I_w[-1,1]T_k. \tag{2.2.43}$$

Essential now is the computation of the <u>modified Chebyshev moments</u> $I_w[-1,1]T_k$. In the Sections (2.2.3.4-6) the computation of these moments is described for three different weight functions, using the recurrence relations of Branders and Piessens (1975), Piessens and Branders (1973, 1975) and Piessens, Mertens and Branders (1974). An important property to be investigated is the numerical stability of such recurrence relations (Gautschi (1967)). In many cases, forward recursion is not stable and Olver's algorithm must be used (Olver (1967)).

2.2.3.4. Computation of the modified moments for the weight functions $\sin(\omega x)$ and $\cos(\omega x)$

Define

$$I_{\sin}[c_1,c_2]f := \int_{c_1}^{c_2} \sin(\omega x) f(x) dx \tag{2.2.44}$$

and

$$I_{\cos}[c_1,c_2]f := \int_{c_1}^{c_2} \cos(\omega x) f(x) dx. \tag{2.2.45}$$

In a first step, the original integrals are transformed onto the range [-1,1]. In the case of the sine weight function we obtain :

$$I_{\sin}[c_1,c_2]f = \tfrac{1}{2}(c_2-c_1)(\cos((c_1+c_2)\omega/2) \int_{-1}^{+1} \sin(\lambda x)\ \psi(x) dx +$$

$$+ \sin((c_1+c_2)\omega/2) \int_{-1}^{+1} \cos(\lambda x)\ \psi(x) dx) \tag{2.2.46}$$

where

$$\psi(x) := f((c_1+c_2)/2 + (c_2-c_1)x/2) \tag{2.2.47}$$

and

$$\lambda := (c_2-c_1)\omega/2. \tag{2.2.48}$$

Then $\psi(x)$ is replaced by its expansion (2.2.30), and we obtain

$$I_{\sin}[c_1,c_2]f \simeq \tfrac{1}{2}(c_2-c_1) \sum_{k=0}^{d}{}'' a_k^{(d)} (\cos((c_1+c_2)\omega/2) S_k(\lambda) +$$

$$+ \sin((c_1+c_2)\omega/2) C_k(\lambda)), \tag{2.2.49}$$

where $S_k(\lambda)$ and $C_k(\lambda)$ denote the sine and cosine Chebyshev moments respectively :

$$S_k(\lambda) := \int_{-1}^{1} \sin(\lambda x)\, T_k(x)\, dx, \qquad k = 0,1,\ldots,d, \tag{2.2.50}$$

$$C_k(\lambda) := \int_{-1}^{1} \cos(\lambda x)\, T_k(x)\, dx, \qquad k = 0,1,\ldots,d. \tag{2.2.51}$$

The integral $I_{cos}[c_1,c_2]f$ is handled in a similar way and also requires the moments $S_k(\lambda)$ and $C_k(\lambda)$ to be calculated. For this calculation we make use of the following recurrence relations (Piessens and Branders (1975)) :

$$\begin{aligned} \lambda^2(k-1)(k-2)S_{k+2}(\lambda) - 2(k^2-4)(\lambda^2-2k^2+2)S_k(\lambda) + \\ + \lambda^2(k+1)(k+2)S_{k-2}(\lambda) = \\ = -8(k^2-4)\sin(\lambda) - 24\lambda\cos(\lambda) \end{aligned} \tag{2.2.52}$$

with initial values

$$S_1(\lambda) = 2(\sin(\lambda)-\lambda\cos(\lambda))\lambda^{-2}$$

$$S_3(\lambda) = \lambda^{-2}\sin(\lambda)(18-48\lambda^{-2}) + \lambda^{-1}\cos(\lambda)(48\lambda^{-2}-2)$$

and

$$\begin{aligned} \lambda^2(k-1)(k-2)C_{k+2}(\lambda) - 2(k^2-4)(\lambda^2-2k^2+2)C_k(\lambda) + \\ + \lambda^2(k+1)(k+2)C_{k-2}(\lambda) = \\ = 24\lambda\sin(\lambda) - 8(k^2-4)\cos(\lambda) \end{aligned} \tag{2.2.53}$$

with initial values

$$C_0(\lambda) = 2\lambda^{-1}\sin(\lambda),$$

$$C_2(\lambda) = 8\lambda^{-2}\cos(\lambda)-\lambda^{-3}(2\lambda^2-8)\sin(\lambda),$$

$$C_4(\lambda) = 32\lambda^{-4}(\lambda^2-12)\cos(\lambda)+2\lambda^{-5}(\lambda^4-80\lambda^2+192)\sin(\lambda).$$

Note that

$$S_{2k}(\lambda) = C_{2k+1}(\lambda) = 0, \qquad k = 0,1,\ldots . \qquad (2.2.54)$$

Furthermore, the $S_k(\lambda)$ can be computed indirectly from the $C_k(\lambda)$ using

$$S_{2k-1}(\lambda) = \begin{cases} -\dfrac{\sin(\lambda)}{2k(k-1)} - \dfrac{\lambda}{4k} C_{2k}(\lambda) + \dfrac{\lambda}{4(k-1)} C_{2k-2}(\lambda), & k \geq 2, \\ \sin(\lambda) - \lambda(C_0(\lambda) + C_2(\lambda))/4, & k = 1. \end{cases} \qquad (2.2.55)$$

When $\lambda \geq 24$ we apply forward recursion for $C_k(\lambda)$ and formula (2.2.55) for $S_k(\lambda)$. For $2 < \lambda < 24$ we solve (2.2.52-53) by means of Olver's algorithm (Olver (1967)), because in that range both forward and backward recursion are unstable and also the indirect computation of the $S_k(\lambda)$ through (2.2.55) causes a loss of significant figures. Together with Olver's algorithm we employ the asymptotic formulae, for $k \to \infty$,

$$\begin{aligned} S_k(\lambda) \sim & -2\sin(\lambda)k^{-2} - 2(3\lambda\cos(\lambda) + \sin(\lambda))k^{-4} + \\ & + 2(-15\lambda\cos(\lambda) + (15\lambda^2-1)\sin(\lambda))k^{-6} + \\ & + 2((105\lambda^3-63\lambda)\cos(\lambda) + (210\lambda^2-1)\sin(\lambda))k^{-8} + \\ & + \ldots \qquad\qquad , k \text{ odd}, \end{aligned} \qquad (2.2.56)$$

and

$$\begin{aligned} C_k(\lambda) \sim & -2\cos(\lambda)k^{-2} - 2(\cos(\lambda) - 3\lambda\sin(\lambda))k^{-4} - \\ & - 2((1-15\lambda^2)\cos(\lambda) - 15\lambda\sin(\lambda))k^{-6} + \\ & + 2((210\lambda^2-1)\cos(\lambda) - (105\lambda^3-63\lambda)\sin(\lambda))k^{-8} + \\ & + \ldots \qquad\qquad , k \text{ even}. \end{aligned} \qquad (2.2.57)$$

For values of $\lambda \leq 2$ there is no need for a method which takes special care with regard to the weight function, because it does not oscillate rapidly. In this case we use Gauss-Legendre and Kronrod formulae for the direct computation of I_{sin} and I_{cos}.

2.2.3.5. Computation of modified moments for a weight function with end-point singularities of algebraic-logarithmic type

A special integration technique is required if the weight function becomes infinite on the integration interval. For example the integral $I_u[c_1,c_2]f$ where

$$u(x) = (x-a)^{\alpha}(b-x)^{\beta} \log^{\mu}(x-a) \log^{\nu}(b-x) \tag{2.2.58}$$

with

$$a = c_1 < c_2 < b \text{ or } a < c_1 < c_2 = b \tag{2.2.59}$$

and

$$\mu, \nu = 0 \text{ or } 1 \tag{2.2.60}$$

has an end-point singularity at $c_1 = a$ or $c_2 = b$. We shall handle these cases separately.

If $c_1 = a$ $(c_2 \neq b)$, $I_u[c_1,c_2]f$ transforms into

$$I_u[a,c_2]f = \left(\frac{c_2-a}{2}\right)^{\alpha+1} \sum_{i=0}^{\mu} \log^{\mu-i}(c_2-a) \int_{-1}^{1} (1+x)^{\alpha} \log^{i}\left(\frac{1+x}{2}\right)\psi(x)\, dx \tag{2.2.61}$$

where

$$\psi(x) := \left(\frac{2b-a-c_2}{2} - \frac{c_2-a}{2}x\right)^{\beta} \log^{\nu}\left(\frac{2b-a-c_2}{2} - \frac{c_2-a}{2}x\right) f\left(\frac{a+c_2}{2} + \frac{c_2-a}{2}x\right). \tag{2.2.62}$$

Thus, when $\psi(x)$ is replaced by its approximation (2.2.30) we have :

$$I_u[a,c_2]f \simeq \left(\frac{c_2-a}{2}\right)^{\alpha+1} \sum_{i=0}^{\mu} \log^{\mu-i}(c_2-a) \sum_{k=0}^{d}{}'' a_k^{(d)} G_{k,i}(\alpha) \tag{2.2.63}$$

where

$$G_{k,i}(\alpha) := \int_{-1}^{1} (1+x)^{\alpha} \log^{i}\left(\frac{1+x}{2}\right) T_k(x)\, dx. \tag{2.2.64}$$

If $c_2 = b$ $(c_1 \neq a)$, the integral $I_u[c_1,c_2]f$ is treated in a similar way. We arrive at an approximation of the form

$$I_u[c_1,b]f \approx \left(\frac{b-c_1}{2}\right)^{\beta+1} \sum_{i=0}^{\nu} \log^{\nu-i}(b-c_1) \sum_{k=0}^{d}{}'' a_k^{(d)} H_{k,i}(\beta), \qquad (2.2.65)$$

with

$$H_{k,i}(\beta) := \int_{-1}^{1} (1-x)^{\beta} \log^{i}\left(\frac{1-x}{2}\right) T_k(x)\, dx. \qquad (2.2.66)$$

Due to the relation

$$H_{k,i}(\beta) = (-1)^k G_{k,i}(\beta), \qquad k = 0,1,\ldots,d, \quad i = 0,1, \qquad (2.2.67)$$

it will be sufficient to provide the recurrence relations only for the modified moments $G_{k,i}(\alpha)$,

$$(k-1)(k+\alpha+1)G_{k,0}(\alpha) + k(k-\alpha-2)G_{k-1,0}(\alpha) = -2^{\alpha+1}$$

with initial values

$$G_{0,0}(\alpha) = 2^{\alpha+1}/(\alpha+1) \qquad (2.2.68)$$

$$G_{1,0}(\alpha) = G_{0,0}(\alpha)/(\alpha+2),$$

and

$$(k-1)(k+\alpha+1)G_{k,1}(\alpha) + k(k-\alpha-2)G_{k-1,1}(\alpha) =$$

$$= kG_{k-1,0}(\alpha) - (k-1)G_{k,0}(\alpha) \qquad (2.2.69)$$

with initial values

$$G_{0,1}(\alpha) = - G_{0,0}(\alpha)/(\alpha+1),$$

$$G_{1,1}(\alpha) = G_{0,1}(\alpha+1) - G_{0,1}(\alpha).$$

In both cases forward recursion is numerically stable.

2.2.3.6. Computation of the modified moments involving the weight function $1/(x-c)$

The expression $I_h[c_1,c_2]f$ with

$$h(x) = 1/(x-c) \tag{2.2.70}$$

is interpreted as a Cauchy principal value if $c \in (c_1,c_2)$, and sometimes called the Hilbert transform of f over $[c_1,c_2]$. By a change of the integration variable we obtain

$$I_h[c_1,c_2]f = \oint_{-1}^{1} \psi(x)/(x-c')dx \tag{2.2.71}$$

where

$$\psi(x) := f((c_1+c_2)/2 + (c_2-c_1)x/2) \tag{2.2.72}$$

and

$$c' := (2c-c_2-c_1)/(c_2-c_1) \tag{2.2.73}$$

By using an approximation for ψ of the form (2.2.30), we get an approximation to $I_h[c_1,c_2]f$, of the form

$$I_h[c_1,c_2]f \approx \sum_{k=0}^{d}{}'' a_k^{(d)} V_k(c') \tag{2.2.74}$$

where

$$V_k(c') := \oint_{-1}^{+1} T_k(x)/(x-c')dx. \tag{2.2.75}$$

The modified moments can be determined by solving the following recurrence relation (Piessens, Van Roy-Branders and Mertens (1976)) :

$$V_{k+1}(c') - 2c'V_k(c') + V_{k-1}(c') = \begin{cases} 0, & k \text{ odd}, \\ 4/(1-k^2), & k \text{ even}, \end{cases} \tag{2.2.76}$$

with initial values

$$V_0(c') = \log|(1-c')/(1+c')|,$$

$$V_1(c') = 2 + c'V_0(c').$$

If $c' \in (-1,1)$ or just outside $[-1,1]$, then forward recursion is sufficiently stable numerically. In practice we shall apply (2.2.76) using forward recursion if

$$c_1 - e \leq c \leq c_2 + e \tag{2.2.77}$$

with

$$e := (c_2-c_1)/20. \tag{2.2.78}$$

Otherwise, we assume that c is lying sufficiently far outside $[c_1,c_2]$ to enable the computation of $I_h[c_1,c_2]f$ by using Gauss-Legendre or Kronrod formulae.

2.2.4. Numerical quadrature and extrapolation

One way of analyzing the error of a quadrature formula is based on approximation theory.

As a consequence of Weierstrass' theorem of uniform approximation (Dieudonné (1960), p. 133), we have

Theorem 2.3

Let the function f *be continuous on* [a,b]. *For a given* $\varepsilon > 0$, *let* d *be the degree of a polynomial* p_d *such that*

$$\max_{x\in[a,b]} |f(x)-p_d(x)| \leq \varepsilon. \tag{2.2.79}$$

Then the absolute error associated with a quadrature sum Q *with positive weights and of degree of precision* d *satisfies*

$$|E[a,b]f| \leq 2(b-a)\varepsilon. \tag{2.2.80}$$

If $\{Q_n\}_n$ is a sequence of positive n-point rules of increasing degree of precision n, Theorem 2.3 implies the convergence of $\{Q_n[a,b]f\}_n$ when f is continuous on [a,b]. For the Gauss quadrature sums it is even possible to prove (Davis and Rabinowitz (1975), pp. 99-102) :

Theorem 2.4
The sequence of n-point Gauss rules satisfies

$$\lim_{n\to\infty} Q_n[a,b]f = I[a,b]f. \tag{2.2.81}$$

whenever f *is bounded and Riemann integrable on* [a,b].

With the use of approximation theory it is furthermore possible to produce bounds on the integration error of a given quadrature sum, which are valid within specific function classes.

Theorem 2.5
Let

$$|f^{(k+1)}(x)| \leq M_{k+1}, \quad x \in [a,b]. \tag{2.2.82}$$

Then the absolute error of a positive quadrature rule of degree of precision d > k *satisfies*

$$|E[a,b]f| \leq c_k \, d^{-k}(b-a)^{k+2} \, M_{k+1}/(d-k) \tag{2.2.83}$$

where

$$c_k = 2e^k \, 3^{k+1}/(k+1). \tag{2.2.84}$$

This is a consequence of Jackson's (1930) estimate of how closely a continuous function can be approximated by a polynomial (Rivlin (1969), pp. 23).

The error bound of (2.2.83) suggests two different types of limiting procedures for improving the accuracy of an integral approximation, one involving an increase of the degree of precision and one involving subdivision of the interval. We have already considered examples of the first procedure. Examples of the second one are the so-called m-panel rules $Q_r^{(m)}[a,b]$, which arise from partitioning [a,b] into m equal panels and applying the correspondingly scaled version of the fixed r-point rule $Q_r[a,b]$ on each panel. The number of subintervals, m is called the <u>mesh ratio</u>. We shall denote the error of $Q_r^{(m)}$ by $E_r^{(m)}$. The m-panel rules converge under quite general conditions on f.

<u>Theorem 2.6</u>

If $Q_r[a,b]$ *has degree of precision* $d > 0$, *then*

$$\lim_{m\to\infty} Q_r^{(m)}[a,b]f = I[a,b]f \tag{2.2.85}$$

for each f *bounded and Riemann integrable on* $[a,b]$.

For functions with a bounded (k+1)st derivative, information about the speed of convergence can be deduced from Theorem 2.5.

<u>Theorem 2.7</u>

If

$$|f^{(k+1)}(x)| \leq M_{k+1}, \; x \in [a,b], \tag{2.2.86}$$

and if $Q_r[a,b]$ *is positive and has degree of precision* $d > k$, *then*

$$|E_r^{(m)}[a,b]f| \leq c_k \, d^{-k}(b-a)^{k+2} \, m^{-(k+1)} \, M_{k+1}/(d-k) \tag{2.2.87}$$

with c_k *defined by* (2.2.84).

It should be noted that the error estimates (2.2.83) and (2.2.87) are very pessimistic in practice. Consider for example the $r(2^\ell)$-point Gauss rule $Q_{r(2^\ell)}[0,1]$ on the one hand and $Q_r^{(2^\ell)}[0,1]$ on the other. For $k = 0$ we obtain

$$|E_{r(2^\ell)}[0,1]f| \leq 6 \, M_1/(2^{\ell+1}r-1) \tag{2.2.88}$$

and

$$|E_r^{(2^\ell)}[0,1]f| \leq 6 \, M_1/(2^{\ell+1}r-2^\ell). \tag{2.2.89}$$

Table 2.8 displays the results of both sequences with $r = 4$, $\ell = 1,2,\ldots,5$, for the integral $I[0,1](5x^{3/2}/2)$.

ℓ	number of points $4(2^\ell)$	$Q_{4(2^\ell)}[0,1](5x^{3/2}/2)$	$Q_4^{(2^\ell)}[0,1](5x^{3/2}/2)$
0	4	0.9998759625294496	
1	8	0.9999952235074709	0.9999780705733770
2	16	0.9999998303586389	0.9999961233722831
3	32	0.9999999943105116	0.9999993147024912
4	64	0.9999999998155494	0.9999998788553750
5	128	0.9999999999943176	0.9999999785844577
	Exact	1.0000000000000000	1.0000000000000000

Table 2.8

The conditions of Theorems 2.4 and 2.6 may be relaxed, to include certain singular functions (which become infinite on [0,1]). Indeed, for certain functions it is possible to <u>ignore the singularity</u> (Davis and Rabinowitz (1975), pp. 144-146). For example if $Q_r[0,1]$ represents the r-point Gauss rule, then (2.2.81) and (2.2.85) are still valid for

$$f(x) = x^\beta \tag{2.2.90}$$

with arbitrary $\beta > -1$. The rate of convergence, however, is very slow for $\beta < 0$. Table 2.9 lists approximations to $(\beta+1)I[0,1]x^\beta$ for $\beta = -9/10$, with the sequence of Gauss formulae $Q_{4(2^\ell)}[0,1]$ and with the sequence of $Q_4^{(2^\ell)}[0,1]$. Note that in the latter case the main contribution to the error $E_4^{(2^\ell)}[0,1]f$ is

$$E_4[0,2^{-\ell}]f = 2^{-\ell(\beta+1)}\ E_4[0,1]f. \tag{2.2.91}$$

So with $\beta = -9/10$ the total absolute error is only approximately halved when $\ell = 10$.

In cases like the one of Table 2.9 it may be advantageous to <u>extrapolate to the limit</u>. Several extrapolation techniques exist. Their possible failure or success and their efficiency when applied to a sequence $\{Q^{(m_\ell)}[0,1]f\}_\ell$ depends on the form of the asymptotic expansion, called the <u>Euler-Maclaurin expansion</u> of the error $E^{(m)}[0,1]f$ as $m \to \infty$.

ℓ	number of points $4(2^\ell)$	$Q_{4(2^\ell)}[0,1](x^{-9/10}/10)$	$Q_4^{(2^\ell)}[0,1](x^{-9/10}/10)$
0	4	0.3456039085893030	0.3456039085893030
1	8	0.4236404760487249	0.3894268014395712
2	16	0.4952135967838964	0.4303150615042657
3	32	0.5592066099192942	0.4684651575991962
4	64	0.6156722904303098	0.5040604558887212
5	128	0.6651629572879993	0.5372720435364531
6	256		0.5682595505131033
7	512		0.5971719168477944
8	1024		0.6241481085014587
9	2048		0.6493177853003457
10	4096		0.6728019241400230
	Exact	1.0000000000000000	1.0000000000000000

Table 2.9

For example if this expansion is an asymptotic series in increasing powers of $1/m$ possibly multiplied by a factor $\log(m)$, the ε-algorithm of Wynn (1956) can be used for extrapolation (Kahaner (1972), Genz (1975)).

The ε-algorithm is called a convergence acceleration technique, because it attempts to replace the original sequence by faster converging sequences.

2.2.4.1. Asymptotic expansion of the integration error

In this section we shall summarize some important results involving the asymptotic expansion of the integration error, starting with the Euler-Maclaurin series for functions with $p \geq 1$ continuous derivatives in the integration interval. Furthermore we will give generalized error expansions for integrals with algebraic-logarithmic singularities and jump discontinuities.

Let us first introduce the m-panel offset trapezoidal rule (Lyness and Puri (1973)) :

$$R^{[m,\alpha]}[0,1]f = \frac{1}{m} \sum_{n=-\infty}^{\infty} \theta_n f((n-1+x)/m) \tag{2.2.92}$$

where θ_n is one when the corresponding abscissa falls inside (0,1), 1/2 when it coincides with an end-point $x = 0$ or $x = 1$ and zero outside $[0,1]$, and where

$$x = (\alpha+1)/2, \qquad |\alpha| \leqslant 1. \tag{2.2.93}$$

We shall denote the quadrature error of the rule (2.2.92) by $E^{[m,\alpha]}[0,1]f$.

Rule $R^{[m,\alpha]}$ has the interesting property that its error functional expansion leads to an error expansion for the m-panel of an arbitrary rule in a quite straightforward way. Let Q_r be a quadrature rule of a non-negative polynomial degree, with nodes $x_k \in [0,1]$ and weights w_k, $k = 1,2,\ldots,r$, where the weights assigned to the end-points are equal (and possibly zero), then

$$Q_r^{(m)} = \sum_{k=1}^{r} w_k \, R^{[m,\alpha_k]}, \qquad \alpha_k = 2x_k-1, \tag{2.2.94}$$

and correspondingly

$$E_r^{(m)} = \sum_{k=1}^{r} w_k \, E^{[m,\alpha_k]}. \tag{2.2.95}$$

Theorem 2.8

If $f \in C^{p+1}[0,1]$, *then with the notations of* (2.2.92-93) *we have*

$$E^{[m,\alpha]}[0,1]f = \sum_{q=1}^{p} \frac{m^{-q}}{q!} \bar{B}_q(x) \int_0^1 f^{(q)}(x)dx + O(m^{-p-1}), \tag{2.2.96}$$

where $\overline{B}_q$ *is the periodic extension of the Bernoulli polynomial of degree* q*, as defined in Abramowitz and Stegun* (1964).
The following theorem is an immediate consequence.

Theorem 2.9

If $f \in C^{p+1}[0,1]$, *and* Q *is a quadrature sum of polynomial degree* d *with* $0 \leqslant d \leqslant p$ *and with equal weights assigned to the end-points, then*

$$E^{(m)}[0,1]f = \sum_{q=d+1}^{p} \frac{m^{-q}}{q!}(E[0,1]\overline{B}_q) \int_0^1 f^{(q)}(x)dx + O(m^{-p-1}). \quad (2.2.97)$$

Note that, because of symmetry of the Bernoulli polynomials with respect to $x = 1/2$, the odd-numbered coefficients vanish, leaving an even expansion if Q is a symmetric rule.

A well-known special case of (2.2.96-97) is the classical Euler-Maclaurin expansion obtained for the trapezoidal rule $R^{[m,1]}$.

We shall now give generalizations of (2.2.96) for functions with singularities (Lyness and Ninham (1967)).

(i) Integrand with algebraic end-point singularity
If

$$f(x) = x^{\beta}\, h(x), \quad (2.2.98)$$

with $0 \geqslant \beta > -1$ and $h \in C^{p+1}[0,1]$, then

$$E^{[m,\alpha]}[0,1]f = \sum_{q=1}^{p} a_q^{(1)} m^{-\beta-q} + \sum_{q=1}^{p} a_q^{(2)} m^{-q} + O(m^{-p-1}) \quad (2.2.99)$$

where $a_q^{(1)}$ and $a_q^{(2)}$ are independent of m.
An important special case of (2.2.99) is that with

$$m = \lambda^{-\ell} = \mu^{\ell} = m_{\ell},\ \ell \geqslant 1 \text{ integer},\ \mu > 1 \text{ integer}. \quad (2.2.100)$$

By arranging the exponents $\beta + q$ and q in an increasing sequence $\{\zeta_q\}_q$ and denoting

$$\lambda_q = \lambda^{\zeta_q} \tag{2.2.101}$$

we obtain from (2.2.99) :

$$E^{[m_\ell,\alpha]}[0,1]f = \sum_{q=1}^{w} u_q(\lambda_q)^\ell + O(\lambda^{\ell(p+1)}) \tag{2.2.102}$$

where $w = p$ if $\beta = 0$ and $w = 2p$ if $\beta \neq 0$, the u_q are independent of ℓ, and

$$1 > \lambda_1 > \lambda_2 > \ldots > \lambda_w > \lambda^{p+1}. \tag{2.2.103}$$

(ii) <u>Integrand with an algebraic-logarithmic singularity</u>
For

$$f(x) = x^\beta \log(x)\ h(x) \tag{2.2.104}$$

with $0 \geqslant \beta > -1$ and $h \in C^{p+1}[0,1]$,
we have

$$E^{[m,\alpha]}[0,1]f = \sum_{q=1}^{p} (b_q^{(1)}+b_q^{(2)}\log(m))m^{-\beta-q} + \sum_{q=1}^{p} b_q^{(3)}m^{-q} + O(m^{-p-1}) \tag{2.2.105}$$

where $b_q^{(1)}$, $b_q^{(2)}$ and $b_q^{(3)}$ are independent of m. The form of (2.2.105) can be obtained by differentiating (2.2.99) with respect to β. When f contains a factor $\log^s(x)$ one can proceed by differentiating (2.2.105) $s - 1$ times with respect to β.

If m has a value given by (2.2.100) it is possible to write the logarithmic terms in (2.2.105) as

$$m_\ell^{-\beta-q}\ b_q^{(2)}\ \log(m_\ell) = -\ \ell b_q^{(2)}\ \log(\lambda)\ \lambda^{\ell(\beta+q)}, \tag{2.2.106}$$

and (2.2.105) becomes

$$E^{[m_\ell,\alpha]}[0,1]f = \sum_{q=1}^{w}\sum_{i=0}^{1} u_{qi}\ \ell^i(\lambda_q)^\ell + O(\lambda^{\ell(p+1)}) \tag{2.2.107}$$

with constant coefficients u_{qi} (not depending on ℓ) and λ_q subject to (2.2.103).

(iii) Integrand with interior algebraic singularities.
Consider

$$f(x) = x^{\beta}(1-x)^{\gamma}|x-\sigma|^{\eta}|x-\theta|^{\phi}\ \mathrm{sgn}(x-\theta)h(x) \qquad (2.2.108)$$

with

$$\beta,\ \gamma,\ \eta \text{ and } \phi > -1,\ h \in C^{p+1}[0,1],\ \sigma \text{ and } \theta \in (0,1). \qquad (2.2.109)$$

It may show singularities not only at the end-points but also at the interior points σ and θ. Theorem 2.10 gives the expansion of $E^{[m,\alpha]}[0,1]f$ with an explicit expression for the coefficients. Note that interior singularities cause terms in the expansion with the coefficient of the $(1/m)$-power depending on m.

Theorem 2.10

Let f(x) *be given by* (2.2.108-109). *Define the functions* ψ_0, ψ_1, ψ_σ, ψ_θ *by*

$$f(x) = x^{\beta}\psi_0(x) = (1-x)^{\gamma}\ \psi_1(x) = |x-\sigma|^{\eta}\ \psi_\sigma(x) = \\ = |x-\theta|^{\phi}\ \mathrm{sgn}(x-\theta)\ \psi_\theta(x). \qquad (2.2.110)$$

Then

$$E^{[m,\alpha]}[0,1]f = \sum_{q=1}^{p} \frac{1}{(q-1)!}\,(\psi_0^{(q-1)}(0)\ c_{q-1}^{[m,\alpha]}(0,\beta) + \\ + \psi_1^{(q-1)}(1)\ d_{q-1}^{[m,\alpha]}(1,\gamma) + \\ +\psi_\sigma^{(q-1)}(\sigma)(c_{q-1}^{[m,\alpha]}(\sigma,\eta) + d_{q-1}^{[m,\alpha]}(\sigma,\eta)) + \\ +\psi_\theta^{(q-1)}(\theta)(c_{q-1}^{[m,\alpha]}(\theta,\phi) - d_{q-1}^{[m,\alpha]}(\theta,\phi)))+ \\ + O(m^{-p-1}), \qquad (2.2.111)$$

with

$$c_{q-1}^{[m,\alpha]}(t,\beta) = \bar{\zeta}(-\beta-q+1,\frac{\alpha+1}{2}-mt)m^{-\beta-q} \qquad (2.2.112)$$

and

$$d_{q-1}^{[m,\alpha]}(t,\beta) = (-1)^{q-1}\,\bar{\zeta}(-\beta-q+1,mt-\frac{\alpha+1}{2})m^{-\beta-q} \qquad (2.2.113)$$

where $\bar{\zeta}$ *denotes the periodic extension of the generalized* ζ *-function* $\zeta(s,\hat{a})$ *over the interval* $[0,1]$ *(Lyness and Ninham* (1967)) :

$$\zeta(s,\hat{a}) = \sum_{n=1}^{\infty} \frac{1}{(n+\hat{a})^s}, \qquad \mathrm{Re}(s) > 1,\ 0 < \hat{a} \leqslant 1, \qquad (2.2.114)$$

with an analytic continuation for $\mathrm{Re}(s) < 1$, *and*

$$\bar{\zeta}(s,a) = \zeta(s,\hat{a}), \qquad (a-\hat{a}) \textit{ integer} . \qquad (2.2.115)$$

Thus when $m = \lambda^{-\ell}$ as given by (2.2.100) we see that a singular factor $|x-\sigma|^\eta$ in $f(x)$ brings terms in $\lambda^{\ell(\eta+q)}$ into the error expansion (2.2.111) with coefficients depending on ℓ through $\bar{\zeta}(-\eta-q+1, \pm(\alpha+1)/2 \mp \lambda^{-\ell}\sigma)$. In connection with our investigation of the applicability of the ε-algorithm (Section 2.2.4.2) it is important to determine the nature of this dependence on ℓ. The $\bar{\zeta}$-function is periodic with period 1 in its second argument. Hence $\bar{\zeta}(A,\pm B \mp \lambda^{-\ell}\sigma)$ is periodic in ℓ with period L if

$$\lambda^{-\ell}(1-\lambda^{-L})\sigma = \text{integer}. \qquad (2.2.116)$$

This relates L, if it exists, to the position σ of the singularity in the integration range. If σ has a representation

$$\sigma = \sum_{i=1}^{\infty} a_i\, \lambda^i \qquad (2.2.117)$$

where a_i is either 0 or 1 and the sequence $\{a_i\}_i$ is periodic with period L, it is easy to verify that (2.2.116) holds for all positive integers . For example, when $\lambda = 1/2$, (2.2.117) gives the binary representation of σ and L is its binary period. For $\lambda = 1/2$: $\sigma = 1/3$ or 2/3 has L = 2, $\sigma = k/7$ with k = 1,2,...,6 has L = 3, etc... .

2.2.4.2. The ε-algorithm

The ε-algorithm of Wynn (1956) is a recursive procedure for calculating a sequence transformation due to Shanks (1955).

For a given sequence $\{T_\ell\}_\ell$ of real numbers, let $H_{\ell,n}(T)$ represent the Hankel determinant

$$H_{\ell,n}(T) := \begin{vmatrix} T_\ell & T_{\ell+1} & \cdots & T_{\ell+n-1} \\ T_{\ell+1} & T_{\ell+2} & \cdots & T_{\ell+n} \\ \cdot & \cdot & & \cdot \\ \cdot & \cdot & & \cdot \\ \cdot & \cdot & & \cdot \\ T_{\ell+n-1} & T_{\ell+n} & \cdots & T_{\ell+2n-2} \end{vmatrix}, \quad n = 1,2,\ldots, \quad \ell = 0,1,\ldots,$$

$$H_{\ell,0}(T) := 1, \quad \ell = 0,1,\ldots \,. \tag{2.2.118}$$

Define

$$T_{\ell,2n} := \frac{H_{\ell,n+1}(T)}{H_{\ell,n}(\Delta^2 T)}$$

$$T_{\ell,2n+1} := \frac{H_{\ell,n}(\Delta^3 T)}{H_{\ell,n+1}(\Delta T)}, \quad n, \ell = 0,1,\ldots, \tag{2.2.119}$$

if $H_{\ell,n}(\Delta^2 T)$ and $H_{\ell,n+1}(\Delta T)$ do not vanish.

Theorem 2.11 formulates the fundamental property of the table $\{T_{\ell,n}\}_{\ell,n}$ defined by (2.2.119).

Theorem 2.11

Let the sequence $\{E_\ell\}_\ell$ *satisfy a homogeneous linear difference equation of order* n *with constant coefficients i.e.*

$$\sum_{i=0}^{n} r_i \, E_{\ell+i} = 0, \qquad \ell = 0,1,\ldots, \tag{2.2.120}$$

then if

$$T_\ell = \tilde{I} + E_\ell, \; \ell = 0,1,\ldots, \tag{2.2.121}$$

transformation (2.2.118-119) *of the sequence* $\{T_\ell\}_\ell$ *gives*

$$T_{\ell,2n} = \tilde{I}, \; \ell = 0,1,\ldots, \tag{2.2.122}$$

whenever $T_{\ell,2n}$ *exists.*

The transforms $\{T_{\ell,n}\}_{\ell,n}$ given by (2.2.119) can be arranged in a triangular table :

$$\begin{array}{lll} T_{0,0} & & \\ T_{1,0} & T_{0,1} & \\ T_{2,0} & T_{1,1} & \\ \cdot & \cdot & \cdot \\ \cdot & \cdot & \cdot \\ \cdot & \cdot & \cdot \\ T_{\ell,0} & T_{\ell-1,1} & \\ T_{\ell+1,0} & & \end{array} \tag{2.2.123}$$

which can be constructed recursively, without computing the Hankel determinants, by means of the ε-algorithm of Wynn (1956) :

$$T_{\ell,-1} := 0,$$

$$T_{\ell,0} := T_\ell, \qquad \ell = 0,1,\ldots, \tag{2.2.124}$$

$$T_{\ell,n+1} := T_{\ell+1,n-1} + \frac{1}{T_{\ell+1,n}-T_{\ell,n}} .$$

Theorem 2.11 relates the efficiency of the ε-algorithm to the existence of a linear difference equation with constant coefficients for the sequence $\{T_\ell-\tilde{I}\}_\ell$. Theorem 2.12 below gives the order of the difference equation for a particular type of sequence which is of interest for our application to numerical integration.

Theorem 2.12

The sequence $\{E_\ell\}_\ell$ *with*

$$E_\ell = u(\ell)\,\ell^i\lambda^\ell, \qquad \ell = 0,1,\ldots, \tag{2.2.125}$$

for a non-negative integer i *and a real* λ *and where the function* $u(\ell)$ *is periodic with integer period* L :

$$u(\ell+L) = u(\ell), \qquad \ell = 0,1,\ldots, \tag{2.2.126}$$

satisfies a linear difference equation with constant coefficients of order

$$n = (i+1)L. \tag{2.2.127}$$

As a consequence the sequence $\{E_\ell\}_\ell$ with

$$E_\ell = \sum_{q=1}^{w}\sum_{i=0}^{s} u_{qi}(\ell)\,\ell^i(\lambda_q)^\ell, \qquad \ell = 0,1,\ldots, \tag{2.2.128}$$

for some integers $s \geq 0$ and $w \geq 1$, real λ_q, and the u_{qi} periodic functions of ℓ with period L, satisfies a linear difference equation with constant coefficients. Theorem 2.13 gives information about the rate of convergence of the columns in the extrapolation table for a simplified form of the sequence (2.2.128).

Theorem 2.13

If $\{T_\ell\}_\ell$ *is given by*

$$T_\ell = \tilde{I} + \sum_{q=1}^{w} u_q(\lambda_q)^\ell \qquad (2.2.129)$$

with u_q *independent of* ℓ *and*

$$1 > |\lambda_1| > \ldots > |\lambda_w| > 0, \qquad (2.2.130)$$

then

$$T_{\ell,2n} = \tilde{I} + O((\lambda_{n+1})^\ell), \qquad n < w, \quad \textit{as } \ell \to \infty. \qquad (2.2.131)$$

2.2.4.3. Application of the ε-algorithm for extrapolation in numerical integration

The quadrature error expansions for singular functions in the foregoing section, have been formulated for the offset trapezoidal rule. However, through (2.2.94), the expansions for $Q^{(m)}$ where Q is a quadrature sum exact for constants and with equal weights at the end-points, are analogous, except for the fact that some terms may vanish because of symmetry in Q or because of its polynomial degree. So we can consider a sequence of rules $\{R^{[m_\ell,\alpha]}\}_\ell$ without loss of generality. For our applications we may also restrict ourselves to a geometric sequence of mesh ratios $m_\ell = \lambda^{-\ell}$.

If $f \in C^{p+1}(0,1)$ but has possibly algebraic end-point singularities, its quadrature error expansion can be written in the form (2.2.129), with

$$\tilde{I} = I[0,1]f + O(\lambda^{\ell(p+1)}) \qquad (2.2.132)$$

and

$$1 > \lambda_1 > \ldots > \lambda_w > \lambda^{p+1}. \qquad (2.2.133)$$

So the convergence of the (2n)-th column of the ε-table ($n < w$) is as specified by (2.2.131). Note that with modified Romberg integration the same convergence would be observed in the n-th column.

In Tables 2.10-11 we treat the integral $(1/100)\ I[0,1]x^{-99/100}$, which is approximated with 2^ℓ-panels of the 2-point Gauss rule. The values of $|T_{\ell,2n}-I|$ are listed in Table 2.10.

Table 2.11 gives the ratios

$$q_{\ell,2n} = \frac{|T_{\ell+1,2n}-I|}{|T_{\ell,2n}-I|}. \tag{2.2.134}$$

In view of the asymptotic expansion

$$T_\ell \sim I + u_1\lambda^{\ell/100} + u_2\lambda^{4\ell} + u_3\lambda^{6\ell} + u_4\lambda^{8\ell} + \dots \tag{2.2.135}$$

with $\lambda = 1/2$, we have that $q_{\ell,0} \to (1/2)^{1/100}$, $q_{\ell,2} \to (1/2)^4, \dots$ as $\ell \to \infty$.

ℓ \ n	0	1	2
0	0.970		
1	0.964	0.187	
2	0.957	$0.135\ 10^{-1}$	$0.250\ 10^{-3}$
3	0.950	$0.929\ 10^{-3}$	$0.404\ 10^{-5}$
4	0.944	$0.610\ 10^{-4}$	
5	0.937		

Values of $|T_{\ell,2n}-I|$

Table 2.10

ℓ \ n	0	1	2
0	0.993		
1	0.99309	0.07	
2	0.9930925	0.068	0.0161
3	0.99309249	0.065	↓
4	0.99309295	↓	↓
	↓		
	$(1/2)^{1/100}$	1/16	1/64
	(≈ 0.99309295)	(=0.0625)	(=0.015625)

Values of $q_{\ell,2n}$

Table 2.11

If logarithmic end-point singularities or algebraic-logarithmic interior singularities are present, then the integration error is asymptotically of the form (2.2.128). Consider for example $I[0,1](c|x-1/7|^{1/2})$ where c is a normalization constant such that the integral is one. The integrand has a singularity at x = 1/7 which has binary period L = 3. If we use 2^{ℓ}-panels of the mid-point rule we have

$$T_{\ell} \sim I + u_1(\ell)\lambda^{3\ell/2} + u_2\lambda^{2\ell} + u_3\lambda^{4\ell} + u_4\lambda^{8\ell} + \dots \qquad (2.2.136)$$

where u_1 is periodic in ℓ with period 3. This periodicity can be observed in Table 2.12, where the entry sequence is given with

$$q^*_{\ell,0} = \frac{|T_{\ell+3}-I|}{|T_{\ell}-I|}, \qquad \ell = 0,3,6. \qquad (2.2.137)$$

Note that $q^*_{\ell,0} \to \lambda^{3/2+3} = (1/2)^{9/2} \simeq 4.42\ 10^{-2}$.

ℓ	$R^{[2^{\ell},0]}[0,1]f$ $T_{\ell} = T_{\ell,0}$	$E^{[2^{\ell},0]}[0,1]f$	$q^*_{\ell,0}$
0	1.057656	$5.77\ 10^{-2}$	
1	0.979158	$-2.08\ 10^{-2}$	
2	0.958106	$-4.19\ 10^{-2}$	
3	1.005464	$5.46\ 10^{-3}$	$9.46\ 10^{-2}$
4	0.999934	$-6.61\ 10^{-5}$	
5	0.998381	$-1.62\ 10^{-3}$	
6	1.000302	$3.02\ 10^{-4}$	$5.53\ 10^{-2}$
7	1.000012	$1.23\ 10^{-5}$	
8	0.999932	$-6.77\ 10^{-5}$	
9	1.000014	$1.43\ 10^{-5}$	$4.74\ 10^{-2}$

Table 2.12

Table 2.13 displays the absolute errors of the $T_{\ell,2n}$ in the ε-table.

ℓ \ n	0	1	2	3	4
0	$5.77\ 10^{-2}$				
1	$2.08\ 10^{-2}$	$4.96\ 10^{-2}$			
2	$4.19\ 10^{-2}$	$2.73\ 10^{-2}$	$1.06\ 10^{-2}$		
3	$5.46\ 10^{-3}$	$5.12\ 10^{-4}$	$1.15\ 10^{-3}$	$4.96\ 10^{-6}$	
4	$6.61\ 10^{-5}$	$2.22\ 10^{-3}$	$5.15\ 10^{-4}$	$5.67\ 10^{-7}$	$2.31\ 10^{-8}$
5	$1.62\ 10^{-3}$	$7.60\ 10^{-4}$	$1.77\ 10^{-4}$	$3.82\ 10^{-8}$	$1.40\ 10^{-9}$
6	$3.02\ 10^{-4}$	$5.03\ 10^{-5}$	$3.35\ 10^{-5}$	$5.12\ 10^{-9}$	
7	$1.23\ 10^{-5}$	$9.83\ 10^{-5}$	$1.33\ 10^{-5}$		
8	$6.77\ 10^{-5}$	$2.72\ 10^{-5}$			
9	$1.43\ 10^{-5}$				

Values of $|T_{\ell,2n}-I|$

Table 2.13

Theorem 2.12 tells us that $u_1(\ell)\lambda^{3\ell/2}$ satisfies a linear difference equation with constant coefficients and of order $n = 3$. We give the $q_{\ell,2n} = q_{\ell,6}$ in Table 2.14. It emerges that $q_{\ell,6} \to 1/4$ as ℓ increases.

ℓ	$q_{\ell,6}$
3	0.979
4	0.430
5	0.345
6	0.313
7	0.293
	↓
	1/4

Table 2.14

III. Algorithm Descriptions

3.1. QUADPACK contents

We shall use the term integrator to indicate either the algorithm or the integration routine. If we want to state explicitly that the double precision version of the routine is meant, we shall put a D in front of the single precision name.

In QUADPACK the following naming conventions have been used :

- first letter of an integrator name denotes
  - Q - Quadrature routine
- second letter of an integrator name denotes
  - N - Non-adaptive integrator
  - A - Adaptive integrator
- third letter of an integrator name denotes
  - G - General purpose user defined integrand
  - W - Weight function, e.g. certain explicit integrand forms
- fourth letter of an integrator name indicates the type of integrals or integrands the routine is especially designed for :
  - S - Singularities can be more readily integrated
  - P - Points of special difficulty (like singularities) can be input
  - I - Infinite interval of integration
  - O - Oscillatory trigonometric weight functions $\cos(\omega x)$ and $\sin(\omega x)$
  - F - Fourier integrals
  - C - Cauchy principal value integrals

E - Extended parameter list for an increased flexibility (compared with the integrator without the letter E in its name).

The following list gives an overview of the QUADPACK routines :

- QNG — is a simple non-adaptive automatic integrator, based on a sequence of rules with increasing degree of algebraic precision.

- QAG,QAGE — are simple globally adaptive integrators. It is possible to choose between 6 pairs of quadrature formulae for the rule evaluation component.
  The program QAG only invokes a call to routine QAGE which has a number of input and output parameters in addition to those of QAG. The output parameters of QAGE return information about the quadrature process, e.g. lists of the subintervals and their local integral and error contributions.

- QAGS — is an integrator based on globally adaptive interval subdivision in connection with extrapolation, which will eliminate the effects of integrand singularities of several types.

- QAGP — serves the same purposes as QAGS, but also allows the user to provide explicit information about the location and type of trouble-spots i.e. the abscissae of internal singularities, discontinuities and other difficulties of the integrand function.

- QAGI — handles integration over infinite intervals. The infinite range is mapped onto a finite interval and subsequently the same strategy as in QAGS is applied.

- QAWO — is an integrator for the evaluation of

  $$\int_a^b \left\{ \begin{matrix} \cos(\omega x) \\ \sin(\omega x) \end{matrix} \right\} f(x)dx$$ over a finite interval [a,b],

  where ω and f are specified by the user. The rule evaluation component is based on the modified Clenshaw-Curtis technique.

An adaptive subdivision scheme is used in connection with an extrapolation procedure, which is a modification of that in QAGS and allows the algorithm to deal with singularities in f(x).

- QAWF calculates the Fourier transform $\int_a^\infty \left\{ \begin{matrix} \cos(\omega x) \\ \sin(\omega x) \end{matrix} \right\} f(x)dx$ for user-provided ω and f. The procedure of QAWO is applied on successive finite intervals, and convergence acceleration by means of the ε-algorithm is applied to the series of integral approximations.

- QAWS,QAWSE approximate $\int_a^b w(x)f(x)dx$, with a < b finite,

$$w(x) = (x-a)^\alpha (b-x)^\beta v(x), \quad \alpha , \beta > -1$$

where v may be one of the following functions

$$v(x) = \begin{cases} 1 \\ \log(x-a) \\ \log(b-x) \\ \log(x-a)\log(b-x) \end{cases} \tag{3.1.1}$$

The user specifies α, β and the type of the function v. A globally adaptive subdivision strategy is applied, with modified Clenshaw-Curtis integration on those subintervals which contain a or b.
The program QAWS is a driver routine for QAWSE which has an extended parameter list (compared with QAWS). The additional parameters return information about the quadrature process, like for instance lists of the subintervals and their local integral contributions and error estimates.

- QAWC,QAWCE compute $⨍_a^b f(x)/(x-c)dx$ for user-specified c and f -

the notation $⨍$ indicates that the integral must be interpreted as a Cauchy principal value integral -. The strategy is globally adaptive. Modified Clenshaw-Curtis integration is used on those intervals containing the point x = c.
The program QAWC is a driver routine for QAWCE which has an extended parameter list (compared with QAWC). These parameters return information about the quadrature process, e.g. lists of the subintervals and their local integral contributions and error estimates.

3.2. Prototype of algorithm description

Let us first make some notational conventions. The integrators compute an approximation RESULT to the integral I of a function F over an interval [A,B] :

$$I = \int_A^B F(X)\,dX, \tag{3.2.1}$$

hopefully satisfying the accuracy requirement

$$|I\text{-RESULT}| \leq TOL = \max\{EPSABS, EPSREL|I|\} \tag{3.2.2}$$

where EPSABS and EPSREL are absolute and relative error tolerances specified by the user. Moreover, every integrator attempts to provide an estimate ABSERR of the absolute error |I-RESULT|. In most practical cases the following inequalities will hold :

$$|I\text{-RESULT}| \leq ABSERR \leq TOL. \tag{3.2.3}$$

Throughout this chapter we shall use the notation E(J)F for an estimate of an integral approximation over a subinterval $J \subseteq [A,B]$.

The following scheme of a quadrature algorithm can be considered to be the top level description of a top-down design which will be refined in the following sections :

Initialize : RESULT, ABSERR, TOL.
While ABSERR > TOL : attempt to reduce ABSERR and update intermediate results.

Return.

Scheme 3.1

Here, '<u>attempt to reduce ABSERR</u>' represents the basic strategy of the algorithm. We make use of three basic strategies : non-adaptive, globally adaptive, and globally adaptive with extrapolation. The first is employed in QNG, the second one in QAG-QAGE, QAWS-QAWSE and QAWC-QAWCE and the third one in QAGS, QAGP, QAGI and QAWO.

3.3. Algorithm-schemes

3.3.1. QNG (quadrature, non-adaptive, general-purpose)

Consider the sequence $\{Q_{n_k}F\}$ where Q_{n_1} is the 10-point Gauss rule (n_1=10) and the Q_{n_k}, k = 2,3,4 are the Patterson rules related to Q_{10} (see Section 2.2.2), so that $n_k = 2n_{k-1}+1$, $2 \leq k \leq 4$. The respective error estimates E_{10}, E_{21}, E_{43} and E_{87} are computed according to Section 3.4.1. Due to the non-adaptive character, the part '<u>attempt to reduce ABSERR</u>' of the general Scheme 3.1 is replaced by '<u>compute the next quadrature sum in the sequence provided k < 4</u>'. A representation of QNG is given in Scheme 3.2.

Initialize : RESULT = Q_{21}, ABSERR = E_{21}, TOL = max {EPSABS, EPSREL |RESULT|}, n = 21.
While ABSERR > TOL and n < 87 : RESULT = Q_{2n+1}, ABSERR = E_{2n+1}, update TOL, n = 2n+1.

Return.

Scheme 3.2

3.3.2. QAG (quadrature, adaptive, general-purpose) QAGE (quadrature, adaptive, general-purpose, extended)

Let us denote the integral approximation over $J \subseteq [A,B]$ by $Q(J)F$, where Q represents the $(2k+1)$-point Kronrod rule. The user can choose between $k = 7,10,15,20,25$ or 30. The error estimate $E(J)F$ for $Q(J)F$ is normally calculated according to Section 3.4.1, as the modulus of the difference of $Q(J)F$ and the k-point Gauss quadrature sum for F. The local integral contributions and error estimates over a partition $\mathbb{P}[A,B]$ of $[A,B]$ are in each step summed to

$$\text{RESULT} = \sum_{J \in \mathbb{P}[A,B]} Q(J)F \tag{3.3.1}$$

and

$$\text{ABSERR} = \sum_{J \in \mathbb{P}[A,B]} E(J)F. \tag{3.3.2}$$

The globally adaptive strategy requires to replace '<u>attempt to reduce ABSERR</u>' in Scheme 3.1 by '<u>subdivide the interval with the largest error estimate</u>'. In QAG all subdivisions are bisections. The criteria for abnormal termination and the corresponding tests are clarified in Section 3.4.3.1-2. Scheme 3.3 gives an outline of the algorithm :

```
Initialize : RESULT = Q[A,B]F, ABSERR = E[A,B]F,
             TOL = max {EPSABS,EPSREL|RESULT|}.
While ABSERR > TOL, and the interval with the largest error
                    estimate is not too small,
                    and the maximum number of subdivision has not
                    yet been achieved,
                    and no roundoff error is detected,
      subdivide the interval with the largest error estimate,
                    and update RESULT, ABSERR, TOL.

Return.
```

Scheme 3.3

3.3.3. QAWS (quadrature, adaptive, weight function, singularities) QAWSE (quadrature, adaptive, weight function, singularities, extended)

The integrand is W(X) F(X), where the function W(X) is given by (3.1.1). The strategy is an adaptation of that in QAG. We start by bisecting the original interval [A,B] and applying modified Clenshaw-Curtis integration (using 13 and 25 points) on both halves. In all further steps modified Clenshaw-Curtis integration is used for those subintervals which have A or B as one of their end-points, and Gauss-Kronrod integration (using 7 and 15 points) is applied to all other subintervals.

3.3.4. QAWC (quadrature, adaptive, weight function, Cauchy principal value) QAWCE (quadrature, adaptive, weight function, Cauchy principal value, extended)

The integrand is W(X) F(X) where

$$W(X) = 1/(X-C) \qquad (3.3.3)$$

If $A < C < B$ the integral is to be interpreted in the sense of a Cauchy principal value.

The strategy is a modified version of that in QAG. In each step the subinterval with the largest error estimate is subdivided. Assuming that this is the interval $[C_1,C_2]$, $C_1 < C_2$, the subdivision occurs as follows :

- if $C \notin [C_1,C_2]$ then $[C_1,C_2]$ is bisected;
- if $C_1 < C \leq (C_1+C_2)/2$ then $[C_1,C_2]$ is subdivided at the point $(C+C_2)/2$;
- otherwise subdivision occurs at the point $(C+C_1)/2$.

This method is used to avoid a subdivision occurring at the abscissa C of the non-integrable singularity.

Concerning the basic rules, modified Clenshaw-Curtis integration (using 13 and 25 points) is performed whenever

$$C_1 - D < C < C_2 + D, \qquad (3.3.4)$$

where

$$D = (C_2 - C_1)/20 \qquad (3.3.5)$$

Otherwise Gauss-Kronrod integration (using 7 and 15 points) is applied.

3.3.5. QAGS (quadrature, adaptive, general-purpose, singularities)

The local integral approximations Q(J)F where J belongs to a partition $\mathbb{P}[A,B]$ of [A,B], are provided by the 21-point Kronrod rule Q. The local error estimates E(J)F are computed according to Section 3.4.1, by using Q together with the 10-point Gauss rule. Define

$$SQ := \sum_{J \in \mathbb{P}[A,B]} Q(J)F, \qquad (3.3.6)$$

$$SE := \sum_{J \in \mathbb{P}[A,B]} E(J)F. \qquad (3.3.7)$$

We will say that the algorithm has reached level ℓ if the smallest interval in the partition of [A,B] is of length $|B-A|2^{-\ell}$. During level ℓ all intervals of this particular length are called small intervals. An interval is called large when it is not small. When a small interval (with respect to the current level) is selected for bisection we say that a new level is introduced. This selection does not necessarily imply immediate bisection of the current interval, but may rather lead to a new selection, from the set of large intervals.
With the following definitions

SBE := sum of the error estimates over all large intervals, (3.3.8)
EXQ := integral approximation obtained by extrapolation, (3.3.9)
EXE := error estimate with respect to EXQ, (3.3.10)
EXTOL := max {EPSABS, EPSREL|EXQ|}, (3.3.11)

algorithm QAGS satisfies the description of Scheme 3.4.
The extrapolation is carried out by means of the ε-algorithm. For the position of the result EXQ in the extrapolation table, and the derivation of EXE from the table, see Section 3.4.2.

```
Initialize : SQ = Q[A,B]F, SE = E[A,B]F, SBE = SE,
             TOL = max{EPSABS,EPSREL|SQ|},
             EXE = ∞, EXTOL = TOL, level = 0.
While SE > TOL
      and EXE > EXTOL,
      and the interval selected for subdivision is not too small,
      and the maximum number of subdivisions has not yet been
      reached,
      and no roundoff error is detected,
      and further calculations are still expected to yield
      improvement,
   select the interval with the largest error estimate for subdivi-
   sion :
   - if no new level is introduced, subdivide the current interval
     and update SQ, SE, SBE, TOL;
   - otherwise :
     * while SBE > EXTOL, and no roundoff error is detected over the
             large intervals, subdivide that large interval which
             has largest error, and update SQ, SE, SBE, TOL;
     * extrapolate, and update EXQ, EXE, EXTOL, SBE(=SE),
             level = level+1.
IF EXE/|EXQ| > SE/|SQ|, set RESULT = SQ, ABSERR = SE;
        otherwise test for divergence of the integral
        and set RESULT = EXQ, ABSERR = EXE.

Return.
```

Scheme 3.4

3.3.6. QAGI (quadrature, adaptive, general-purpose, infinite interval)

The infinite range of integration is transformed to (0,1]

$$\int_A^{\pm\infty} F(X)dX = \pm \int_0^1 F(A \pm (1-T)/T)T^{-2}\, dT, \qquad (3.3.12)$$

if A is finite.

Otherwise the following substitution is used

$$\int_{-\infty}^{\infty} F(X)dX = \int_{0}^{\infty} (F(X) + F(-X))dX$$

$$= \int_{0}^{1} (F((1-T)/T) + F((-1+T)/T))T^{-2} \, dT. \qquad (3.3.13)$$

For integrating over the unit range the strategy of QAGS is used. However, the basic rules are different : the integral approximation $Q(J)F$ over subinterval $J \subseteq (0,1]$ is now obtained with the 15-point Kronrod rule. The error estimate $E(J)F$ is obtained by using Q together with the 7-point Gauss rule. In QAGI rules of lower degrees are used than in QAGS, since it is likely, due to the transformation that a singularity occurs in the transformed integrand at the point 0, so little would be gained from using a formula of higher degree.

3.3.7. QAGP (quadrature, adaptive, general-purpose, points of singularity)

The algorithm is similar to QAGS. The definition of level is, however, slightly different. Assume that N break-points of the interval $[A,B]$, $A < B$, are provided by the user :

$$A < P_1 < P_2 < \dots < P_N < B \qquad (3.3.14)$$

and define

$$P_0 = A, \; P_{N+1} = B. \qquad (3.3.15)$$

The first integral approximation is computed by splitting up the interval $[A,B]$ and summing the contributions from all subintervals $[P_k,P_{k+1}]$, $k = 0,1,\dots,N$. From now on any subdivision is a bisection. The definition of small and large subintervals, however, is different. During the execution of level $\ell \geq 0$ (see Section 3.3.5, QAGS), the smallest interval in the partition of $[P_k,P_{k+1}]$ has length $(P_{k+1}-P_k)2^{-\ell}$. Subintervals of this length are called small, the other ones large.

3.3.8. QAWO (quadrature, adaptive, weight function, oscillating)

The integrand is $W(X)\ F(X)$, where $W(X)$ is $\cos(\omega X)$ or $\sin(\omega X)$, and ω is specified by the user, as well as F on [A,B].

An extrapolation strategy is followed which is a modification of that in QAGS. Assume that a subinterval has length

$$L = |B-A|2^{-\ell}. \tag{3.3.16}$$

If

$$L\omega > 4, \tag{3.3.17}$$

the integration over this subinterval is performed by means of a modified 25-point Clenshaw-Curtis procedure, and the error estimate is computed from this approximation together with the result of the 13-point formula (see Section 3.4.1). If condition (3.3.17) is not satisfied the 7/15-point Gauss-Kronrod integration is carried out.

In the QAWO algorithm level (see Section 3.3.5, QAGS) is initialized to be zero, and a new level is introduced such as in QAGS but only with respect to the intervals where Gauss-Kronrod integration is performed.

3.3.9. QAWF (quadrature, adaptive, weight function, Fourier)

The integrand is $W(X)\ F(X)$ where $W(X)$ is $\cos(\omega X)$ or $\sin(\omega X)$, ω and F being specified by the user. The integration interval is semi-infinite : $[A,\infty)$ where A is a finite, user-specified limit.

Over successive intervals

$$C_k = [A+(k-1)c,\ A+kc],\ k = 1,2,\ldots, \tag{3.3.18}$$

integration is performed by means of the integrator QAWO.
The intervals C_k are of constant length

$$c = (2[|\omega|]+1)\ \pi/|\omega|, \tag{3.3.19}$$

where $[|\omega|]$ represents the largest integer which is still smaller than or equal to $|\omega|$. Since c equals an odd number of half periods, the integral contributions over succeeding intervals will alternate in sign when the function F is positive and monotonically decreasing over $[A,\infty)$. For convergence acceleration of the series formed by the integral contributions we use the ε-algorithm of Wynn (1956), which is indeed suitable for summing alternating series (Smith and Ford (1979)).

In contrast to the other codes, the integrator QAWF works only with a user-specified absolute error tolerance (EPSABS). Over the interval C_k it attempts to satisfy the accuracy requirement

$$TOL_k := u_k \text{ EPSABS} \tag{3.3.20}$$

with

$$u_k := (1-p)p^{k-1}, \qquad k = 1,2,\ldots, \tag{3.3.21}$$

and p = 9/10. However, when difficulties occur during the integration within some subinterval and are flagged by QAWF, the accuracy requirement within each of the following intervals C_i is relaxed, i.e. the tolerance over C_i is set to

$$TOL_i := u_i \max \{\text{EPSABS}, \max_{1\leqslant k\leqslant i-1} E(C_k)F\}. \tag{3.3.22}$$

3.4. Heuristics used in the algorithms

3.4.1. Local error estimates

The estimate E(J)F of the absolute error on a subinterval $J \subseteq [A,B]$ is based on D(J)F, the modulus of the difference of Q(J)F and some integral approximation of lower degree.

- If the modified Clenshaw-Curtis method is used for the integration, the absolute error estimate on this subinterval is set to

$$E(J)F := D(J)F \tag{3.4.1}$$

- If the integration is performed using the Gauss-Kronrod procedure we also need

$$\text{RESABS} :\simeq \int_J |F(X)|\,dX \tag{3.4.2}$$

and

$$\text{RESASC} :\approx \int_J |F(X)-M(J)F|\, dX, \tag{3.4.3}$$

where M(J)F is an approximation to the mean value of the integrand on J :

$$M(J)F = Q(J)F/\text{length of } J. \tag{3.4.4}$$

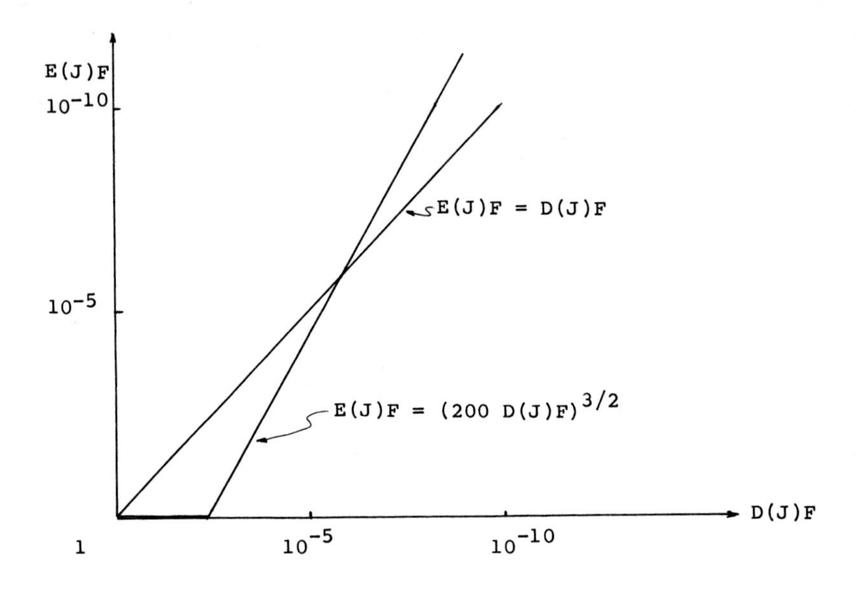

Figure 3.1

The error estimate is then set according to the following formula

$$E(J)F = \text{RESASC}\times\min\,\{1,(200D(J)F/\text{RESASC})^{3/2}\}, \tag{3.4.5}$$

but is increased if it appears to be below the attainable accuracy limit imposed by the relative machine precision.

The power 3/2 has been determined experimentally. In Figure 3.1 the error estimate (3.4.5) for an integral with RESASC = 1 is compared with the classical estimate E(J)F = D(J)F. The advantage of (3.4.5) is that E(J)F is less pessimistic for small values of D(J)F and more reliable for large D(J)F.

3.4.2. Implementation of the ε-algorithm : subroutine QEXT

The triangular ε-table is computed from the input sequence of quadrature results positioned in the first column, numbered zero. Whenever a new element is added to the first column, a new lower diagonal can be computed, from which a new extrapolated integral approximation is derived.
Instead of carrying out the full algorithm, a condensed version of the ε-table is computed. Only the elements needed for the computation of the next lower diagonal are preserved, and only the even-numbered columns are explicitly calculated (Rutishauser (1967)).

Let ε_0, ε_1, ε_2, ε_3 be the 4 elements in the table on which the computation of a new element ε is based :

$$\begin{array}{ccc} & \varepsilon_0 & \\ \varepsilon_3 & \varepsilon_1 & \varepsilon \\ & \varepsilon_2 & \end{array}$$

Scheme 3.5

For each element ε in the new lower diagonal an error estimate

$$e := |\varepsilon_0 - \varepsilon_1| + |\varepsilon_1 - \varepsilon_2| + |\varepsilon_2 - \varepsilon|$$

is computed. The element ε in the lower diagonal with the smallest error estimate e is returned as the new extrapolated value

$$EXQ := \varepsilon, \tag{3.4.6}$$

together with the error estimate

$$EXE := |\varepsilon - r_1| + |\varepsilon - r_2| + |\varepsilon - r_3| \tag{3.4.7}$$

where r_1 is the last, r_2 the 2nd last and r_3 the 3rd last value of EXQ.

In order to avoid low-accuracy problems no extrapolated values are accepted until at least 5 elements are present in the entry sequence.

Caution is also taken with regard to the possible occurrence of irregular behaviour in the table (Wynn (1966)). If a large element (pole) is detected then all upper diagonals including the one with the pole are deleted (see Figure 3.2). Furthermore, if the input sequence exceeds limexp = 50 elements, then the first upper diagonal is deleted at each call, so that the table is kept of constant size from then on.

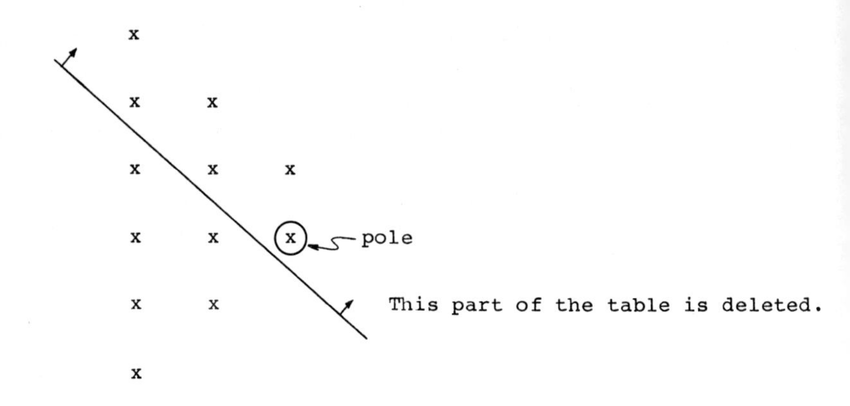

Figure 3.2

3.4.3. Criteria for abnormal termination, and test for divergence of the integral

In the adaptive algorithms some or all of the following termination criteria are implemented : the algorithm terminates whenever

- roundoff error is detected,
- a too small interval must be subdivided,
- further calculations are not expected to yield any improvement,
- an upper bound for subdivisions has been reached.

Only the first three termination criteria need further explanation, which will be given in the following subsections. A test for divergence of the integral is only carried out in algorithms which use extrapolation (see 3.4.3.4).

3.4.3.1. Roundoff error

We denote the given integration interval by [A,B], the absolute and relative error tolerances by EPSABS and EPSREL respectively, and the relative machine accuracy by e_m.

The first circumstance under which an adaptive routine will flag roundoff error and stop, occurs if the inequality

$$100e_m Q[A,B]\,|F| \geq E[A,B]F > \max\{EPSABS, EPSREL|Q[A,B]F|\} \qquad (3.4.8)$$

is satisfied after the initial step of the integration process which is performed over the whole interval [A,B]. Using this test we terminate early in cases where the exact integral is zero and a pure relative accuracy is requested (EPSABS=0). Consequently such problems are handled with a minimum number of integrand evaluations.

Let us assume now that the (sub)interval J is subdivided into J' and J". Basically we test for the occurrence of

(1) $E(J')F + E(J")F > (1-\delta_1)E(J)F$ (3.4.9)
together with
$1-\delta_2 < Q(J)F/(Q(J')F + Q(J")F) < 1+\delta_2$ (3.4.10)

or

(2) $E(J')F + E(J")F > (1-\delta_1)E(J)F$ (3.4.11)
together with total number of subdivisions $\geq$ 10. (3.4.12)

The variables δ_1 and δ_2 have heuristically chosen values : $\delta_1 = 0.01$ and $\delta_2 = 0.0001$.
The occurrence of 10 times (1) or 20 times (2) results into flagging roundoff, and termination of the computations. Note however that, in the algorithms QAWS and QAWC, (1) and (2) are only tested in the case where Gauss-Kronrod integration is applied on both subintervals J' and J".

In all integrators using extrapolation an additional test is performed. If (1) is satisfied 5 times after the introduction of a new level, (cf. Scheme 3.4) the algorithm decides that roundoff error is present on the large intervals. This decision will result in a change of the basic strategy of the algorithm : in all further steps the interval with the largest error estimate is selected for the next subdivision.

3.4.3.2. Too small subinterval

After subdividing the interval $J = [A_1,B_2]$ into $J' = [A_1,A_2]$ and $J'' = [A_2,B_2]$ it is tested if the following inequality is satisfied :

$$\max\{|A_1|,|B_2|\} \leqslant (1+100e_m)(|A_2|+10^3u), \tag{3.4.13}$$

where e_m denotes the relative machine accuracy and u is the smallest positive machine number.
If (3.4.13) is satisfied the routines flag local difficulties and terminate, because the length of interval J is getting too small for further subdivision.

3.4.3.3. Roundoff in extrapolation table

If the extrapolated result EXQ provided by the extrapolation routine QEXT (Section 3.4.2) cannot be improved by 6 successive extrapolations, although after the 6th extrapolation step the error estimate EXE satisfies

$$\text{EXE} < 10^{-3} \sum_J E(J)F, \tag{3.4.14}$$

it is decided that further calculations would not yield any improvement. This situation is flagged and the computation is terminated.

In the Fourier integrator QAWF, routine QEXT is used for convergence acceleration of the sequence formed by the partial sums of the integral contributions over successive intervals. A flag indicating roundoff in the extrapolation table is set if the extrapolated result EXQ is not improved after 20 successive extrapolations even though after the 20th extrapolation step

$$\text{EXE} < 10^{-3} \left(\sum_{i=1}^{k} E(C_i)F + 50|Q(C_i)F| \right) \tag{3.4.15}$$

is satisfied, where the C_i, $i = 1,2,\ldots,k$ are the intervals over which integration has been carried out.

3.4.3.4. Test for divergence

Since the ε-algorithm has the property of making certain divergent sequences convergent, it is necessary to test for divergence of the integral in the algorithms which use extrapolation.
After the initial integration over the given interval [A,B], it is tested whether the integrand changes sign within this interval.
It is assumed that it does not change sign if the following inequality is satisfied

$$|Q[A,B]F| \geq (1-50e_m)Q[A,B]\,|F|. \qquad (3.4.16)$$

The test for divergence is executed if the extrapolation converges (i.e. an element ε of the extrapolation table is going to be returned as the approximation to the integral), and one of the following conditions is satisfied :

either
the integrand does not change sign over [A,B]
or

$$\max\{|\varepsilon|, |\sum_J Q(J)F|\} > 10^{-2}\, Q[A,B]|F|. \qquad (3.4.17)$$

Divergence or extremely slow convergence is assumed if
either

$$10^{-2} > \varepsilon/\sum_J Q(J)F \qquad (3.4.18)$$

or

$$\varepsilon/\sum_J Q(J)F > 10^{2} \qquad (3.4.19)$$

or

$$\sum_J E(J)F > |\sum_J Q(J)F| \qquad (3.4.20)$$

and accordingly the error flag is set.
It must be noted that for a divergent integral the algorithm may also terminate with any other type of flag (e.g. indicating roundoff, or local difficulties, or roundoff in the ε-table).

If the divergence flag is set, the result from the ε-table and its error estimate are returned, because these values may still be acceptable in the case of very slowly convergent improper integrals like for instance

$$\int_0^1 x^\alpha dx = 1/(\alpha+1) \tag{3.4.21}$$

for $\alpha = -0.999$. For a non-integer $\alpha < -1$, however, the ε-table converges to $1/(\alpha+1)$.

IV. Guidelines for the Use of QUADPACK

4.1. General remarks

Before providing guidelines for the use of the automatic integration package in a given situation, we shall first eliminate some occasions where the application of an automatic quadrature procedure is too expensive or wasteful or even impossible.

(i) If only one or a few integrals are involved, not much is lost by trying to apply an automatic integrator anyway and see what comes out. If, however, a problem turns out to be too difficult for an automatic solution within the required error tolerance, ways must be found for simplifying it analytically. Several techniques for doing so are demonstrated by Abramowitz (1954). In Davis and Rabinowitz' (1975) section on improper integrals special methods for dealing with singularities are illustrated, such as variable substitution and subtracting out singularities. Also Krylov (1962, pp. 202-206) discusses methods for weakening the singularity. In each case the method to be used depends heavily on the character of the integrand. A general transformation technique, for eliminating end-point singularities regardless of the type, is given by Iri, Moriguti and Takasawa (1970), and Takahasi and Mori (1974). We also found extrapolation by the ε-algorithm as applied in the algorithm QAGS (Section 3.3.5) to be successful in many circumstances. Nevertheless one still has to take care of possible roundoff error which may influence the integrand evaluation in the neighbourhood of a singularity.
Consider for example the integral

$$I = \int_0^{\pi/2} \log(1-x \, \mathrm{cotan}(x))dx, \qquad (4.1.1)$$

where the subtraction in the argument of the logarithm causes a considerable loss of significant figures in the neighbourhood of $x = 0$. In order to avoid this cancellation one could make use of the power series expansion of $x\,\text{cotan}(x)$ and, for small values of x, replace the integrand in (4.1.1) by

$$f(x) = 2\log(x) + \log(\frac{1}{3} + \frac{x^2}{45} + \frac{2x^4}{945} + \frac{x^6}{4725} + \ldots) \qquad (4.1.2)$$

(ii) Special considerations are also needed when a large number of integrals has to be computed, even if they are easy to approximate numerically. Lyness (1969) and Lyness and Kaganove (1976) discuss the implication of the expense and of the jagged performance profile of an automatic quadrature routine when applied in iterative calculations. We want to stress the expense problem, because some efficiency is always sacrificed for the sake of reliability.

If many integrals of the same type are to be computed, one could construct a suitable quadrature formula. For example, if all integrals have the same fixed weight function w which does not change sign on the integration interval, computation of the corresponding Gauss formula could be taken into consideration (Gautschi (1968), Golub and Welsch (1969)). If a weight function w depending on one or more parameters is present and the integral must be calculated for many values of these parameters, modified Clenshaw-Curtis integration is suitable, provided the interval is finite and the modified Chebyshev moments $I_w T_k$ can be determined with sufficient accuracy.

Here is not the place, however, to investigate the question whether automatic quadrature should be used or not. In the rest of this chapter we shall rather attempt to help users who already decided to use QUADPACK, with the selection of an appropriate routine or a combination of several routines for handling their particular problems.

In any case one of the first questions to be answered by the user is related to the amount of computer time he wants to spend, versus his own time he is willing to spend, for example, for a manual subdivision of the integration interval or for analytic manipulations of the integrand :

(i) The user may not care about computer time, or equivalently not be willing to do any analysis of the problem. Especially when only a very low number of integrals has to be calculated, this attitude can be perfectly reasonable. In this case it is clear that either the most sophisticated routine for finite intervals, QAGS, or its counterpart for infinite ranges, QAGI has to be used. These routines are able to cope with rather difficult (even with singular) integrand functions. This way of proceeding may be expensive. But the integrator is supposed to give a fairly reliable answer, with additional information in the case of a failure, (communicating through its error estimate and flag). Of course the programs (like any other routines based on numerical rather than symbolic integration) cannot be totally reliable, but they have passed severe tests and represent the state of the art in numerical quadrature.

(ii) The user may want to examine the integrand function :
If local difficulties occur, such as discontinuities, singularities of the integrand or its derivatives, high peaks at one or more points within the interval, the user is well advised to split up the interval at these points. A further examination of the integrand is needed on all subintervals to select a suitable integrator for each of them. If this should yield problems due to relative accuracies to be imposed on -finite- subintervals one can make use of QAGP, which must be provided (through its parameter list) with the positions of the local trouble spots. However, if strong singularities are present and a high accuracy is requested, separate applications of QAGS on the subintervals may yield a better result.
Furthermore, if the integrand is really troublesome and its analytic examination complicated, the extended routine QAGE may prove helpful for locating trouble spots of the integrand. In addition to the global integral approximation and the corresponding error estimate, QAGE returns the list of subintervals which it has produced in the subdivision process while adapting to the integrand, together with the according integral and error contributions.
For infinite intervals only the general-purpose routine QAGI is supplied. It is based on an application of the QAGS algorithm after a transformation of the original interval onto (0,1] (see 3.3.12-13). It may turn out, however, that another type of

transformation is more appropriate, or the user might prefer to break up the original interval and use QAGI only on the infinite part. These kinds of actions suggest a combined use of different QUADPACK integrators. Note that, when the only difficulty is an integrand singularity at the finite integration limit, it will in general not be necessary to break up the interval, as QAGI deals with several types of singularity at the finite boundary-point of the integration interval. It also handles slowly convergent improper integrals, provided the integrand does not oscillate over the entire infinite range. If it does it might help to sum successive positive and negative contributions to the integral -e.g. integrate between the zeroes - with one of the finite-range integrators, and apply convergence acceleration possibly by means of the routine QEXT (Section 3.4.2) which implements the ε-algorithm. The combined use of QEXT with a QUADPACK integrator is illustrated in Section 5.2. Such quadrature problems include the Fourier transform as a special case. For Fourier transforms the automatic integrator QAWF is provided in QUADPACK.

In summary, we dispose of

- QNG for well-behaved integrands,

- QAG-QAGE for functions without singularities or discontinuities, which are too difficult for QNG, and, in particular, for functions with oscillating behaviour of a non-specific type,

- QAGS, which is the general-purpose integrator,

- QAWO for functions, possibly singular, containing either the factor $\cos(\omega x)$ or $\sin(\omega x)$ with a known value of ω,

- QAWS-QAWSE for integrands with algebraic and/or logarithmic end-point singularities of known type,

- QAWC-QAWCE for Cauchy principal values,

- QAGI for integrals over an infinite interval,

- QAWF for Fourier transform integrals.

4.2. Decision tree for finite-range integration

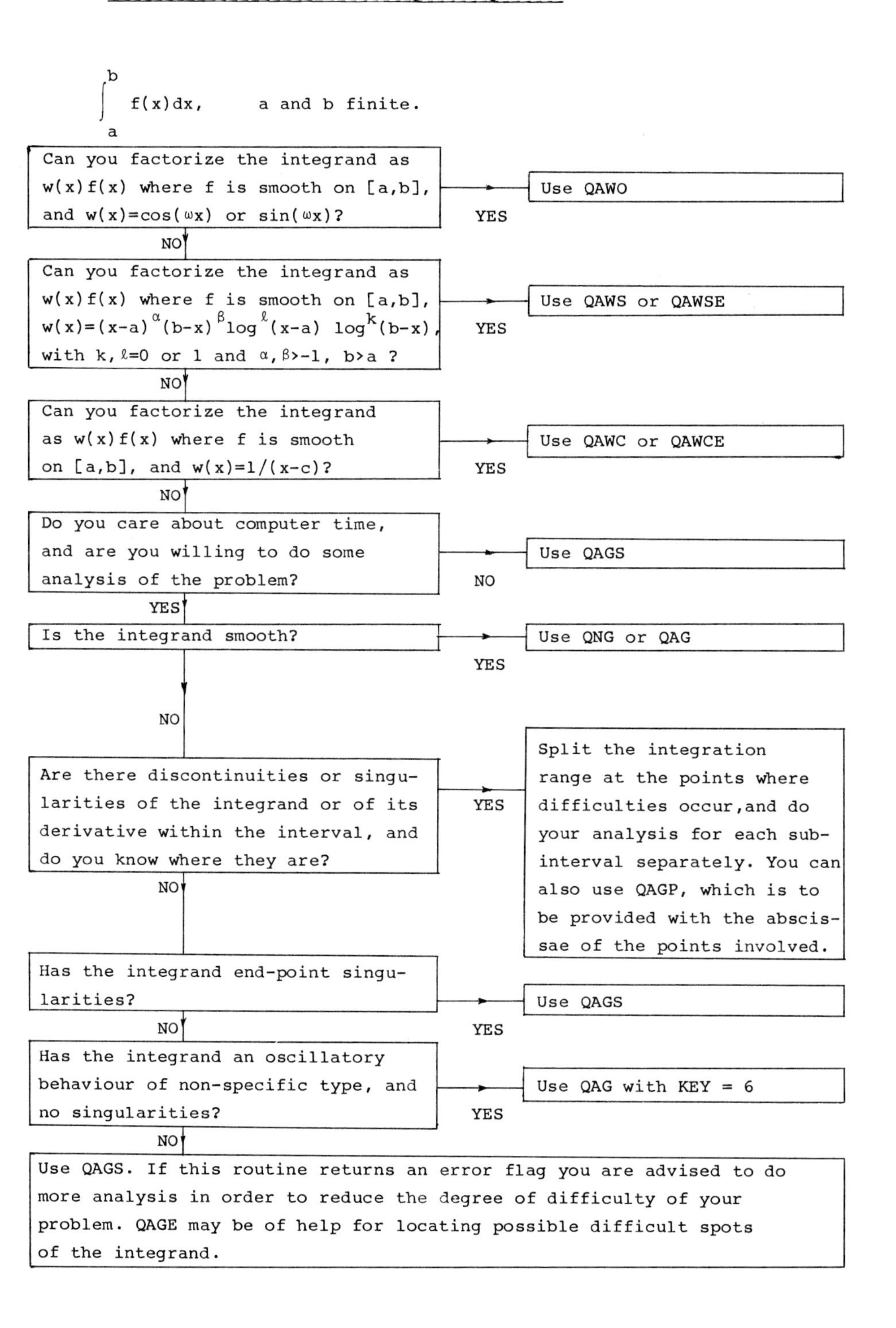

4.3. Decision tree for infinite-range integration

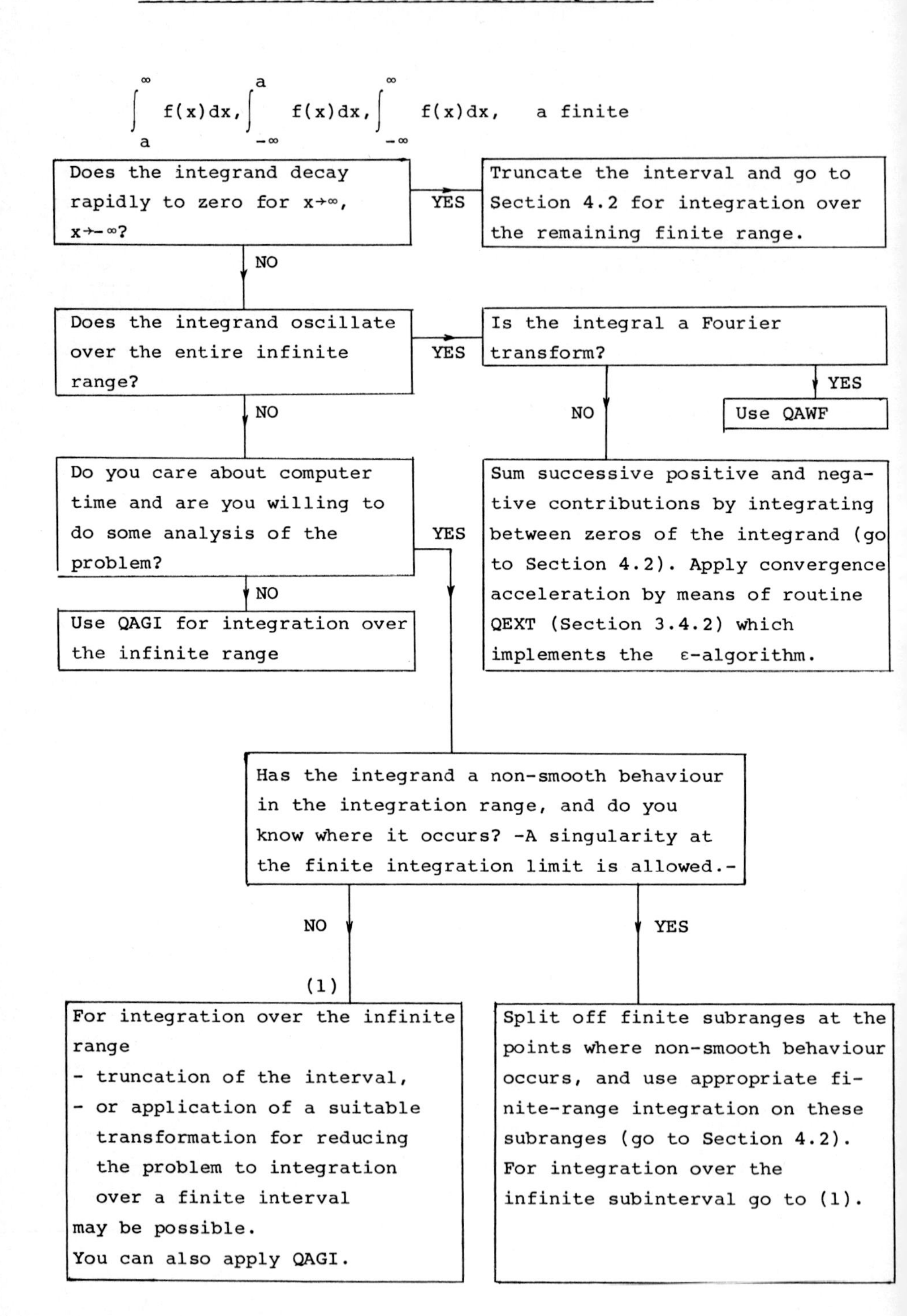

4.4. Numerical examples

The primary purpose of this section, is to give an idea of the range of applications of the QUADPACK integrators for some specific classes of problems. To this end some parameter studies will be carried out, based on the integrals listed in Table 4.1.

All examples were run on an IBM 3033, in double precision (16 digit mantissa). For the first set of test problems we required a relative accuracy of 10^{-8}, except for integral 13 where an absolute accuracy of 10^{-8} was requested, since (D)QAWF works with an absolute error tolerance only. Figures 4.1-16a,b show the actual error and the number of integrand evaluations used, as functions of the parameter . A brief interpretation of the results is given in Table 4.2, completed by Table 4.3 which explains the meaning of the error flags IER returned by the integrators. Those computations which terminated with IER$\neq$0 are indicated in the figures by means of ●, ■ or ▲.

The first three integrals are included to demonstrate, for the program DQAG, the appropriate values of the input parameter KEY, in case of a singular, peaked or oscillating integrand. KEY takes (integer) values between 1 and 6 where higher values of KEY correspond to a higher degree of accuracy of the local quadrature rule to be used. When a singularity of a peak is present, the best choice appears to be a low degree of accuracy (KEY=1) whereas a high degree of accuracy (KEY=6) is appropriate for an oscillating integrand. Any adaptive algorithm makes many subdivisions in the vicinity of a local difficulty. In the oscillating case, however, subdivisions are spread uniformly over the entire interval, and only a small number of subdivisions is needed when the quadrature rule has a fairly high degree of accuracy. The three previous examples are now repeated to compare DQAG with the best KEY value found, with DQNG and DQAGS. Considering the entire parameter set, these problems are clearly too difficult for DQNG, but it almost always yields the most efficient choice in the range where it succeeds. DQAGS is by far superior for the test function with the algebraic-logarithmic end-point singularity. However, DQAG copes much better with a high peak and is more efficient for the oscillating function. For the peaked integrand, DQAGS breaks down for $\alpha>10$ with an error code IER=5 (erroneously indicating that the integral is almost divergent).

The results for integrals 7 and 8 show that, although DQAGS deals very well with the internal singularity at a point with small binary period, DQAGP deserves preference if the singularity is situated at a point with large binary period. DQAWS is better than DQAGS for integral 9, where the integrand consists of a factor with algebraic end-point singularities and a factor which increases the effect of the singularity at $x = -1$ for increasing . Note that DQAWS is provided explicitly with the values of the powers (-1/2) characterizing the singularities at both end-points.

For integral 10 the minor manipulation of dividing out the singular factor, as indicated in Table 4.1, tailors the function for integration by DQAWS, so that DQAWS does the job not only more accurately but also more efficiently than the general-purpose DQAGS. This is due to the very smooth behaviour of the factor with which the singular part is multiplied.

If the procedure of dividing out a singularity leaves us with a factor which is still singular at the other end-point, or which shows any other difficult behaviour, we gain considerably less by using DQAWS, depending on the severity of the integrand difficulty on which no information can be supplied to the routine. This happens for integral 11.

With integral 12 we compare the performance of DQAG with KEY=6 and of DQAWO, when a factor $\sin(\omega x)$ occurs in the integrand. Naturally, the larger ω becomes the larger the gain with DQAWO.

With its extrapolation technique, DQAWO also copes with the algebraic end-point singularities involved in problem 13, without having been informed of their presence. In this circumstance the use of DQAWS is an alternative where not the oscillatory but the singular nature of the function is described to the routine. DQAWS performs best as long as $\omega = 2^{\alpha}$ is reasonably small. Then there is an α-range where DQAWO, DQAWS and DQAGS are of comparable efficiency. Finally, α gets too large to be dealt with by means of DQAWS or DQAGS, and DQAWO becomes superior.

The 14-th integration is one over an infinite interval. However, truncation of this interval is possible, because of the rapid decline of the function as x increases. The presence of cos(x) and of the begin-point singularity requires the application of DQAWF or DQAWO. DQAWF appears to be more efficient when the size of the integration interval for DQAWO ($b = 20\times 2^{\alpha}$) becomes sufficiently large.

The possibility to truncate the interval and the absence of difficult integrand behaviour make integral 15 suitable for DQAG as well as for DQAGI. In view of the length of the interval, DQAG with KEY=6 is a very good choice. Example 16 illustrates the performance of DQAGI for a rather slowly convergent improper integral with a singularity at the left end-point.

Finally, with problem 17 a Cauchy principal value integral of the Hilbert transform type is included.

So far each numerical example was treated for only one error tolerance. We now repeat problem 1 with $\alpha=-1/2$, problem 2 with $\alpha=5$ and problem 3 with $\alpha=6$, and plot the actual relative error and the number of evaluations for DQAG and for DQAGS, as functions of the tolerated relative error (Fig. 4.17, 4.18 and 4.19).

1. $\int_0^1 x^\alpha \log(1/x)dx = (\alpha+1)^{-2}$

2. $\int_0^1 4^{-\alpha}/((x-\pi/4)^2+16^{-\alpha})dx = \arctan((4-\pi)4^{\alpha-1}) + \arctan(\pi 4^{\alpha-1})$

3. $\int_0^\pi \cos(2^\alpha \sin(x))dx = \pi\, J_0(2^\alpha)$

4. see integral 1.

5. see integral 2.

6. see integral 3.

7. $\int_0^1 |x-1/3|^\alpha\, dx = ((2/3)^{\alpha+1} + (1/3)^{\alpha+1})/(\alpha+1)$

8. $\int_0^1 |x-\pi/4|^\alpha\, dx = ((1-\pi/4)^{\alpha+1} + (\pi/4)^{\alpha+1})/(\alpha+1)$

9. $\int_{-1}^1 (1-x^2)^{-1/2}/(x+1+2^{-\alpha})dx = \pi((1+2^{-\alpha})^2-1)^{-1/2}$

10. $\int_0^{\pi/2} (\sin(x))^{\alpha-1}\, dx = \int_0^{\pi/2} x^{\alpha-1}(\sin(x)/x)^{\alpha-1}\, dx = 2^{\alpha-2}(\Gamma(\alpha/2))^2/\Gamma(\alpha)$

Table 4.1

11. $\int_0^1 (\log(1/x))^{\alpha-1}\, dx = \int_0^1 (1-x)^{\alpha-1} (\log(1/x)/(1-x))^{\alpha-1}\, dx = \Gamma(\alpha)$

12. $\int_0^1 \exp(20(x-1)) \sin(2^\alpha x)dx = (20 \sin(2^\alpha) - 2^\alpha \cos(2^\alpha) +$

$+ 2^\alpha \exp(-20))/(400+4^\alpha)$

13. $\int_0^1 (x(1-x))^{-1/2} \cos(2^\alpha x)dx = \pi \cos(2^{\alpha-1})\, J_0(2^{\alpha-1})$

14. $\int_0^b x^{-1/2} \exp(-2^{-\alpha}x) \cos(x)dx = (1+\varepsilon)\sqrt{\pi}\, (1+4^{-\alpha})^{-1/4} \cos(\arctan(2^\alpha)/2)$

where $\varepsilon=0$ if $b=\infty$.
and $|\varepsilon| < 10^{-16}$ if $b=20\times 2^\alpha$

15. $\int_0^b x^2 \exp(-2^{-\alpha}x)dx = (1+\varepsilon)\, 2^{3\alpha+1}$

where $\varepsilon=0$ if $b=\infty$
and $|\varepsilon| < 10^{-14}$ if $b=40\times 2^\alpha$

16. $\int_0^\infty x^{\alpha-1}/(1+10x)^2\, dx = 10^{-\alpha}(1-\alpha)\pi/(\sin(\pi\alpha)),\ \alpha\neq 1$

$= 1/10 \qquad \alpha=1$

17. $⨍_0^5 2^{-\alpha}(((x-1)^2+4^{-\alpha})(x-2))^{-1}\, dx = (2^{-\alpha}\ln(3/2)-2^{-\alpha-1} \ln((16+4^{-\alpha})/(1+4^{-\alpha}))$

$- \arctan(2^{\alpha+2}) - \arctan(2^\alpha))/(1+4^{-\alpha})$

Table 4.1 (cont'd)

inte-gral	values of parameter α	characteristics of integral	integrators used	best integrator for this problem	failures and flags (IER) for α-values within range
1	-0.9(+0.1)0 (+0.2)2.6	end-point singularity (at x=0) of integrand or derivative	DQAG with KEY = 1,3,6	DQAG with KEY = 1 more efficient for pronounced singularity	KEY = 1,3,6: α=-0.9 (IER=3)
2	0(+1)20	integrand peak of height 4^{α} at $x = \pi/4$	DQAG with KEY = 1,3,6	DQAG with KEY = 1 more efficient	KEY=1: $\alpha \geqslant 18$ (IER=2) KEY=3,6: $\alpha \geqslant 19$ (IER=2)
3	0(+1)10	integrand oscillates more strongly for increasing α	DQAG with KEY = 1,3,6	DQAG with KEY=6 more efficient when integrand oscillates	
4	see integral 1		DQNG DQAGS DQAG (KEY=1)	DQAGS without failures, more efficient and accurate than DQAG	DQNG: $\alpha \leqslant 1.0$ (IER=1) DQAG: $\alpha = -0.9$ (IER=3)
5	see integral 2		DQNG DQAGS DQAG (KEY=1)	DQAG applicable in relatively large α-range	DQNG: $\alpha \geqslant 2$ (IER=1) DQAGS: $\alpha \geqslant 10$ (IER=5) DQAG: $\alpha \geqslant 18$ (IER=2)

Table 4.2

integral	values of parameter α	characteristics of integral	integrators used	best integrator for this problem	failures and flags (IER) for α-values within range
6	see integral 3		DQNG DQAGS DQAG (KEY=6)	DQAG more efficient than DQAGS	DQNG : $\alpha \geqslant 7$ (IER=1)
7	-0.8(+0.1)2.1	integrand singularity at internal point with small binary period (x=1/3)	DQAGS DQAGP (point of sing. supplied)	DQAGS more efficient	
8	-0.8(+0.1)2.1	integrand singularity at internal point $x = \pi/4$	DQAGS DQAGP (point of sing. supplied)	DQAGP wider applicable, more efficient and accurate	DQAGS $\alpha \leqslant -0.5$ (IER=3)
9	1(+1)20	singular integrand factor (sing. at end-points) × factor which increases the effect of the sing. as α increases.	DQAGS DQAWS (type of sing. at end-points supplied)	DQAWS wider applicable, more efficient and accurate	DQAGS: $\alpha \geqslant 17$ (IER=4 or 5)

Table 4.2 (cont'd)

integral	values of parameter α	characteristics of integral	integrators used	best integrator for this problem	failures and flags (IER) for α-values within range
10	0.1(+0.1)2	integrand factor with algebraic end-point singularity (at $x=0$) $\times$ well-behaved factor	DQAGS DQAWS (end-point sing. supplied)	DQAWS more efficient and accurate	
11	0.1(+0.1)2	integrand factor with algebraic end-point singularity (at $x=1$) $\times$ singular factor	DQAGS DQAWS (algebraic end-point sing. at $x=1$ supplied)	efficiency and accuracy comparable	
12	0(+1)9	integrand factor $\sin(\omega x)$, $\omega=2^{\alpha}$, $\times$ well-behaved factor	DQAG (KEY=6) DQAWO	DQAWO more efficient for large α	
13	0(+1)8	integrand factor $\cos(\omega x)$, $\omega=2^{\alpha}$, $\times$ factor with algebraic end-point singularity	DQAGS DQAWO (ω supplied) DQAWS (end-point sing. supplied)	DQAWS more efficient and accurate (efficiency of all 3 integrators comparable for $\alpha=7,8$)	DQAGS: $\alpha=4$ (IER=5)

Table 4.2 (cont'd)

integral	values of parameter α	characteristics of integral	integrators used	best integrator for this problem	failures and flags (IER) for α-values within range
14	0(+1)6	infinite interval, integrand factor cos(x) × factor with end-point singularity × factor which tends to zero rapidly	DQAWF (ω=1 supplied) DQAWO	DQAWF more efficient for $\alpha \geqslant 6$ and more accurate	
15	0(+1)5	infinite interval, well-behaved integrand which tends to zero rapidly	DQAG (KEY=6) DQAGI	DQAG more efficient	
16	0.1(+0.1)1.9	slowly convergent integral over infinite interval, integrand with end-point singularity	DQAGI		
17	0(+1)10	Cauchy principal value integral	DQAWC		

Table 4.2 (cont'd)

flag IER	meaning of flag
0	normal exit
1	maximum number of function evaluations has been achieved; for the adaptive routines this number is determined by a limit on the number of interval subdivisions, set within the integrator.
2	occurrence of roundoff error (makes further improvements of the already reached accuracy impossible)
3	local difficulty in integrand behaviour
4	roundoff error on extrapolation
5	divergent integral (or slowly convergent integral)
6	invalid input parameters
7	limiting number of cycles attained ((D)QAWF only)

Table 4.3

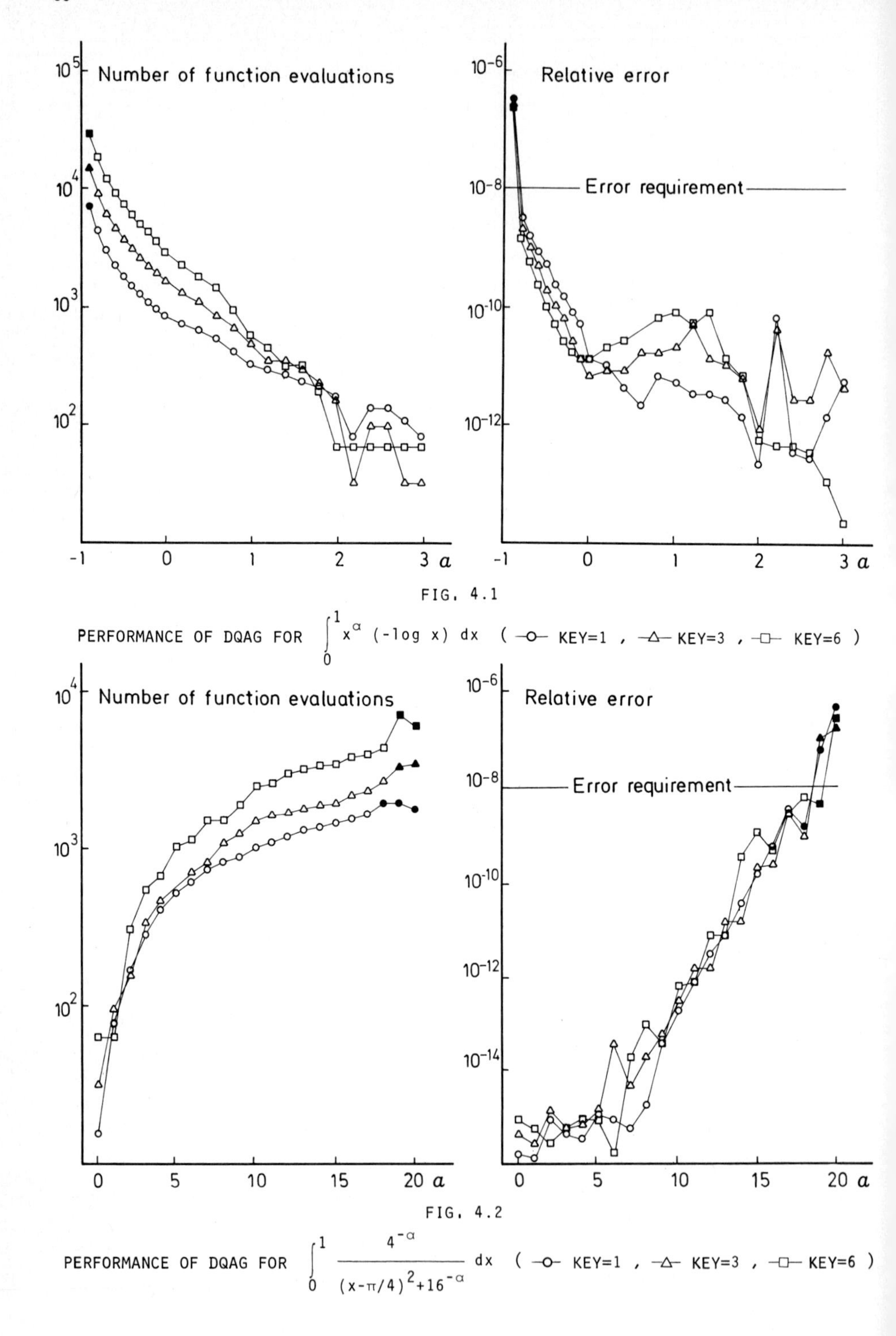

FIG. 4.1

PERFORMANCE OF DQAG FOR $\int_0^1 x^{\alpha} (-\log x)\, dx$ (—○— KEY=1 , —△— KEY=3 , —□— KEY=6)

FIG. 4.2

PERFORMANCE OF DQAG FOR $\int_0^1 \frac{4^{-\alpha}}{(x-\pi/4)^2+16^{-\alpha}}\, dx$ (—○— KEY=1 , —△— KEY=3 , —□— KEY=6)

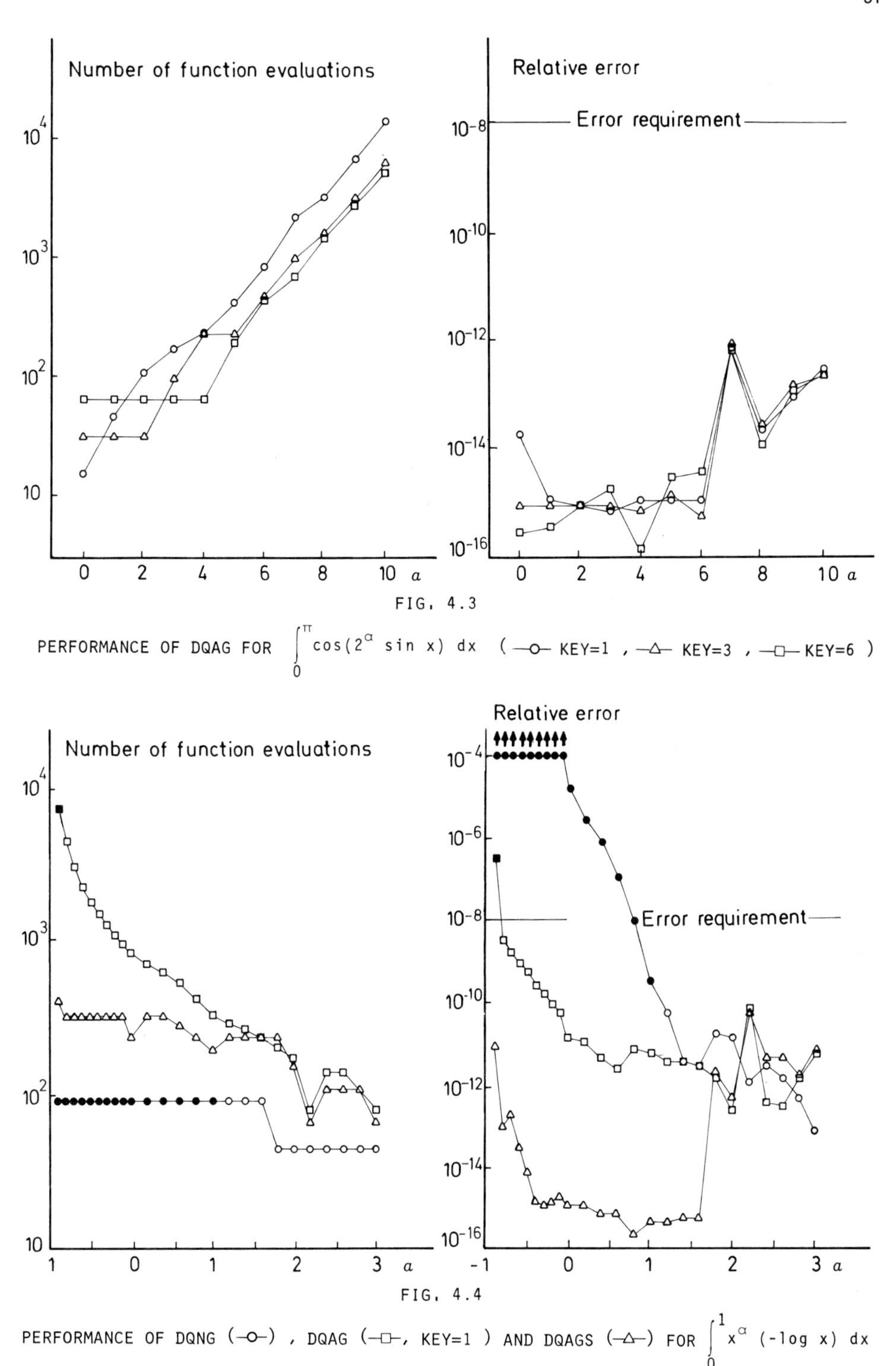

FIG. 4.3

PERFORMANCE OF DQAG FOR $\int_0^{\pi} \cos(2^{\alpha} \sin x)\, dx$ (—○— KEY=1 , —△— KEY=3 , —□— KEY=6)

FIG. 4.4

PERFORMANCE OF DQNG (—○—) , DQAG (—□—, KEY=1) AND DQAGS (—△—) FOR $\int_0^1 x^{\alpha} (-\log x)\, dx$

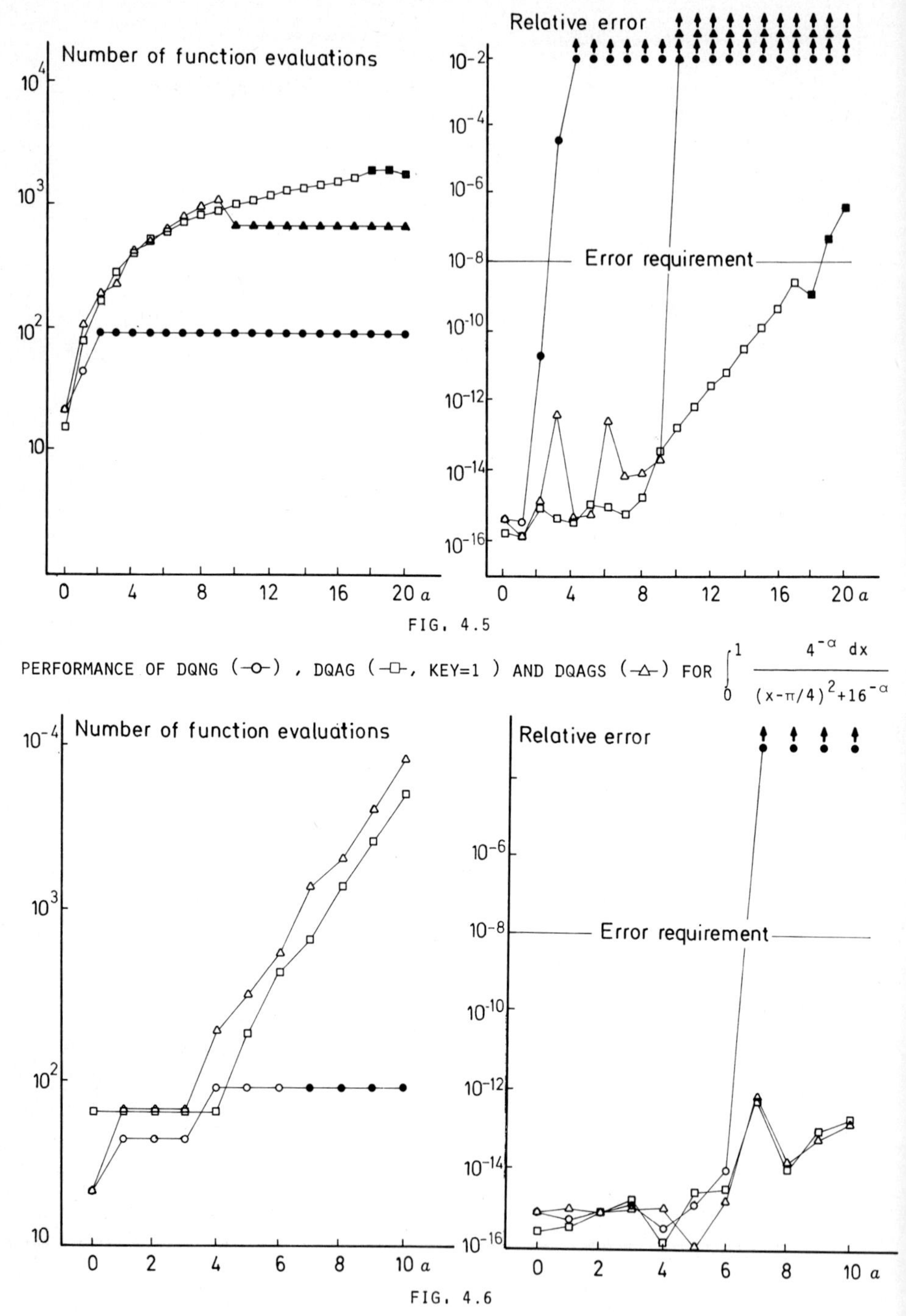

FIG. 4.5

PERFORMANCE OF DQNG (—○—), DQAG (—□—, KEY=1) AND DQAGS (—△—) FOR $\int_0^1 \frac{4^{-\alpha}\,dx}{(x-\pi/4)^2+16^{-\alpha}}$

FIG. 4.6

PERFORMANCE OF DQNG (—○—), DQAG (—□—, KEY=6) AND DQAGS (—△—) FOR $\int_0^{\pi} \cos(2^{\alpha} \sin x)\,dx$

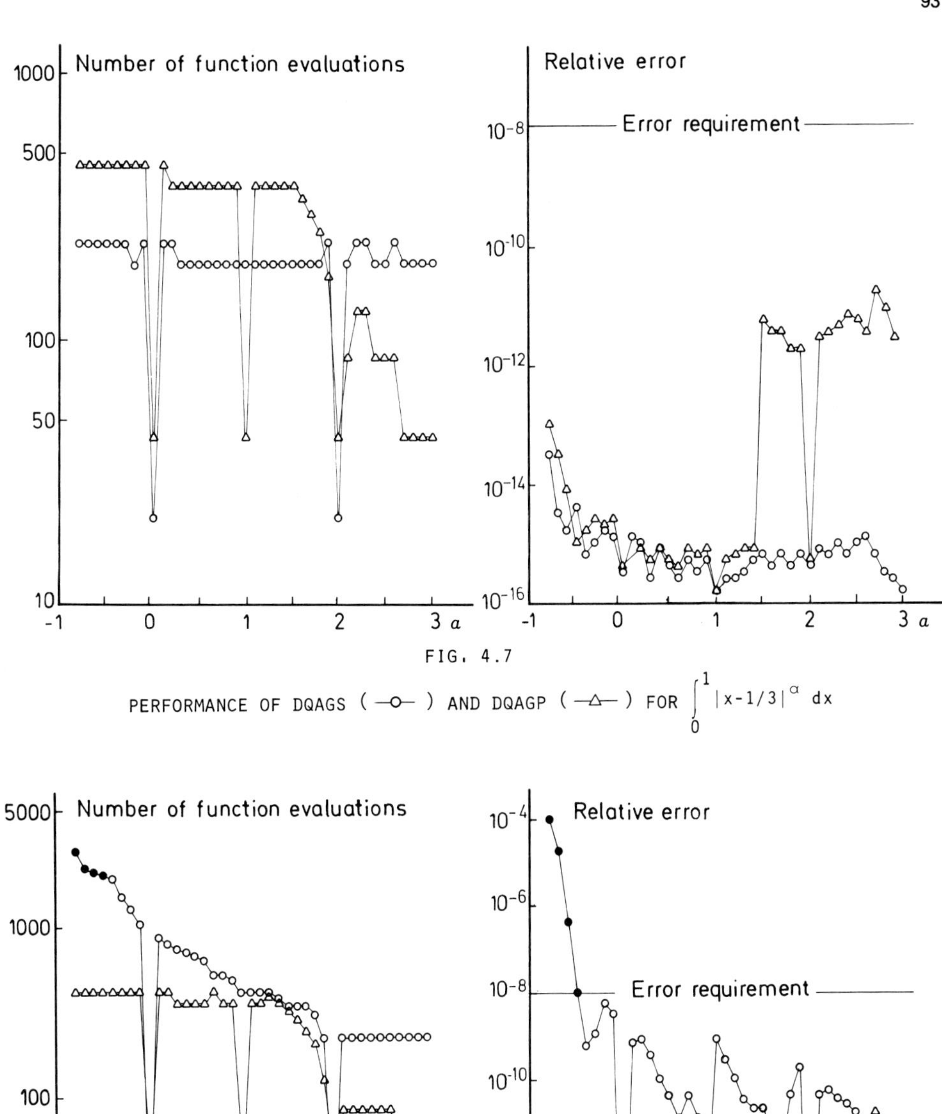

FIG. 4.7

PERFORMANCE OF DQAGS (—○—) AND DQAGP (—△—) FOR $\int_0^1 |x-1/3|^{\alpha}\, dx$

FIG. 4.8

PERFORMANCE OF DQAGS (—○—) AND DQAGP (—△—) FOR $\int_0^1 |x-\pi/4|^{\alpha}\, dx$

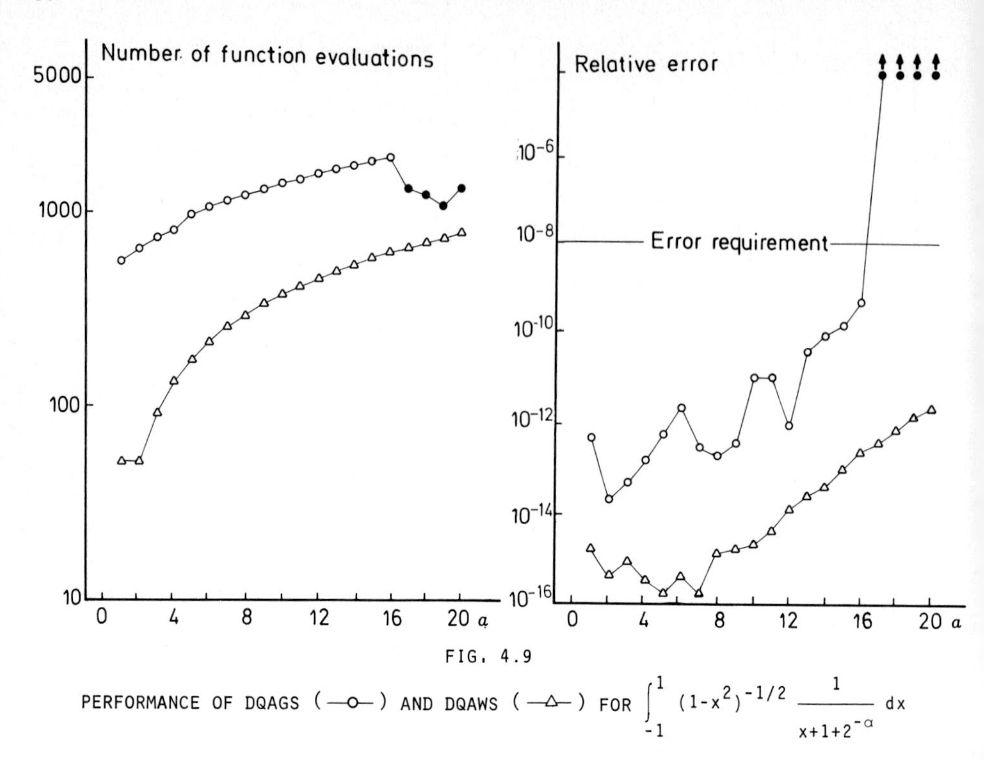

FIG. 4.9

PERFORMANCE OF DQAGS (—o—) AND DQAWS (—△—) FOR $\int_{-1}^{1} (1-x^2)^{-1/2} \frac{1}{x+1+2^{-\alpha}} dx$

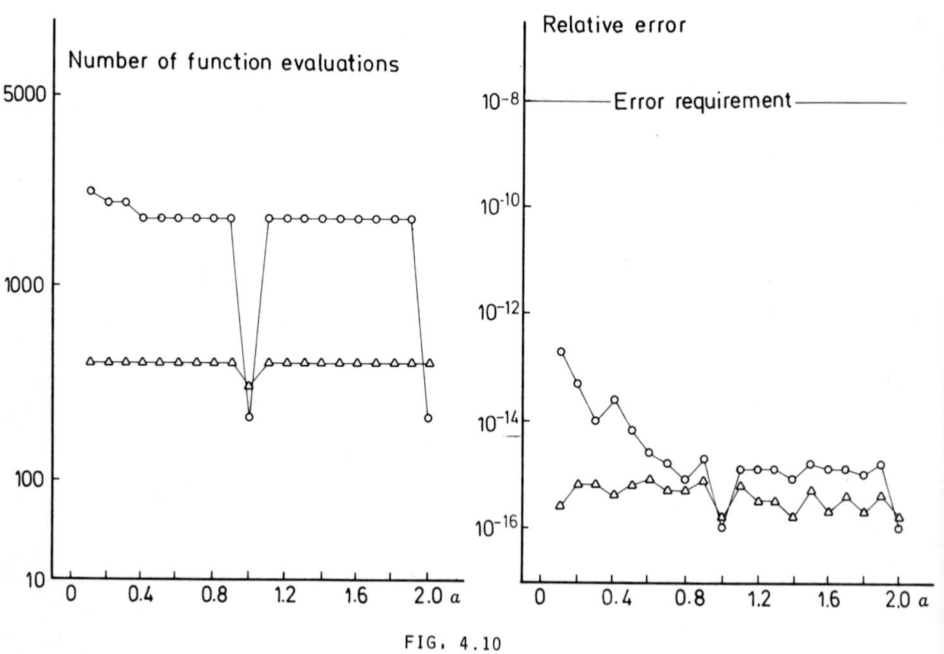

FIG. 4.10

PERFORMANCE OF DQAGS (—o—) AND DQAWS (—△—) FOR $\int_{0}^{\pi/2} (\sin x)^{\alpha-1} dx$

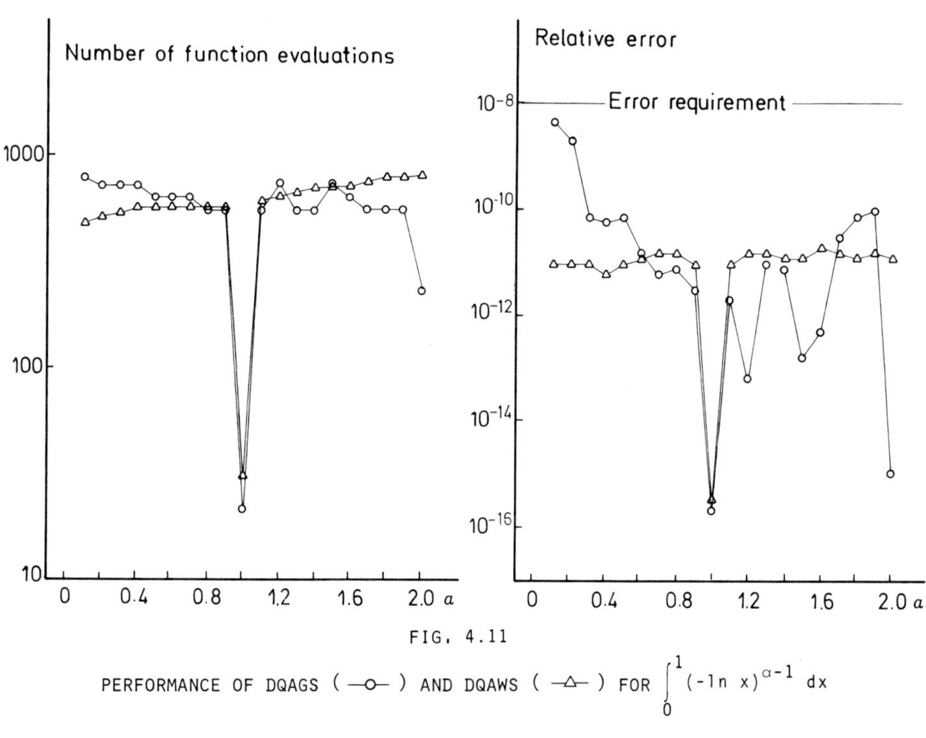

FIG. 4.11

PERFORMANCE OF DQAGS (—o—) AND DQAWS (—△—) FOR $\int_0^1 (-\ln x)^{\alpha-1}\, dx$

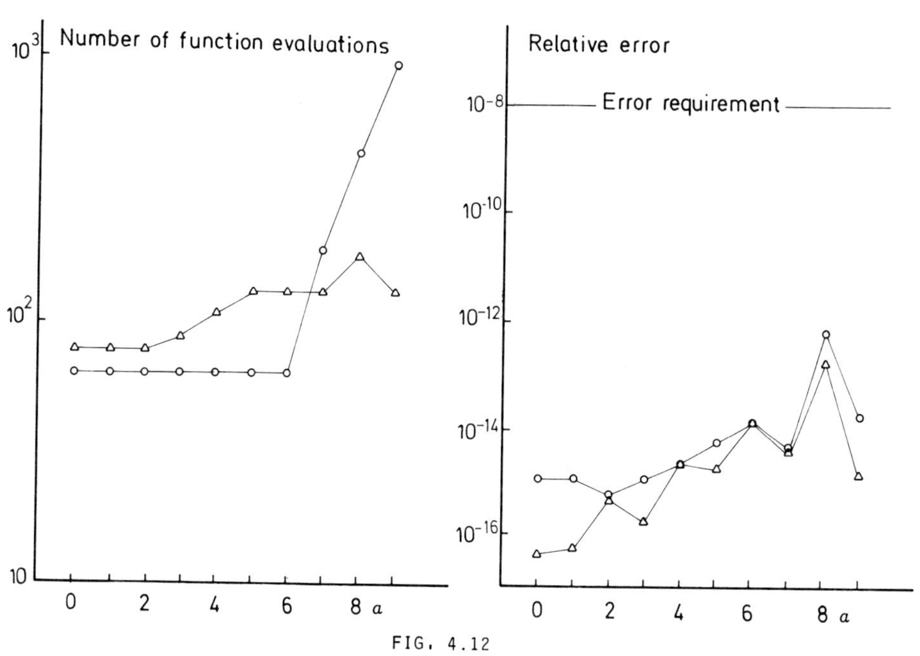

FIG. 4.12

PERFORMANCE OF DQAG (—o—, KEY=6) AND DQAWO (—△—) FOR $\int_0^1 \exp\,(20x-20)\,\sin(2^{\alpha}x)\, dx$

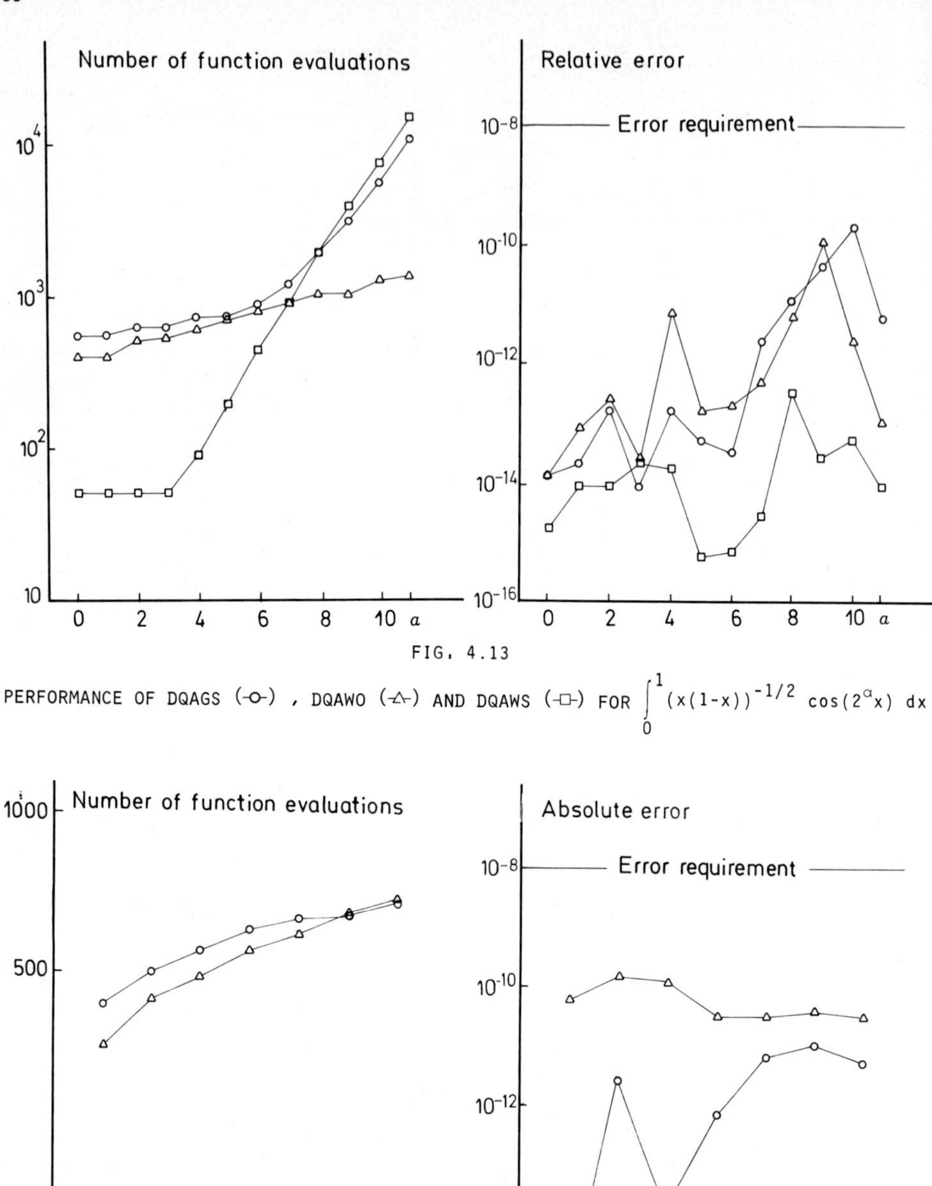

FIG. 4.13

PERFORMANCE OF DQAGS (-○-), DQAWO (-△-) AND DQAWS (-□-) FOR $\int_0^1 (x(1-x))^{-1/2} \cos(2^\alpha x)\,dx$

Number of function evaluations

Absolute error

Error requirement

FIG. 4.14

PERFORMANCE OF DQAWF (-○-) AND DQAWO (-△-) FOR $\int_0^\infty x^{-1/2} \exp(-2^\alpha x) \cos x\,dx$

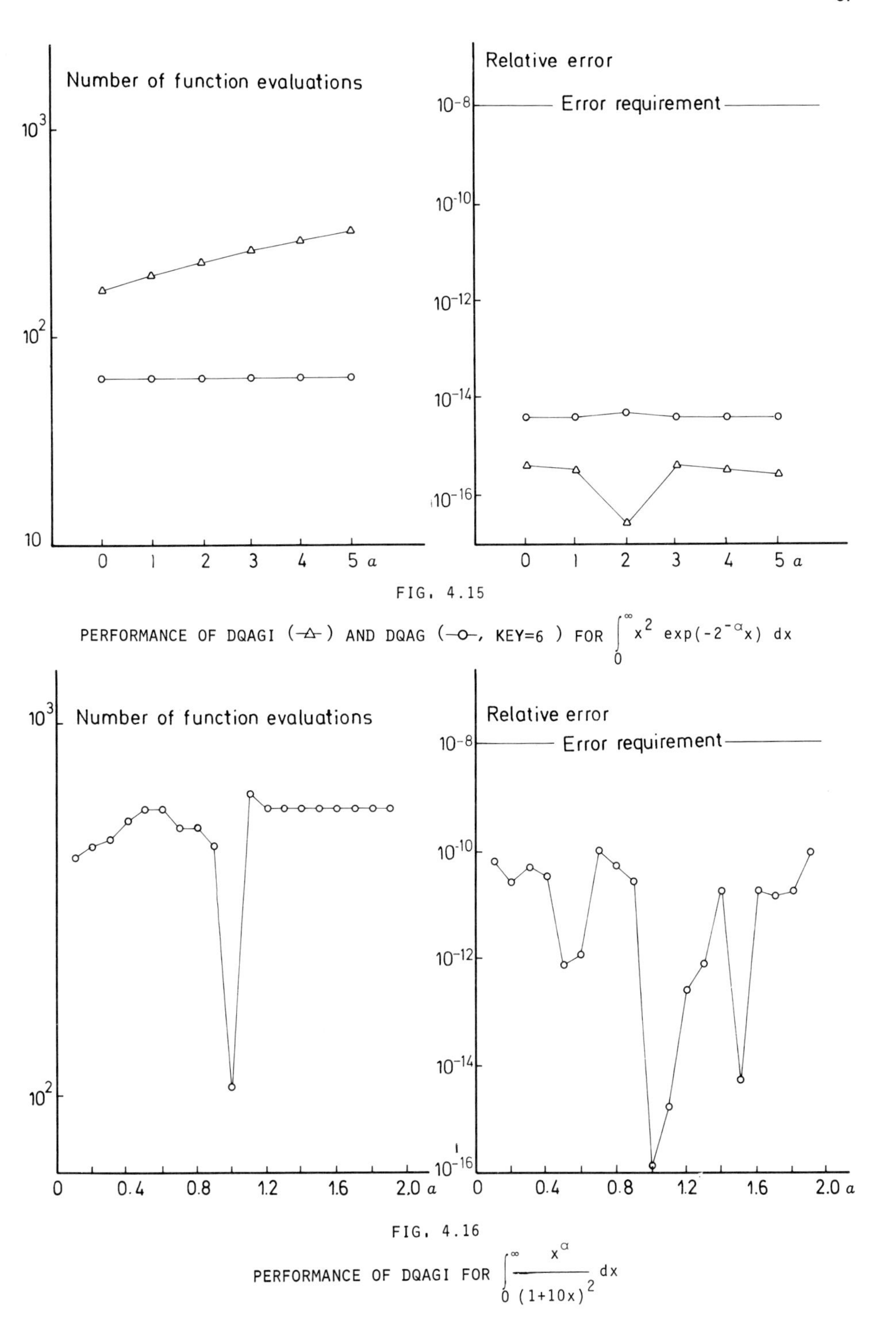

FIG. 4.15

PERFORMANCE OF DQAGI (—△—) AND DQAG (—○—, KEY=6) FOR $\int_0^\infty x^2 \exp(-2^{-\alpha}x)\,dx$

FIG. 4.16

PERFORMANCE OF DQAGI FOR $\int_0^\infty \frac{x^\alpha}{(1+10x)^2}\,dx$

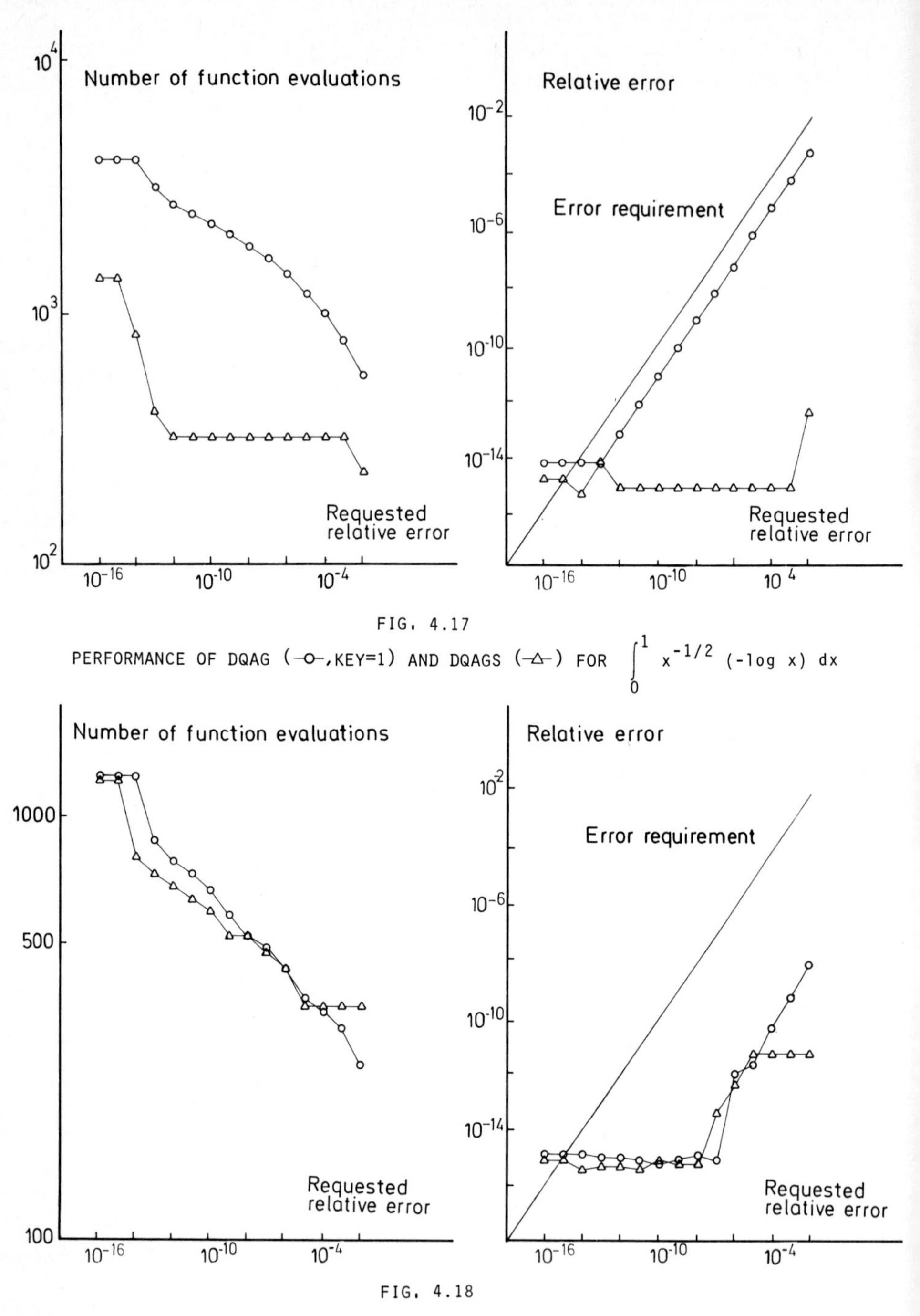

FIG. 4.17

PERFORMANCE OF DQAG (–o–, KEY=1) AND DQAGS (–△–) FOR $\int_0^1 x^{-1/2}\,(-\log x)\,dx$

FIG. 4.18

PERFORMANCE OF DQAG (–o–, KEY=1) AND DQAGS (–△–) FOR $\int_0^1 \frac{4^{-5}}{(x-\pi/4)^2+16^{-5}}\,dx$

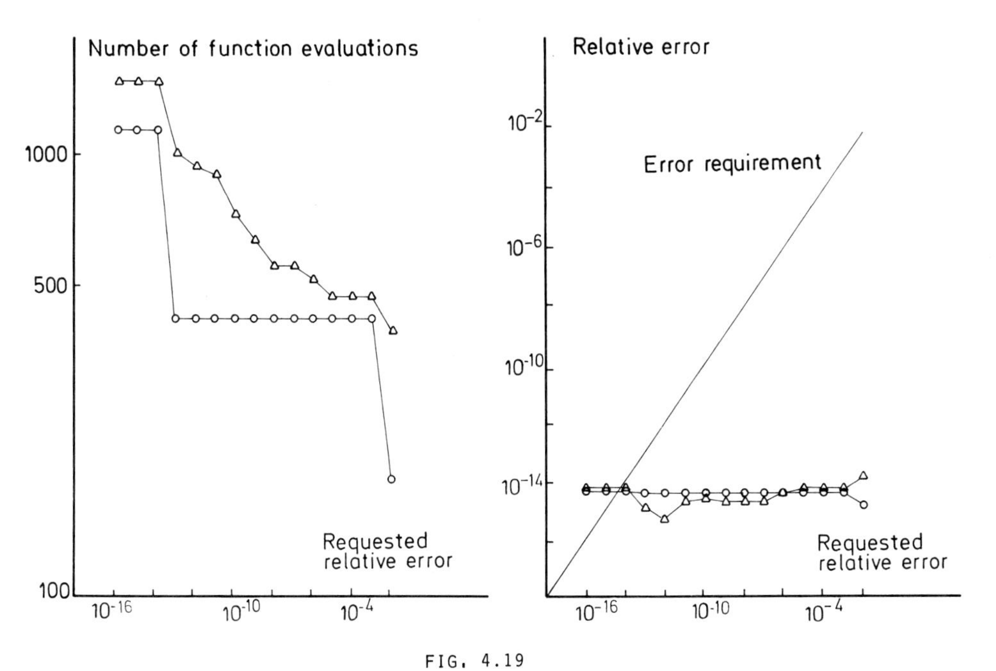

FIG. 4.19

PERFORMANCE OF DQAG (-o-, KEY=6) AND DQAGS (-Δ-) FOR $\int_0^1 \cos(2^6 \sin x)dx$

4.5. Sample programs illustrating the use of the QUADPACK integrators

Preliminary note

If the function f in $\int_a^b w(x)f(x)dx$ is not defined at one or more points of the integration interval (e.g. because of the presence of a singularity), care should be taken with regard to the coding of the FUNCTION-subprogram. Preferably the function value should be set equal to the limit of the function. Whenever this limit does not exist or is infinite, the value zero should be assigned. To this respect, the subroutines QNG, QAG, QAGE, QAGS, QAGP and QAGI do not evaluate the integrand function at the end-points of the integration interval, and QAGP does not evaluate the function at the user-provided interval break-points. However, the Clenshaw-Curtis rules in QAWO, QAWS, QAWSE, QAWC and QAWCE do require the integrand values at the end-points and so does QAWF at the finite boundary-point.

4.5.1. Sample program illustrating the use of QNG

$$\int_0^1 \sqrt{x}\,\log(x)\,dx = 4/9$$

```
      REAL A,ABSERR,B,EPSABS,EPSREL,F,RESULT
      INTEGER IER,NEVAL
      EXTERNAL F
C   A AND B ARE THE INTEGRATION LIMITS
      A = 0.0E0
      B = 1.0E0
C   EPSABS AND EPSREL DETERMINE THE ACCURACY REQUIREMENT
      EPSABS = 0.0E0
      EPSREL = 1.0E-3
C
      CALL QNG(F,A,B,EPSABS,EPSREL,RESULT,ABSERR,NEVAL,IER)
C
C   PRINT THE RESULT
      WRITE(6,900) RESULT
      WRITE(6,901) ABSERR
      WRITE(6,902) NEVAL
      WRITE(6,903) IER
  900 FORMAT(26H INTEGRAL APPROXIMATION = ,E15.8)
  901 FORMAT(30H ESTIMATE OF ABSOLUTE ERROR = ,E9.2)
  902 FORMAT(34H NUMBER OF FUNCTION EVALUATIONS = ,I5)
  903 FORMAT(14H ERROR CODE = ,I1)
      STOP
      END


      REAL FUNCTION F(X)
      REAL X
      F = SQRT(X)*ALOG(X)
      RETURN
      END
```

Output:

```
 INTEGRAL APPROXIMATION = -0.44444460E+00
 ESTIMATE OF ABSOLUTE ERROR =  0.22E-04
 NUMBER OF FUNCTION EVALUATIONS =    87
 ERROR CODE = 0
```

4.5.2. Sample program illustrating the use of QAG

$$\int_0^{\pi} \cos(100\sin(x))dx = \pi J_0(100) \simeq 0.06278740$$

```
      REAL A,ABSERR,B,EPSABS,EPSREL,F,RESULT
      INTEGER IER,KEY,NEVAL
      EXTERNAL F
C   A AND B ARE THE INTEGRATION LIMITS
      A = 0.0E0
      B = 3.14159265E0
C   EPSABS AND EPSREL DETERMINE THE ACCURACY REQUIREMENT
      EPSABS = 0.0E0
      EPSREL = 1.0E-3
C   KEY DETERMINES THE INTEGRATION RULE USED IN QAG
      KEY = 6
C
      CALL QAG(F,A,B,EPSABS,EPSREL,KEY,RESULT,ABSERR,NEVAL,
     *  IER)
C
C   PRINT THE RESULT
      WRITE(6,900) RESULT
      WRITE(6,901) ABSERR
      WRITE(6,902) NEVAL
      WRITE(6,903) IER
  900 FORMAT(26H INTEGRAL APPROXIMATION = ,E15.8)
  901 FORMAT(30H ESTIMATE OF ABSOLUTE ERROR = ,E9.2)
  902 FORMAT(34H NUMBER OF FUNCTION EVALUATIONS = ,I5)
  903 FORMAT(14H ERROR CODE = ,I1)
      STOP
      END


      REAL FUNCTION F(X)
      REAL X
      F = COS(1.0E2*SIN(X))
      RETURN
      END
```

Output:

```
 INTEGRAL APPROXIMATION =  0.62787168E-01
 ESTIMATE OF ABSOLUTE ERROR =  0.10E-04
 NUMBER OF FUNCTION EVALUATIONS =   427
 ERROR CODE = 0
```

4.5.3. Sample program illustrating the use of QAGS

$$\int_0^1 x^{-1/2} \log(x)dx = -4$$

```
      REAL A,ABSERR,B,EPSABS,EPSREL,F,RESULT
      INTEGER IER,NEVAL
      EXTERNAL F
C    A AND B ARE THE INTEGRATION LIMITS
      A = 0.0E0
      B = 1.0E0
C    EPSABS AND EPSREL DETERMINE THE ACCURACY REQUIREMENT
      EPSABS = 0.0E0
      EPSREL = 1.0E-3
C
      CALL QAGS(F,A,B,EPSABS,EPSREL,RESULT,ABSERR,NEVAL,IER)
C
C    PRINT THE RESULT
      WRITE(6,900) RESULT
      WRITE(6,901) ABSERR
      WRITE(6,902) NEVAL
      WRITE(6,903) IER
  900 FORMAT(26H INTEGRAL APPROXIMATION = ,E15.8)
  901 FORMAT(30H ESTIMATE OF ABSOLUTE ERROR = ,E9.2)
  902 FORMAT(34H NUMBER OF FUNCTION EVALUATIONS = ,I5)
  903 FORMAT(14H ERROR CODE = ,I1)
      STOP
      END


      REAL FUNCTION F(X)
      REAL X
      F = ALOG(X)/SQRT(X)
      RETURN
      END
```

Output:

```
 INTEGRAL APPROXIMATION = -0.40000243E+01
 ESTIMATE OF ABSOLUTE ERROR =  0.25E-03
 NUMBER OF FUNCTION EVALUATIONS =   315
 ERROR CODE = 0
```

4.5.4. Sample program illustrating the uses of QAGP

$$\int_0^3 x^3 \log|(x^2-1)(x^2-2)|dx = 61 \log(2) + 77 \log(7)/4 - 27$$

$$\simeq 52.740748$$

```
      REAL A,ABSERR,B,EPSABS,EPSREL,F,POINTS,RESULT
      INTEGER IER,NEVAL,NPTS2
      DIMENSION POINTS(4)
      EXTERNAL F
C   A AND B ARE THE INTEGRATION LIMITS
      A = 0.0E0
      B = 3.0E0
C   POINTS(I),I=1,...,NPTS2-2 ARE THE ABSCISSAE WHERE A DIFFICULTY
C   OF THE INTEGRAND IS SITUATED
      NPTS2 = 4
      POINTS(1) = 1.0E0
      POINTS(2) = SQRT(2.0E0)
C   EPSABS AND EPSREL DETERMINE THE ACCURACY REQUIREMENT
      EPSABS = 0.0E0
      EPSREL = 1.0E-3
C
      CALL QAGP(F,A,B,NPTS2,POINTS,EPSABS,EPSREL,RESULT,ABSERR,
     *  NEVAL,IER)
C
C   PRINT THE RESULT
      WRITE(6,900) RESULT
      WRITE(6,901) ABSERR
      WRITE(6,902) NEVAL
      WRITE(6,903) IER
  900 FORMAT(26H INTEGRAL APPROXIMATION = ,E15.8)
  901 FORMAT(30H ESTIMATE OF ABSOLUTE ERROR = ,E9.2)
  902 FORMAT(34H NUMBER OF FUNCTION EVALUATIONS = ,I5)
  903 FORMAT(14H ERROR CODE = ,I1)
      STOP
      END


      REAL FUNCTION F(X)
      REAL X
      F = X**3*ALOG(ABS((X**2-1.0E0)*(X**2-2.0E0)))
      RETURN
      END
```

Output:

```
 INTEGRAL APPROXIMATION =  0.52740822E+02
 ESTIMATE OF ABSOLUTE ERROR =  0.16E-03
 NUMBER OF FUNCTION EVALUATIONS =   777
 ERROR CODE = 0
```

4.5.5. Sample program illustrating the use of QAGI

$$\int_0^{\infty} \log(x)/(1+100x^2)dx = - \log(10)/20 \simeq -0.3616892.$$

```
      REAL ABSERR,BOUND,EPSABS,EPSREL,F,RESULT
      INTEGER IER,INF,NEVAL
      EXTERNAL F
C   BOUND AND INF DETERMINE THE INTEGRATION LIMITS
      BOUND = 0.0E0
      INF = 1
C   EPSABS AND EPSREL DETERMINE THE ACCURACY REQUIREMENT
      EPSABS = 0.0E0
      EPSREL = 1.0E-3
C
      CALL QAGI(F,BOUND,INF,EPSABS,EPSREL,RESULT,ABSERR,NEVAL,
     *  IER)
C
C   PRINT THE RESULT
      WRITE(6,900) RESULT
      WRITE(6,901) ABSERR
      WRITE(6,902) NEVAL
      WRITE(6,903) IER
  900 FORMAT(26H INTEGRAL APPROXIMATION = ,E15.8)
  901 FORMAT(30H ESTIMATE OF ABSOLUTE ERROR = ,E9.2)
  902 FORMAT(34H NUMBER OF FUNCTION EVALUATIONS = ,I5)
  903 FORMAT(14H ERROR CODE = ,I1)
      STOP
      END


      REAL FUNCTION F(X)
      REAL X
      F = ALOG(X)/(1.0E0+1.0E2*X**2)
      RETURN
      END
```

Output:

```
 INTEGRAL APPROXIMATION = -0.36168936E+00
 ESTIMATE OF ABSOLUTE ERROR =  0.55E-05
 NUMBER OF FUNCTION EVALUATIONS =   285
 ERROR CODE = 0
```

4.5.6. Sample program illustrating the use of QAWO

$$\int_0^1 \log(x)\sin(10\pi x)\,dx = -(\gamma+\log(10\pi)-\mathrm{Ci}(10\pi))/(10\pi)$$

$$\approx -0.1281316$$

```
      REAL A,ABSERR,B,EPSABS,EPSREL,F,OMEGA,PI,RESULT
      INTEGER IER,INTEGR,NEVAL
      EXTERNAL F
C   A AND B ARE THE INTEGRATION LIMITS
      A = 0.0E0
      B = 1.0E0
C   OMEGA AND INTEGR DETERMINE THE WEIGHT FUNCTION
      PI = 3.14159265E0
      OMEGA = 10.0E0*PI
      INTEGR = 2
C   EPSABS AND EPSREL DETERMINE THE ACCURACY REQUIREMENT
      EPSABS = 0.0E0
      EPSREL = 1.0E-3
C
      CALL QAWO(F,A,B,OMEGA,INTEGR,EPSABS,EPSREL,RESULT,ABSERR,
     *  NEVAL,IER)
C
C   PRINT THE RESULT
      WRITE(6,900) RESULT
      WRITE(6,901) ABSERR
      WRITE(6,902) NEVAL
      WRITE(6,903) IER
  900 FORMAT(26H INTEGRAL APPROXIMATION = ,E15.8)
  901 FORMAT(30H ESTIMATE OF ABSOLUTE ERROR = ,E9.2)
  902 FORMAT(34H NUMBER OF FUNCTION EVALUATIONS = ,I5)
  903 FORMAT(14H ERROR CODE = ,I1)
      STOP
      END


      REAL FUNCTION F(X)
      REAL X
      F = 0.0E0
      IF(X.GT.0.0E0) F = ALOG(X)
      RETURN
      END
```

Output:

```
 INTEGRAL APPROXIMATION = -0.12813687E+00
 ESTIMATE OF ABSOLUTE ERROR =  0.75E-04
 NUMBER OF FUNCTION EVALUATIONS =   215
 ERROR CODE = 0
```

4.5.7. Sample program illustrating the use of QAWF

$$\int_0^\infty x^{-1/2} \cos(\pi x/2)dx = 1$$

```
      REAL A,ABSERR,EPSABS,F,OMEGA,PI,RESULT
      INTEGER IER,INTEGR,NEVAL
      EXTERNAL F
C   A IS THE FINITE INTEGRATION LIMIT
      A = 0.0E0
C   OMEGA AND INTEGR DETERMINE THE WEIGHT FUNCTION
      PI = 3.14159265E0
      OMEGA = 0.5E0*PI
      INTEGR = 1
C   EPSABS IS THE REQUESTED ABSOLUTE ACCURACY
      EPSABS = 1.0E-3
C
      CALL QAWF(F,A,OMEGA,INTEGR,EPSABS,RESULT,ABSERR,NEVAL,
     *  IER)
C
C   PRINT THE RESULT
      WRITE(6,900) RESULT
      WRITE(6,901) ABSERR
      WRITE(6,902) NEVAL
      WRITE(6,903) IER
  900 FORMAT(26H INTEGRAL APPROXIMATION = ,E15.8)
  901 FORMAT(30H ESTIMATE OF ABSOLUTE ERROR = ,E9.2)
  902 FORMAT(34H NUMBER OF FUNCTION EVALUATIONS = ,I5)
  903 FORMAT(14H ERROR CODE = ,I1)
      STOP
      END


      REAL FUNCTION F(X)
      REAL X
      F = 0.0E0
      IF(X.GT.0.0E0) F = 1.0E0/SQRT(X)
      RETURN
      END
```

Output:

```
 INTEGRAL APPROXIMATION =  0.99999696E+00
 ESTIMATE OF ABSOLUTE ERROR =  0.70E-03
 NUMBER OF FUNCTION EVALUATIONS =   380
 ERROR CODE = 0
```

4.5.8. Sample program illustrating the use of QAWS

$$\int_0^1 \log(x)/(1+(\ln(x))^2)^2\,dx = (\mathrm{Ci}(1)\sin(1)+(\pi/2-\mathrm{Si}(1))\cos(1))/\pi$$

$$\simeq 0.1892752$$

```
      REAL A,ABSERR,ALFA,B,BETA,EPSABS,EPSREL,F,RESULT
      INTEGER IER,INTEGR,NEVAL
      EXTERNAL F
C   A AND B ARE THE INTEGRATION LIMITS
      A = 0.0E0
      B = 1.0E0
C   ALFA, BETA AND INTEGR DETERMINE THE WEIGHT FUNCTION
      ALFA = 0.0E0
      BETA = 0.0E0
      INTEGR = 2
C   EPSABS AND EPSREL DETERMINE THE ACCURACY REQUIREMENT
      EPSABS = 0.0E0
      EPSREL = 1.0E-3
C
      CALL QAWS(F,A,B,ALFA,BETA,INTEGR,EPSABS,EPSREL,RESULT,
     *  ABSERR,NEVAL,IER)
C
C   PRINT THE RESULT
      WRITE(6,900) RESULT
      WRITE(6,901) ABSERR
      WRITE(6,902) NEVAL
      WRITE(6,903) IER
  900 FORMAT(26H INTEGRAL APPROXIMATION = ,E15.8)
  901 FORMAT(30H ESTIMATE OF ABSOLUTE ERROR = ,E9.2)
  902 FORMAT(34H NUMBER OF FUNCTION EVALUATIONS = ,I5)
  903 FORMAT(14H ERROR CODE = ,I1)
      STOP
      END


      REAL FUNCTION F(X)
      REAL X
      F = 0.0E0
      IF (X.GT.0.0E0) F = 1.0E0/(1.0E0+ALOG(X)**2)**2
      RETURN
      END
```

Output:

```
 INTEGRAL APPROXIMATION = -0.18927363E+00
 ESTIMATE OF ABSOLUTE ERROR =  0.16E-05
 NUMBER OF FUNCTION EVALUATIONS =    40
 ERROR CODE = 0
```

4.5.9. Sample program illustrating the use of QAWC

$$\int_{-1}^{5} dx/(x(5x^3+6)) = \log(125/631)/18$$

$$\approx -0.08994401$$

```
      REAL A,ABSERR,B,C,EPSABS,EPSREL,F,RESULT
      INTEGER IER,NEVAL
      EXTERNAL F
C   A AND B ARE THE INTEGRATION LIMITS
      A = -1.0E0
      B = 5.0E0
C   C IS THE PARAMETER OF THE WEIGHT FUNCTION
      C = 0.0E0
C   EPSABS AND EPSREL DETERMINE THE ACCURACY REQUIREMENT
      EPSABS = 0.0E0
      EPSREL = 1.0E-3
C
      CALL QAWC(F,A,B,C,EPSABS,EPSREL,RESULT,ABSERR,NEVAL,
     *  IER)
C
C   PRINT THE RESULT
      WRITE(6,900) RESULT
      WRITE(6,901) ABSERR
      WRITE(6,902) NEVAL
      WRITE(6,903) IER
  900 FORMAT(26H INTEGRAL APPROXIMATION = ,E15.8)
  901 FORMAT(30H ESTIMATE OF ABSOLUTE ERROR = ,E9.2)
  902 FORMAT(34H NUMBER OF FUNCTION EVALUATIONS = ,I5)
  903 FORMAT(14H ERROR CODE = ,I1)
      STOP
      END


      REAL FUNCTION F(X)
      REAL X
      F = 1.0E0/(5.0E0*X**3+6.0E0)
      RETURN
      END
```

Output:

```
 INTEGRAL APPROXIMATION = -0.89944020E-01
 ESTIMATE OF ABSOLUTE ERROR =  0.21E-05
 NUMBER OF FUNCTION EVALUATIONS =   215
 ERROR CODE = 0
```

4.5.10. Sample program illustrating the use of QAGE

Integration of

$$\int_0^1 |x^2+2x-2|^{-1/2}\, dx = \pi/2 - \arctan(2^{-1/2}) + \log(3)/3$$

$$\simeq 1.504622$$

using QAG or QAGS terminates with an error flag IER=3, which indicates that entremely bad integrand behaviour occurs at one or more points of the integration interval. In order to get information about the position of these points, we can use QAGE, which returns a list of the subintervals (obtained by the adaptive subdivision algorithm) and their local integral and error contributions. In the following program the subintervals are printed out in decreasing order of their error estimate. This output strongly indicates that there is a local difficulty in the interval (0.73199,0.73206). Indeed, in this example, a simple analysis shows that the integrand tends to infinity for $x \to -1+\sqrt{3}$. This information can be used for splitting up the integration range and integrating over $(0,-1+\sqrt{3})$ and $(-1+\sqrt{3},1)$ using QAGS, or for using QAGP, provided with the position of the singularity $-1+\sqrt{3}$.

```
      REAL A,ABSERR,ALIST,B,BLIST,ELIST,EPSABS, EPSREL,F,RESULT
      INTEGER IER,IORD,K,KEY,LAST,LIMIT,NEVAL
      DIMENSION ALIST(100),BLIST(100),ELIST(100),IORD(100),RLIST(100)
      EXTERNAL F
C   A AND B ARE THE INTEGRATION LIMITS
      A = 0.0E+00
      B = 1.0E+00
C   EPSABS AND EPSREL DETERMINE THE ACCURACY REQUIREMENT
      EPSABS = 0.0E+00
      EPSREL = 1.0E-04
C   KEY DETERMINES THE DEGREE OF THE LOCAL INTEGRATION RULE
      KEY = 1
C   LIMIT IS A BOUND ON THE NUMBER OF SUBINTERVALS
      LIMIT = 100
C
      CALL QAGE(F,A,B,EPSABS,EPSREL,KEY,LIMIT,RESULT, ABSERR,
     * NEVAL,IER,ALIST,BLIST,RLIST,ELIST,IORD,LAST)
C
C   PRINT THE RESULT
      WRITE(6,900) RESULT
      WRITE(6,901) ABSERR
      WRITE(6,902) NEVAL
      WRITE(6,903) IER
      WRITE(6,904)
C   PRINT THE SUBINTERVALS AND THEIR INTEGRAL AND
C   ERROR CONTRIBUTIONS
```

```
      DO 10 K=1,LAST
      WRITE(6,905) K,ALIST(IORD(K)),BLIST(IORD(K)),RLIST(IORD(K)),
     *  ELIST(IORD(K))
   10  CONTINUE
  900 FORMAT(26H INTEGRAL APPROXIMATION = ,E15.8)
  901 FORMAT(30H ESTIMATE OF ABSOLUTE ERROR = ,E9.2)
  902 FORMAT(34H NUMBER OF FUNCTION EVALUATIONS = ,I5)
  903 FORMAT(14H ERROR CODE = ,I1)
  904 FORMAT(//10X,22HLIMITS OF SUBINTERVALS,9X,13HINTEGRAL OVER,
     *  3X,10HERROR OVER/41X,11HSUBINTERVAL,5X,11HSUBINTERVAL/)
  905 FORMAT(I3,3(3X,E14.7),3X,E9.2)
      STOP
      END



      FUNCTION F(X)
      F = 1.0E+00/SQRT(ABS(X*X+2.0E+00*X-2.0E+00))
      RETURN
      END
```

Output:

```
 INTEGRAL APPROXIMATION =  0.15038227E+01
 ESTIMATE OF ABSOLUTE ERROR =  0.90E-02
 NUMBER OF FUNCTION EVALUATIONS =    435
 ERROR CODE = 3


          LIMITS OF SUBINTERVALS         INTEGRAL OVER   ERROR OVER
                                         SUBINTERVAL     SUBINTERVAL

  1    0.7319946E+00    0.7320557E+00    0.9622502E-02    0.53E-02
  2    0.7500000E+00    0.1000000E+01    0.4054651E+00    0.24E-02
  3    0.7320557E+00    0.7321777E+00    0.9737561E-02    0.12E-02
  4    0.7343750E+00    0.7500000E+00    0.9204194E-01    0.87E-05
  5    0.7324219E+00    0.7343750E+00    0.3109964E-01    0.26E-05
  6    0.0000000E+01    0.5000000E+00    0.4317178E+00    0.22E-05
  7    0.7321777E+00    0.7324219E+00    0.8592905E-02    0.14E-05
  8    0.5000000E+00    0.6250000E+00    0.1701778E+00    0.85E-06
  9    0.6250000E+00    0.6875000E+00    0.1261218E+00    0.63E-06
 10    0.6875000E+00    0.7187500E+00    0.1032905E+00    0.52E-06
 11    0.7304688E+00    0.7314453E+00    0.1630192E-01    0.30E-06
 12    0.7319336E+00    0.7319946E+00    0.3579979E-02    0.25E-06
 13    0.7187500E+00    0.7265625E+00    0.4438016E-01    0.22E-06
 14    0.7265625E+00    0.7304688E+00    0.3688416E-01    0.18E-06
 15    0.7314453E+00    0.7319336E+00    0.1480879E-01    0.74E-07
```

V. Special Applications of QUADPACK

5.1. Two-dimensional integration

5.1.1. General approach

Like all one-dimensional quadrature codes, the QUADPACK quadrature routines can be used to compute multiple integrals. We will restrict, however, our considerations to double integration, because all QUADPACK routines require at least 15, 21 or 25 function evaluations. Moreover we shall apply different integrators in each dimension, in order to avoid problems from the recursive use of the routines. We consider only quadrature problems which can be reduced to computing an integral or a sum of integrals of the form

$$I = \int_a^b dx \int_{\psi_\ell(x)}^{\psi_u(x)} dy\, F(x,y) \qquad (5.1.1)$$

$$= I[a,b]\,(I[\psi_\ell,\psi_u]F).$$

We denote the approximation obtained for the inner integral by $Q_2(x)$ and its error by $E_2(x)$, the approximation to $I[a,b]Q_2$ by Q_1 and its error by E_1. We shall call the inner integral $I_2(x)$ and the outer integral I_1.

In this manner, the integral can be written as

$$I = \int_a^b (Q_2(x) + E_2(x))\, dx$$

$$= Q_1 + E_1 + \int_a^b E_2(x)dx. \qquad (5.1.2)$$

So we shall try to satisfy an absolute tolerance t by imposing the condition

$$|I-Q_1| \leqslant t_1 + |b-a|t_2 = t \qquad (5.1.3)$$

on the absolute accuracies t_1 and t_2 requested for Q_1 and Q_2 respectively. This is a modification of the method proposed by Patterson (1980).
We let

$$t_1 = \frac{9}{10}t \quad \text{and} \quad |b-a|t_2 = \frac{1}{10}t. \qquad (5.1.4)$$

Note that a thorough investigation of the interface problem involving the absolute tolerances to be requested for the inner and outer integrations, was done by Fritsch, Kahaner and Lyness (1979) in relation to the use of (one-dimensional) adaptive Newton-Cotes quadrature routines.
The absolute error can be estimated as the sum of the error estimate for I_1 plus $|b-a|$ times the maximum of the inner error estimates.

Example

For the computation of

$$I = \int_0^1 dx \int_0^\infty dy\, \frac{e^{-(x+1)y}}{1+e^{-y}}, \qquad (5.1.5)$$

let us apply DQAGI for I_2, DQNG for I_1.
Table 5.1 gives the actual absolute errors, the error estimates and the

corresponding numbers of function evaluations for tolerances $t = 10^{-3}$, 10^{-6} and 10^{-9}.

integral (5.1.5)	t = absolute error tolerance		
	10^{-3}	10^{-6}	10^{-9}
act.abs.error	$8.8\ 10^{-11}$	$1.9\ 10^{-14}$	$6.7\ 10^{-16}$
est.abs.error	$9.2\ 10^{-5}$	$6.8\ 10^{-8}$	$6.9\ 10^{-11}$
# of F-eval.	1605	2355	2955

Table 5.1

5.1.2. Transformation to polar coordinates

Transformation to polar coordinates (r,θ) is useful in several circumstances, for instance when the integrand has a singularity of the form $r^{\alpha}\log(r)$.

If the integral can be written as

$$I = \int_a^b d\theta \int_0^{\phi(\theta)} dr\ r\ F(r\cos(\theta), r\sin(\theta)) \tag{5.1.6}$$

where F has an r^{α}- or $r^{\alpha}\log(r)$-singularity but is otherwise smooth, we can handle the inner integral with DQAWS if $\phi(\theta)$ is finite.

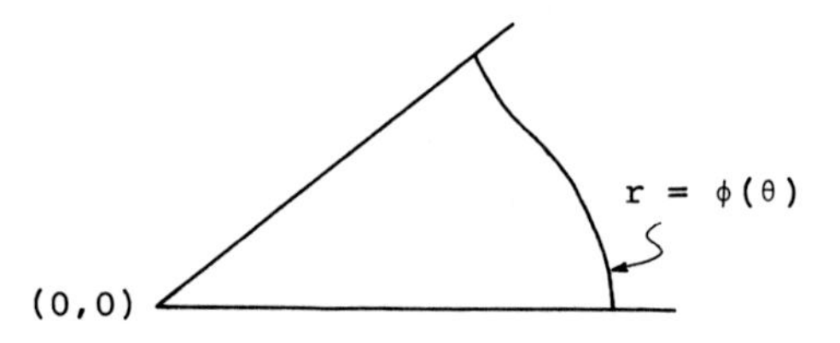

Figure 5.1

In the case of an infinite sector ($\phi(\theta) = \infty$ in (5.1.6)), DQAGI can be used for I_2.

However, DQAWS and DQAGI should be replaced by DQAWO and DQAWF respectively if the integrand contains an oscillatory factor $\cos(\omega r)$ or $\sin(\omega r)$.

Examples

(1) $$\int_{\Delta_1} (x+y)^{-1/2} (1+x+y)^{-2} \, dy \, dx \tag{5.1.7}$$

$$= \int_0^{\pi/2} d\theta \int_0^{\phi(\theta)} dr (\sin(\theta)+\cos(\theta))^{-1/2} \, r^{1/2} \, (1+r\sin(\theta)+r\cos(\theta))^{-2}$$

where

$$\Delta_1 = \{(x,y) \mid 0 \leqslant y \leqslant 1-x, \; 0 \leqslant x \leqslant 1\}. \tag{5.1.8}$$

and

$$\phi(\theta) = (\sin(\theta) + \cos(\theta))^{-1}.$$

Results are given in Table 5.2.

integral (5.1.7)	t = absolute error tolerance		
	10^{-3}	10^{-6}	10^{-9}
act.abs.error	$5.1 \; 10^{-15}$	$5.1 \; 10^{-15}$	$5.1 \; 10^{-15}$
est.abs.error	$5.9 \; 10^{-12}$	$5.9 \; 10^{-12}$	$5.9 \; 10^{-12}$
# of F-eval.	840	840	840

Table 5.2

(2) $$\int_{\mathbb{R}_+^2} (x+y)^{-1/2} \exp(-x-y) \, dy \, dx$$

$$= \int_0^{\pi/2} d\theta \, (\sin(\theta)+\cos(\theta))^{-1/2} \int_0^{\infty} dr \, r^{1/2} \exp(-r\sin(\theta)-r\cos(\theta)) \tag{5.1.9}$$

where $\mathbb{R}_+^2$ is the positive quadrant.

Results are given in Table 5.3.

integral (5.1.9)	t = absolute error tolerance 10^{-3}	10^{-6}	10^{-9}
act.abs.error	$2.7\ 10^{-10}$	$3.5\ 10^{-12}$	$6.8\ 10^{-16}$
est.abs.error	$2.0\ 10^{-5}$	$5.3\ 10^{-8}$	$3.3\ 10^{-11}$
# of F-eval	4275	6015	7425

Table 5.3

3) The integrand of the following problem has a singularity on a curve $r = \rho(\theta)$ within the triangle

$$\Delta_2 = \{(x,y) \mid 0 \leqslant y \leqslant x \leqslant 1\} : \tag{5.1.10}$$

$$\int_{\Delta_2} (x^2+y^2-1/4)^{1/2}\, G(x,y)\, dy\, dx =$$

$$= \int_0^{\pi/4} d\theta \int_0^{1/2} dr\ r\ \ (r^2-1/4)^{1/2}\, G(r\cos(\theta),\ r\sin(\theta)) + \tag{5.1.11}$$

$$+ \int_0^{\pi/4} d\theta \int_{1/2}^{\sec(\theta)} dr\ r\ \ (r^2-1/4)^{1/2}\, G(r\cos(\theta),\ r\sin(\theta)).$$

If G(x,y) is well-behaved, the inner integral in both terms of (5.1.11) can be computed efficiently by means of DQAWS, the outer one by means of DQNG, DQAG or DQAGS.

5.1.3. Cauchy principal values

Quadrature problems of the type

$$I = \int_0^1 dx \, \fint_{\psi_\ell(x)}^{\psi_u(x)} dy \, \frac{G(x,y)}{y-\phi(x)}, \qquad (5.1.12)$$

where the integrand has a non-integrable singularity along the curve $y = \phi(x)$ within the integration domain, have received a certain amount of attention in the literature (Song (1969), Monegato and Lyness (1979)).
For the automatic calculation of (5.1.12) we shall take care of the special form of the integrand by using DQAWC.

Example

$$I = \int_0^1 dx \, \frac{x}{e^x-1} \fint_{-1}^{1} y \quad \frac{(1-y^2)^{-1/2}}{y-\frac{1}{4}\sin(x)} \, dy \qquad (5.1.13)$$

We calculate I_2 by applying DQAWC on the subinterval [-1/2,1/2], and DQAWS on the two remaining intervals, in view of the algebraic singularities at $y = \pm 1$. For I_1 we use DQNG. The results obtained in this way are given in Table 5.4.

integral (5.1.13)	t = absolute error tolerance		
	10^{-3}	10^{-6}	10^{-9}
act.abs.error	$2.2 \; 10^{-16}$	$1.3 \; 10^{-15}$	$1.1 \; 10^{-16}$
est.abs.error	$4.3 \; 10^{-8}$	$2.6 \; 10^{-8}$	$4.1 \; 10^{-12}$
# of F-eval	2205	2565	4165

Table 5.4

5.2. Hankel transform

The Hankel transform $H_n(p;g)$ of the function g is defined as

$$H_n(p;g) = \int_0^\infty xg(x)\, J_n(px)\, dx \qquad (5.2.1)$$

where $J_n(x)$ is the Bessel function of order n. Because of the oscillatory nature of the Bessel functions, the integral (5.2.1) cannot be computed by means of the general purpose integrator QAGI.

We shall now illustrate the use of the QUADPACK routines for the computation of (5.2.1).
The Hankel transform can be written as

$$H_n(p;g) = \sum_{\ell=0}^{\infty} \int_{j_{n,\ell}}^{j_{n,\ell+1}} f(x)\, J_n(x)\, dx \qquad (5.2.2)$$

where $j_{n,\ell}$, $\ell=1,2...$ is the ℓ-th positive zero of $J_n(x)$, $j_{n,0} := 0$, and $f(x) = xg(x/p)/p^2$. The integrals occuring in (5.2.2) can be computed using QNG, QAG or QAGS.
The series in (5.2.2) is an alternating series, and the summation can be carried out using the ε-algorithm of Wynn (1956), which is very effective if g is a smooth, monotonically decreasing function.

The following program illustrates how (5.2.2) can be evaluated using DQAGS for the integration and DQEXT for accelerating the convergence of the series.

Table 5.5 gives the output of this program for

$$\int_0^\infty (1-e^{-x})\, [x \log(1+\sqrt{2})]^{-1}\, J_0(x)\, dx. \qquad (5.2.3)$$

The exact value of this integral is 1.

```
C
C MAIN PROGRAM FOR ILLUSTRATING THE USE OF DQAGS AND DQEXT
C FOR INTEGRATING OVER AN INFINITE INTERVAL USING CONVERGENCE
C ACCELERATION
C
      DOUBLE PRECISION ABSEPS,ABSERR,ALIST,AREA,A,BLIST,
     *        B,DABS,ELIST,EPSABS,ERR,F,PSUM,RESEPS,RESULT,PSUM(52),
     *        RES3LA(3),RLIST,ZEROJN
      INTEGER IER,IORD,LAST,L,NEV,NEVAL,NRES,NUMRL2
C
      EXTERNAL F
C
C INITIALIZE
      EPSABS = 1.0D-12
      RESULT = 0.0D+00
      A = 0.0D+00
      NRES = 0
      NEV = 0
      NUMRL2 = 0
      WRITE(6,800)
C MAIN DO-LOOP
      DO 100 L=1,50
C COMPUTE THE L TH POSITIVE ZERO  OF THE BESSEL FUNCTION JN
        B = ZEROJN(0,L)
C INTEGRATE FROM A TO B
C
        CALL DQAGS(F,A,B,EPSABS,0.0D+00,AREA,ABSERR,NEVAL,IER)
C
        A = B
        NEV = NEV+NEVAL
C EXTRAPOLATE
        NUMRL2 = NUMRL2+1
        RESULT = RESULT+AREA
        PSUM(NUMRL2) = RESULT
        IF(NUMRL2.LT.3) GO TO 100
C
        CALL DQEXT(NUMRL2,PSUM,RESEPS,ABSEPS,RES3LA,NRES)
C
C AREA = L TH TERM OF SERIES
C RESULT = L TH PARTIAL SUM
C RESEPS = EXTRAPOLATED RESULT
C ABSEPS = ESTIMATE OF ABSOLUTE ERROR OF RESEPS
C NEV = NUMBER OF FUNCTION EVALUATIONS ALREADY USED
        WRITE(6,900) L,AREA,RESULT,RESEPS,ABSEPS,NEV
C TEST ON ACCURACY
C***JUMP OUT OF DO-LOOP
        IF(ABSEPS.LE.EPSABS) GO TO 999
100   CONTINUE
999   STOP
800   FORMAT(1H1,5X,1HL,4X,19HL TH TERM OF SERIES,6X,
     *        16HL TH PARTIAL SUM,6X,19HEXTRAPOLATED RESULT,
     *        3X,12HABS.ERR.EST.,1X,9HNUMBER OF/
     *        1X,60X,12H(FROM DQEXT),6X,12H(FROM DQEXT),2X,7HF-EVAL.//)
900   FORMAT(1X,I6,3D24.16,D11.3,I6)
      END
```

1

L	L TH TERM OF SERIES	L TH PARTIAL SUM	EXTRAPOLATED RESULT (FROM DQEXT)	ABS.ERR.EST. (FROM DQEXT)	NUMBER OF F-EVAL.
3	0.9720941305303759D-01	0.1033977523819508D+01	0.1005427556534740D+01	0.170D+39	63
4	-0.5576180665385378D-01	0.9782157171656541D+00	0.9985422798175024D+00	0.170D+39	84
5	0.3721563729882483D-01	0.1015431354464479D+01	0.1000158347761019D+01	0.170D+39	105
6	-0.2707995888016404D-01	0.9883513955843149D+00	0.9999611039893594D+00	0.708D-02	126
7	0.2083678193263091D-01	0.1009188177516946D+01	0.1000004566593879D+01	0.166D-02	147
8	-0.1667240713749006D-01	0.9925157703794558D+00	0.9999989276016102D+00	0.203D-03	168
9	0.1373183762719216D-01	0.1006247608006648D+01	0.1000000132378945D+01	0.447D-04	189
10	-0.1156459972171845D-01	0.9946830082849295D+00	0.9999999700786973D+00	0.580D-05	210
11	0.9913100336013039D-02	0.1004596108620943D+01	0.1000000003873796D+01	0.124D-05	231
12	-0.8620467426260281D-02	0.9959756411946822D+00	0.9999999991586137D+00	0.167D-06	252
13	0.7586267440049880D-02	0.1003561908634732D+01	0.1000000000114305D+01	0.348D-07	273
14	-0.6743526959569978D-02	0.9968183816751621D+00	0.9999999999761460D+00	0.485D-08	294
15	0.6046048020082958D-02	0.1002864429695245D+01	0.1000000000003387D+01	0.983D-09	315
16	-0.5461039724356489D-02	0.9974033899708886D+00	0.9999999999993175D+00	0.142D-09	336
17	0.4964651093302690D-02	0.1002368041064191D+01	0.1000000000000100D+01	0.280D-10	357
18	-0.4539162411976654D-02	0.9978288786522146D+00	0.9999999999999804D+00	0.419D-11	378
19	0.4171163015661070D-02	0.1002000041667876D+01	0.1000000000000003D+01	0.806D-12	399

Table 5.5

5.3. Numerical inversion of the Laplace transform

5.3.1. Introduction

For a given image function $F(p)$ let us approximate its inverse Laplace transform

$$f(t) = \mathcal{L}^{-1}\{F(p)\}. \tag{5.3.1}$$

The original function f is determined by means of the Bromwich integral

$$f(t) = \frac{1}{2\pi i}\int_{c-i\infty}^{c+i\infty} e^{pt}\, F(p)\, dp \tag{5.3.2}$$

where c is a real number such that F(p) is analytic for $Re(p) > c$.

If $|F(p)| \to 0$ uniformly with respect to arg(p) as $p \to \infty$ in the region $Re(p) < c$, Talbot (1976) proposes to replace (5.3.2) by

$$f(t) = \frac{\lambda}{2\pi i}\int_L e^{(\lambda p+\sigma)t}\, \Phi(p)\, dp, \tag{5.3.3}$$

where $\Phi(p) = F(\lambda p+\sigma)$, and λ and σ are chosen so that the contour

$$L = \{p \in \mathbf{C} \mid p = \theta\,\mathrm{cotan}(\theta) + i\theta,\ -\pi < \theta < \pi\} \tag{5.3.4}$$

encloses the singularities of Φ.

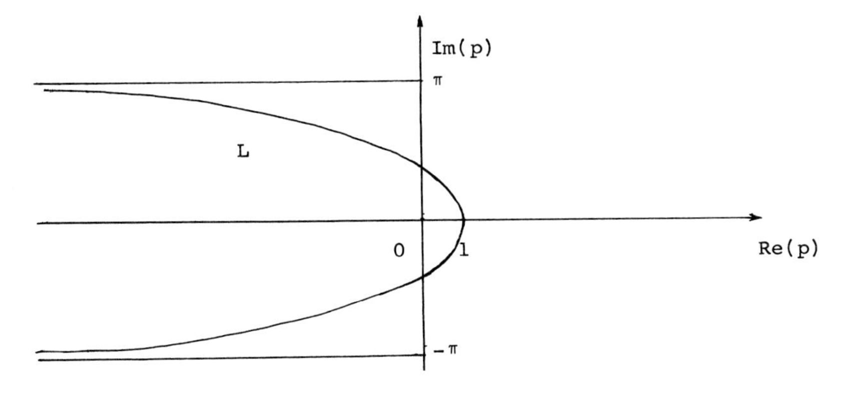

Figure 5.2

The resulting integral,

$$f(t) = \frac{\lambda e^{\sigma t}}{2\pi i} \int_{-\pi}^{\pi} e^{\lambda pt} \, \Phi(p) \, \frac{dp}{d\theta} \, d\theta \tag{5.3.5}$$

can be calculated using the trapezoidal rule. Excellent convergence is assured by the property that the integrand and all its derivatives vanish at the end-points $\pm\pi$ of the integration interval.

5.3.2. Modification of Talbot's method

An alternative contour in (5.3.2) is (Figure 5.3)

$$R = \{p \in \mathbb{C} \mid p = \alpha + iy, \ -\beta \leqslant y \leqslant \beta \ \text{or} \ p = x \pm i\beta, \ -\infty < x \leqslant \alpha\} \tag{5.3.6}$$

where α and β are chosen so that R encloses the singularities of F.

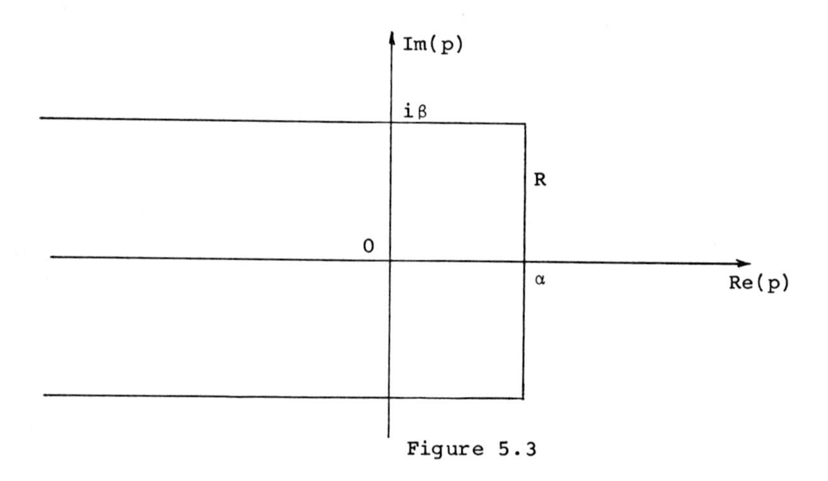

Figure 5.3

For

$$\begin{aligned} G(p) &= \mathrm{Re}(F(p)), \\ H(p) &= \mathrm{Im}(F(p)), \end{aligned} \tag{5.3.7}$$

the integral over R can be written as

$$f(t) = (e^{\alpha t} (I_1+I_2) + I_3)/\pi, \tag{5.3.8}$$

with

$$I_1 = \int_0^\beta G(\alpha+iy)\cos(ty)dy, \tag{5.3.9}$$

$$I_2 = -\int_0^\beta H(\alpha+iy)\sin(ty)dy, \tag{5.3.10}$$

$$I_3 = -t^{-1}\int_{-\infty}^{\alpha t} e^u(G(u/t+i\beta)\sin(\beta t) + H(u/t+i\beta)\cos(\beta t))du. \tag{5.3.11}$$

For these integrals the trapezoidal rule is not an appropriate method. We shall deal with I_1 and I_2 by means of DQAWO, and with I_3 by means of DQAGI.
It is preferable to keep the values of α and β small, especially for large t-values. On the other hand the integration process might suffer from the probable effect of singularities which occur too close to the contour R. A compromise between these alternatives is the choice

$$\alpha = a + c_1/t,$$
$$\beta = b + c_2/t, \qquad c_1 \text{ and } c_2 \text{ constant}, \tag{5.3.12}$$

where a is the smallest value for which F(p) is analytic in $|Re(p)| < a$, and b is the smallest value for which F(p) is analytic in $|Im(p)| > b$. As to the constants in (5.3.12), $1 \leqslant c_1 = c_2 \leqslant 5$ seem to be appropriate in most cases.
In Table 5.6 we give numerical results for the example

$$F(p) = (p^2+1)^{-1},$$
$$\mathcal{L}^{-1}\{F(p)\} = \sin(t), \tag{5.3.13}$$

for

$$t = 10^k, \qquad k = 0,1,\ldots,5. \tag{5.3.14}$$

We tabulate the number of integrand evaluations used for the computation of I_1, I_2 and I_3 with a relative accuracy requirement of 10^{-8}. The relative accuracy in the approximate f(t) is also given. We let $c_1 = c_2 = 2$.

a=0,b=1	number of integrand eval. I_1	I_2	I_3	rel. accuracy in approx. f(t)
t	DQAWO	DQAWO	DQAGI	
1	45	45	135	$9.7\ 10^{-16}$
10	165	135	165	$5.6\ 10^{-16}$
10^2	315	315	135	$4.0\ 10^{-15}$
10^3	465	435	165	$2.6\ 10^{-15}$
10^4	635	635	165	$2.2\ 10^{-12}$
10^5	785	815	165	$3.1\ 10^{-10}$

Table 5.6

Although Talbot's method is more efficient for small t, the present procedure is found to be superior to any other method for large values of t.

We note that, when F(p) has branch points, one must choose the branch for which G and H are continuous along the contour. This is for example the case for the function

$$F(p) = (p^2+1)^{-1/2}$$
$$\mathcal{L}^{-1}\{F(p)\} = J_0(t). \tag{5.3.15}$$

Results are given in Table 5.7.

a=0,b=1	number of integrand eval.			rel. accuracy in approx. f(t)
	I_1	I_2	I_3	
t	DQAWO	DQAWO	DQAGI	
1	45	15	135	$1.6\ 10^{-13}$
10	135	135	135	$1.7\ 10^{-16}$
10^2	285	315	165	$2.1\ 10^{-14}$
10^3	435	435	105	$3.6\ 10^{-13}$
10^4	635	635	135	$8.0\ 10^{-13}$
10^5	785	785	135	$1.9\ 10^{-11}$

Table 5.7

5.3.3. Image function with an infinite number of singularities along a vertical line in the complex plane

Let f be given by the Bromwich integral (5.3.2), and assume that F has an infinite number of singularities along the vertical line $p = c^*$ in the complex plane. Thus the methods of the previous sections cannot be applied, because no appropriate contour of type L (5.3.4) or R (5.3.6) can be found. In this case we can only evaluate the Bromwich integral by integrating over a vertical line. Transforming $p = c + iv$ in (5.3.2), where $c > c^*$ yields

$$f(t) = \frac{2e^{ct}}{\pi} \int_0^\infty \mathrm{Re}(F(c+iv))\cos(vt)dv. \tag{5.3.16}$$

Example

$$F(p) = \frac{1}{p\cosh(p)} \tag{5.3.17}$$

has an infinite number of singularities along the imaginary axis. The original function is discontinuous (Figure 5.4).

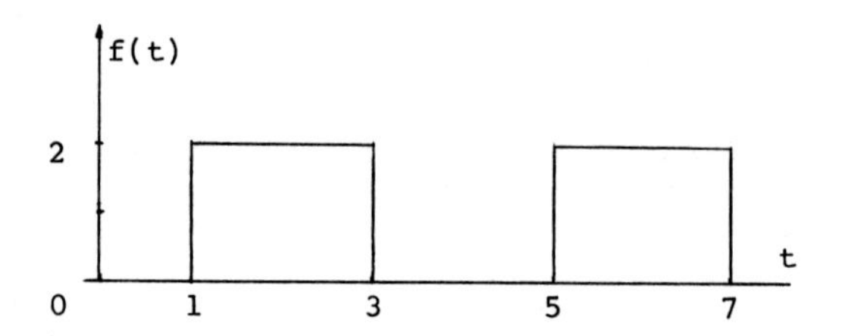

Figure 5.4

By putting

$$c = 2/t \tag{5.3.18}$$

and calculating the resulting integral (5.3.16) by means of DQAWF, we obtain the results given in Table 5.8. The required (absolute) accuracy is 10^{-6}. Although this remarkably high accuracy is reached for a t-value as large as t = 502.5, it emerges that difficulties are experienced especially for integer t-values.
Note, however, that Re(F(c+iv)) in the integrand of (5.3.16) is, in this case, not monotonic but oscillates over the entire interval $(0,\infty)$, so that DQAWF is in fact not a suitable choice for this problem (cf. Section 3.3.9).
When t is an integer, DQAWF integrates over intervals of length $2\pi+\pi/t$ (close to 2π especially when t is large). Because Re(F(p)) oscillates with period 2π we can expect that the ε-algorithm will accelerate the convergence if we integrate over intervals of length

$$\text{an odd number of half periods} = (2k+1)\pi,\ k = 0,1,\ldots\ . \tag{5.3.19}$$

Indeed, by integrating over intervals of length 3π using DQAGS and applying convergence acceleration using the ε-algorithm routine DQEXT (see Section 3.4.2), we obtain the accuracies listed in Table 5.9 (an absolute tolerance of 10^{-6} was requested).

t	actual abs. error	estim. abs. error	# of integrand evaluations
2.0	3.6 10^{-8}	6.8 10^{-7}	5100
5.5	9.2 10^{-10}	8.5 10^{-7}	4325
7.5	1.0 10^{-9}	9.6 10^{-7}	4800
10.0	5.3 10^{-5}	6.0 10^{-5}	22590 (*)
10.5	3.9 10^{-9}	9.2 10^{-7}	4900
100.0	8.6 10^{-4}	4.1 10^{-2}	50800 (*)
100.5	1.1 10^{-9}	3.8 10^{-7}	14300
502.5	2.0 10^{-10}	8.3 10^{-7}	21135

Table 5.8

(*) Maximum number of cycles (50) was reached.

t	actual abs. error	estim. abs. error	# of integrand evaluations
10.0	5.8 10^{-10}	8.8 10^{-7}	8510
20.0	9.1 10^{-11}	9.3 10^{-7}	13105
50.0	7.5 10^{-10}	7.4 10^{-7}	15610
100.0	6.1 10^{-10}	7.6 10^{-7}	18110
308.0	6.6 10^{-12}	9.9 10^{-7}	24580
502.0	9.7 10^{-11}	9.1 10^{-7}	25405
1000.0	1.5 10^{-10}	9.2 10^{-7}	28055

Table 5.9

VI. Implementation Notes and Routine Listings

6.1. Implementation notes

All the programs of QUADPACK are written in Standard FORTRAN and have been checked with the PFORT verifier of Bell Labs.

Before implementing the package on the computer, two types of machine dependencies are to be checked and eventually adjusted :

(i) a single precision and two double precision versions are available. In the single precision version and in one of the double precision versions the abscissae and weights of the quadrature rules are given to 16 decimal digits, and in the other double precision version these constants are accurate to 33 places. The proper version must be chosen according to the precision of the computer. In Section 6.2 only the single precision subroutines are listed.

(ii) three machine dependent constants are extensively used throughout the package : the relative machine precision, the smallest non-underflowing positive number and the largest non-overflowing number. They are given in subroutine (D)QMACO and must be modified depending on the machine characteristics.

For the ease of implementation, the inner structure of the package is pictured in Figure 6.1.

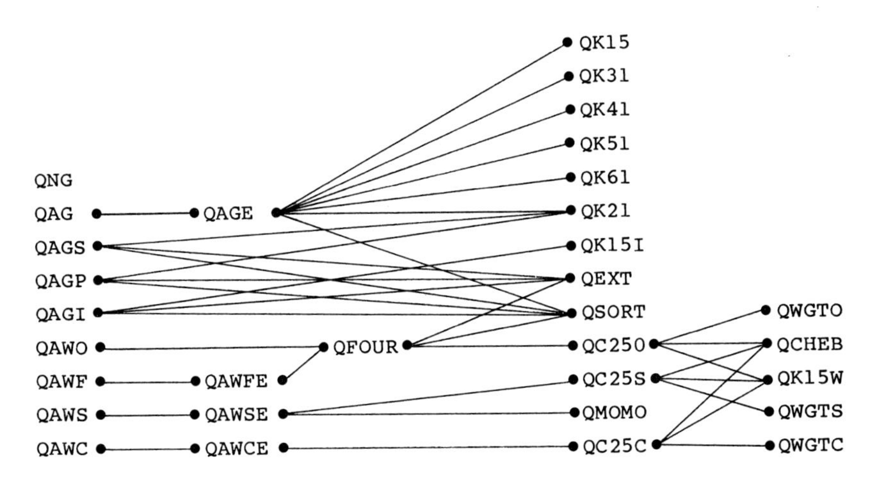

QMACO is used by all the integrators

Fig. 6.1.

6.2. Routine listings


```
      SUBROUTINE QNG(F,A,B,EPSABS,EPSREL,RESULT,ABSERR,NEVAL,IER)
C
C.......................................................................
C
C 1.     QNG
C        NON-ADAPTIVE INTEGRATION
C           STANDARD FORTRAN SUBROUTINE
C
C 2.     PURPOSE
C           THE ROUTINE CALCULATES AN APPROXIMATION  RESULT  TO
C           A GIVEN DEFINITE INTEGRAL  I = INTEGRAL OF  F  OVER (A,B),
C           HOPEFULLY SATISFYING FOLLOWING CLAIM FOR ACCURACY
C           ABS(I-RESULT).LE.MAX(EPSABS,EPSREL*ABS(I)).
C
C
C 3.     CALLING SEQUENCE
C           CALL QNG(F,A,B,EPSABS,EPSREL,RESULT,ABSERR,NEVAL,IER)
C
C        PARAMETERS
C         ON ENTRY
C           F      - REAL
C                    FUNCTION SUBPROGRAM DEFINING THE INTEGRAND
C                    FUNCTION F(X). THE ACTUAL NAME FOR F NEEDS TO BE
C                    DECLARED E X T E R N A L IN THE DRIVER PROGRAM.
C
C           A      - REAL
C                    LOWER LIMIT OF INTEGRATION
C
C           B      - REAL
C                    UPPER LIMIT OF INTEGRATION
C
C           EPSABS - REAL
C                    ABSOLUTE ACCURACY REQUESTED
C           EPSREL - REAL
C                    RELATIVE ACCURACY REQUESTED
C                    IF  EPSABS.LT.0 AND EPSREL.LT.0,
C                    THE ROUTINE WILL END WITH IER = 6.
C
C         ON RETURN
C           RESULT - REAL
C                    APPROXIMATION TO THE INTEGRAL I
C                    RESULT IS OBTAINED BY APPLYING THE 21-POINT
C                    GAUSS-KRONROD RULE (RES21) OBTAINED BY OPTIMAL
C                    ADDITION OF ABSCISSAE TO THE 10-POINT GAUSS RULE
C                    (RES10), OR BY APPLYING THE 43-POINT RULE (RES43)
C                    OBTAINED BY OPTIMAL ADDITION OF ABSCISSAE TO THE
C                    21-POINT GAUSS-KRONROD RULE, OR BY APPLYING THE
C                    87-POINT RULE (RES87) OBTAINED BY OPTIMAL ADDITION
```

```
C                   OF ABSCISSAE TO THE 43-POINT RULE.
C
C           ABSERR - REAL
C                    ESTIMATE OF THE MODULUS OF THE ABSOLUTE ERROR,
C                    WHICH SHOULD EQUAL OR EXCEED ABS(I-RESULT)
C
C           NEVAL  - INTEGER
C                    NUMBER OF INTEGRAND EVALUATIONS
C
C           IER    - IER = 0 NORMAL AND RELIABLE TERMINATION OF THE
C                            ROUTINE. IT IS ASSUMED THAT THE REQUESTED
C                            ACCURACY HAS BEEN ACHIEVED.
C                    IER.GT.0 ABNORMAL TERMINATION OF THE ROUTINE. IT IS
C                            ASSUMED THAT THE REQUESTED ACCURACY HAS
C                            NOT BEEN ACHIEVED.
C                    IER = 1 THE MAXIMUM NUMBER OF STEPS HAS BEEN
C                            EXECUTED. THE INTEGRAL IS PROBABLY TOO
C                            DIFFICULT TO BE CALCULATED BY QNG.
C                        = 6 THE INPUT IS INVALID, BECAUSE
C                            EPSABS.LT.0 AND EPSREL.LT.0,
C                            RESULT, ABSERR AND NEVAL ARE SET TO ZERO.
C
C
C 4.     SUBROUTINES OR FUNCTIONS NEEDED
C              - F (USER-PROVIDED FUNCTION)
C              - QMACO
C              - FORTRAN ABS, AMAX1, AMIN1
C
C.......................................................................
C
      REAL A,ABSC,ABSERR,B,CENTR,DHLGTH,EPMACH,EPSABS,EPSREL,F,FCENTR,
     *  FVAL,FVAL1,FVAL2,FV1,FV2,FV3,FV4,HLGTH,OFLOW,RESULT,RES10,RES21,
     *  RES43,RES87,RESABS,RESASC,RESKH,SAVFUN,UFLOW,W10,W21A,W21B,W43A,
     *  W43B,W87A,W87B,X1,X2,X3,X4
      INTEGER IER,IPX,K,L,NEVAL
C
      DIMENSION FV1(5),FV2(5),FV3(5),FV4(5),X1(5),X2(5),X3(11),X4(22),
     *  W10(5),W21A(5),W21B(6),W43A(10),W43B(12),W87A(21),W87B(23),
     *  SAVFUN(21)
C
C           THE FOLLOWING DATA STATEMENTS CONTAIN THE ABSCISSAE
C           AND WEIGHTS OF THE INTEGRATION RULES USED.
C
C           X1      ABSCISSAE COMMON TO THE 10-, 21-, 43- AND 87-POINT
C                   RULE
C           X2      ABSCISSAE COMMON TO THE 21-, 43- AND 87-POINT RULE
C           X3      ABSCISSAE COMMON TO THE 43- AND 87-POINT RULE
C           X4      ABSCISSAE OF THE 87-POINT RULE
```

```
C           W10      WEIGHTS OF THE 10-POINT FORMULA
C           W21A     WEIGHTS OF THE 21-POINT FORMULA FOR ABSCISSAE X1
C           W21B     WEIGHTS OF THE 21-POINT FORMULA FOR ABSCISSAE X2
C           W43A     WEIGHTS OF THE 43-POINT FORMULA FOR ABSISSAE X1, X3
C           W43B     WEIGHTS OF THE 43-POINT FORMULA FOR ABSCISSAE X3
C           W87A     WEIGHTS OF THE 87-POINT FORMULA FOR ABSCISSAE X1,
C                    X2 AND X3
C           W87B     WEIGHTS OF THE 87-POINT FORMULA FOR ABSCISSAE X4
C
      DATA X1(1),X1(2),X1(3),X1(4),X1(5)/
     *     9.739065285171717E-01,     8.650633666889845E-01,
     *     6.794095682990244E-01,     4.333953941292472E-01,
     *     1.488743389816312E-01/
      DATA X2(1),X2(2),X2(3),X2(4),X2(5)/
     *     9.956571630258081E-01,     9.301574913557082E-01,
     *     7.808177265864169E-01,     5.627571346686047E-01,
     *     2.943928627014602E-01/
      DATA X3(1),X3(2),X3(3),X3(4),X3(5),X3(6),X3(7),X3(8),X3(9),X3(10),
     *  X3(11)/
     *     9.993333609019321E-01,     9.874334029080889E-01,
     *     9.548079348142663E-01,     9.001486957483283E-01,
     *     8.251983149831142E-01,     7.321483889893050E-01,
     *     6.228479705377252E-01,     4.994795740710565E-01,
     *     3.649016613465808E-01,     2.222549197766013E-01,
     *     7.465061746138332E-02/
      DATA X4(1),X4(2),X4(3),X4(4),X4(5),X4(6),X4(7),X4(8),X4(9),X4(10),
     *  X4(11),X4(12),X4(13),X4(14),X4(15),X4(16),X4(17),X4(18),X4(19),
     *  X4(20),X4(21),X4(22)/          9.999029772627292E-01,
     *     9.979898959866787E-01,     9.921754978606872E-01,
     *     9.813581635727128E-01,     9.650576238583846E-01,
     *     9.431676131336706E-01,     9.158064146855072E-01,
     *     8.832216577713165E-01,     8.457107484624157E-01,
     *     8.035576580352310E-01,     7.570057306854956E-01,
     *     7.062732097873218E-01,     6.515894665011779E-01,
     *     5.932233740579611E-01,     5.314936059708319E-01,
     *     4.667636230420228E-01,     3.994248478592188E-01,
     *     3.298748771061883E-01,     2.585035592021616E-01,
     *     1.856953965683467E-01,     1.118422131799075E-01,
     *     3.735212339461987E-02/
      DATA W10(1),W10(2),W10(3),W10(4),W10(5)/
     *     6.667134430868814E-02,     1.494513491505806E-01,
     *     2.190863625159820E-01,     2.692667193099964E-01,
     *     2.955242247147529E-01/
      DATA W21A(1),W21A(2),W21A(3),W21A(4),W21A(5)/
     *     3.255816230796473E-02,     7.503967481091995E-02,
     *     1.093871588022976E-01,     1.347092173114733E-01,
     *     1.477391049013385E-01/
      DATA W21B(1),W21B(2),W21B(3),W21B(4),W21B(5),W21B(6)/
```

```
     *     1.169463886737187E-02,      5.475589657435200E-02,
     *     9.312545458369761E-02,      1.234919762620659E-01,
     *     1.427759385770601E-01,      1.494455540029169E-01/
      DATA W43A(1),W43A(2),W43A(3),W43A(4),W43A(5),W43A(6),W43A(7),
     *  W43A(8),W43A(9),W43A(10)/      1.629673428966656E-02,
     *     3.752287612086950E-02,      5.469490205825544E-02,
     *     6.735541460947809E-02,      7.387019963239395E-02,
     *     5.768556059769796E-03,      2.737189059324884E-02,
     *     4.656082691042883E-02,      6.174499520144256E-02,
     *     7.138726726869340E-02/
      DATA W43B(1),W43B(2),W43B(3),W43B(4),W43B(5),W43B(6),W43B(7),
     *  W43B(8),W43B(9),W43B(10),W43B(11),W43B(12)/
     *     1.844477640212414E-03,      1.079868958589165E-02,
     *     2.189536386779543E-02,      3.259746397534569E-02,
     *     4.216313793519181E-02,      5.074193960018458E-02,
     *     5.837939554261925E-02,      6.474640495144589E-02,
     *     6.956619791235648E-02,      7.282444147183321E-02,
     *     7.450775101417512E-02,      7.472214751740301E-02/
      DATA W87A(1),W87A(2),W87A(3),W87A(4),W87A(5),W87A(6),W87A(7),
     *  W87A(8),W87A(9),W87A(10),W87A(11),W87A(12),W87A(13),W87A(14),
     *  W87A(15),W87A(16),W87A(17),W87A(18),W87A(19),W87A(20),W87A(21)/
     *     8.148377384149173E-03,      1.876143820156282E-02,
     *     2.734745105005229E-02,      3.367770731163793E-02,
     *     3.693509982042791E-02,      2.884872430211531E-03,
     *     1.368594602271270E-02,      2.328041350288831E-02,
     *     3.087249761171336E-02,      3.569363363941877E-02,
     *     9.152833452022414E-04,      5.399280219300471E-03,
     *     1.094767960111893E-02,      1.629873169678734E-02,
     *     2.108156888920384E-02,      2.537096976925383E-02,
     *     2.918969775647575E-02,      3.237320246720279E-02,
     *     3.478309895036514E-02,      3.641222073135179E-02,
     *     3.725387550304771E-02/
      DATA W87B(1),W87B(2),W87B(3),W87B(4),W87B(5),W87B(6),W87B(7),
     *  W87B(8),W87B(9),W87B(10),W87B(11),W87B(12),W87B(13),W87B(14),
     *  W87B(15),W87B(16),W87B(17),W87B(18),W87B(19),W87B(20),W87B(21),
     *  W87B(22),W87B(23)/             2.741455637620724E-04,
     *     1.807124155057943E-03,      4.096869282759165E-03,
     *     6.758290051847379E-03,      9.549957672201647E-03,
     *     1.232944765224485E-02,      1.501044734638895E-02,
     *     1.754896798624319E-02,      1.993803778644089E-02,
     *     2.219493596101229E-02,      2.433914712600081E-02,
     *     2.637450541483921E-02,      2.828691078877120E-02,
     *     3.005258112809270E-02,      3.164675137143993E-02,
     *     3.305041341997850E-02,      3.425509970422606E-02,
     *     3.526241266015668E-02,      3.607698962288870E-02,
     *     3.669860449845609E-02,      3.712054926983258E-02,
     *     3.733422875193504E-02,      3.736107376267902E-02/
C
```

```
C           LIST OF MAJOR VARIABLES
C           -----------------------
C
C           CENTR  - MID POINT OF THE INTEGRATION INTERVAL
C           HLGTH  - HALF-LENGTH OF THE INTEGRATION INTERVAL
C           FCENTR - FUNCTION VALUE AT MID POINT
C           ABSC   - ABSCISSA
C           FVAL   - FUNCTION VALUE
C           SAVFUN - ARRAY OF FUNCTION VALUES WHICH HAVE ALREADY
C                    BEEN COMPUTED
C           RES10  - 10-POINT GAUSS RESULT
C           RES21  - 21-POINT KRONROD RESULT
C           RES43  - 43-POINT RESULT
C           RES87  - 87-POINT RESULT
C           RESABS - APPROXIMATION TO THE INTEGRAL OF ABS(F)
C           RESASC - APPROXIMATION TO THE INTEGRAL OF ABS(F-I/(B-A))
C
C           MACHINE DEPENDENT CONSTANTS
C           ---------------------------
C
C           EPMACH IS THE LARGEST RELATIVE SPACING.
C           UFLOW IS THE SMALLEST POSITIVE MAGNITUDE.
C           OFLOW IS THE LARGEST MAGNITUDE.
C
C***FIRST EXECUTABLE STATEMENT
      CALL QMACO(EPMACH,UFLOW,OFLOW)
C
C           TEST ON VALIDITY OF PARAMETERS
C           ------------------------------
C
      RESULT = 0.0E+00
      ABSERR = 0.0E+00
      NEVAL = 0
      IER = 6
      IF(EPSABS.LT.0.0E+00.AND.EPSREL.LT.0.0E+00) GO TO 999
      HLGTH = 5.0E-01*(B-A)
      DHLGTH = ABS(HLGTH)
      CENTR = 5.0E-01*(B+A)
      FCENTR = F(CENTR)
      NEVAL = 21
      IER = 1
C
C          COMPUTE THE INTEGRAL USING THE 10- AND 21-POINT FORMULA.
C
      DO 70 L = 1,3
      GO TO (5,25,45),L
    5 RES10 = 0.0E+00
      RES21 = W21B(6)*FCENTR
```

```
      RESABS = W21B(6)*ABS(FCENTR)
      DO 10 K=1,5
        ABSC = HLGTH*X1(K)
        FVAL1 = F(CENTR+ABSC)
        FVAL2 = F(CENTR-ABSC)
        FVAL = FVAL1+FVAL2
        RES10 = RES10+W10(K)*FVAL
        RES21 = RES21+W21A(K)*FVAL
        RESABS = RESABS+W21A(K)*(ABS(FVAL1)+ABS(FVAL2))
        SAVFUN(K) = FVAL
        FV1(K) = FVAL1
        FV2(K) = FVAL2
   10 CONTINUE
      IPX = 5
      DO 15 K=1,5
        IPX = IPX+1
        ABSC = HLGTH*X2(K)
        FVAL1 = F(CENTR+ABSC)
        FVAL2 = F(CENTR-ABSC)
        FVAL = FVAL1+FVAL2
        RES21 = RES21+W21B(K)*FVAL
        RESABS = RESABS+W21B(K)*(ABS(FVAL1)+ABS(FVAL2))
        SAVFUN(IPX) = FVAL
        FV3(K) = FVAL1
        FV4(K) = FVAL2
   15 CONTINUE
C
C           TEST FOR CONVERGENCE.
C
      RESULT = RES21*HLGTH
      RESABS = RESABS*DHLGTH
      RESKH = 5.0E-01*RES21
      RESASC = W21B(6)*ABS(FCENTR-RESKH)
      DO 20 K = 1,5
        RESASC = RESASC+W21A(K)*(ABS(FV1(K)-RESKH)+ABS(FV2(K)-RESKH))
     *                  +W21B(K)*(ABS(FV3(K)-RESKH)+ABS(FV4(K)-RESKH))
   20 CONTINUE
      ABSERR = ABS((RES21-RES10)*HLGTH)
      RESASC = RESASC*DHLGTH
      GO TO 65
C
C           COMPUTE THE INTEGRAL USING THE 43-POINT FORMULA.
C
   25 RES43 = W43B(12)*FCENTR
      NEVAL = 43
      DO 30 K=1,10
        RES43 = RES43+SAVFUN(K)*W43A(K)
   30 CONTINUE
```

```
      DO 40 K=1,11
        IPX = IPX+1
        ABSC = HLGTH*X3(K)
        FVAL = F(ABSC+CENTR)+F(CENTR-ABSC)
        RES43 = RES43+FVAL*W43B(K)
        SAVFUN(IPX) = FVAL
   40 CONTINUE
C
C           TEST FOR CONVERGENCE.
C
      RESULT = RES43*HLGTH
      ABSERR = ABS((RES43-RES21)*HLGTH)
      GO TO 65
C
C           COMPUTE THE INTEGRAL USING THE 87-POINT FORMULA.
C
   45 RES87 = W87B(23)*FCENTR
      NEVAL = 87
      DO 50 K=1,21
        RES87 = RES87+SAVFUN(K)*W87A(K)
   50 CONTINUE
      DO 60 K=1,22
        ABSC = HLGTH*X4(K)
        RES87 = RES87+W87B(K)*(F(ABSC+CENTR)+F(CENTR-ABSC))
   60 CONTINUE
      RESULT = RES87*HLGTH
      ABSERR = ABS((RES87-RES43)*HLGTH)
   65 IF(RESASC.NE.0.0E+00.AND.ABSERR.NE.0.0E+00)
     *  ABSERR = RESASC*AMIN1(1.0E+00,(2.0E+02*ABSERR/RESASC)**1.5E+00)
      IF (RESABS.GT.UFLOW/(5.0E+01*EPMACH)) ABSERR = AMAX1
     *  ((EPMACH*5.0E+01)*RESABS,ABSERR)
      IF (ABSERR.LE.AMAX1(EPSABS,EPSREL*ABS(RESULT))) IER = 0
C***JUMP OUT OF DO-LOOP
      IF (IER.EQ.0) GO TO 999
   70 CONTINUE
  999 RETURN
      END
```

```
      SUBROUTINE QAG(F,A,B,EPSABS,EPSREL,KEY,RESULT,ABSERR,NEVAL,IER)
C
C ..................................................................
C
C 1.     QAG
C        COMPUTATION OF A DEFINITE INTEGRAL
C           STANDARD FORTRAN SUBROUTINE
C
C 2.     PURPOSE
C           THE ROUTINE CALCULATES AN APPROXIMATION  RESULT  TO A GIVEN
C           DEFINITE INTEGRAL   I = INTEGRAL OF  F  OVER (A,B),
C           HOPEFULLY SATISFYING FOLLOWING CLAIM FOR ACCURACY
C           ABS(I-RESULT).LE.MAX(EPSABS,EPSREL*ABS(I)).
C
C 3.     CALLING SEQUENCE
C           CALL QAG(F,A,B,EPSABS,EPSREL,KEY,RESULT,ABSERR,NEVAL,IER)
C
C        PARAMETERS
C         ON ENTRY
C            F      - REAL
C                     FUNCTION SUBPROGRAM DEFINING THE INTEGRAND
C                     FUNCTION F(X). THE ACTUAL NAME FOR F NEEDS TO BE
C                     DECLARED E X T E R N A L IN THE DRIVER PROGRAM.
C
C            A      - REAL
C                     LOWER LIMIT OF INTEGRATION
C
C            B      - REAL
C                     UPPER LIMIT OF INTEGRATION
C
C            EPSABS - REAL
C                     ABSOLUTE ACCURACY REQUESTED
C            EPSREL - REAL
C                     RELATIVE ACCURACY REQUESTED
C                     IF  EPSABS.LT.0 AND EPSREL.LT.0,
C                     THE ROUTINE WILL END WITH IER = 6.
C
C            KEY    - INTEGER
C                     KEY FOR CHOICE OF LOCAL INTEGRATION RULE
C                     A GAUSS-KRONROD PAIR IS USED WITH
C                       7 - 15 POINTS IF KEY.LT.2,
C                      10 - 21 POINTS IF KEY = 2,
C                      15 - 31 POINTS IF KEY = 3,
C                      20 - 41 POINTS IF KEY = 4,
C                      25 - 51 POINTS IF KEY = 5,
C                      30 - 61 POINTS IF KEY.GT.5.
C
C         ON RETURN
```

```
C             RESULT - REAL
C                      APPROXIMATION TO THE INTEGRAL
C
C             ABSERR - REAL
C                      ESTIMATE OF THE MODULUS OF THE ABSOLUTE ERROR,
C                      WHICH SHOULD EQUAL OR EXCEED ABS(I-RESULT)
C
C             NEVAL  - INTEGER
C                      NUMBER OF INTEGRAND EVALUATIONS
C
C             IER    - INTEGER
C                      IER = 0 NORMAL AND RELIABLE TERMINATION OF THE
C                              ROUTINE. IT IS ASSUMED THAT THE REQUESTED
C                              ACCURACY HAS BEEN ACHIEVED.
C                      IER.GT.0 ABNORMAL TERMINATION OF THE ROUTINE
C                              THE ESTIMATES FOR RESULT AND ERROR ARE
C                              LESS RELIABLE. IT IS ASSUMED THAT THE
C                              REQUESTED ACCURACY HAS NOT BEEN ACHIEVED.
C                      IER = 1 MAXIMUM NUMBER OF SUBDIVISIONS ALLOWED
C                              HAS BEEN ACHIEVED. ONE CAN ALLOW MORE
C                              SUBDIVISIONS BY INCREASING THE DATA VALUE
C                              OF LIMIT IN QAG (AND TAKING THE ACCORDING
C                              DIMENSION ADJUSTMENTS INTO ACCOUNT).
C                              HOWEVER, IF THIS YIELDS NO IMPROVEMENT IT
C                              IS ADVISED TO ANALYZE THE INTEGRAND IN
C                              ORDER TO DETERMINE THE INTEGRATION
C                              DIFFICULTIES. IF THE POSITION OF A LOCAL
C                              DIFFICULTY CAN BE DETERMINED (E.G.
C                              SINGULARITY, DISCONTINUITY WITHIN THE
C                              INTERVAL) ONE WILL PROBABLY GAIN FROM
C                              SPLITTING UP THE INTERVAL AT THIS POINT
C                              AND CALLING THE INTEGRATOR ON THE
C                              SUBRANGES. IF POSSIBLE, AN APPROPRIATE
C                              SPECIAL-PURPOSE INTEGRATOR SHOULD BE USED
C                              WHICH IS DESIGNED FOR HANDLING THE TYPE OF
C                              DIFFICULTY INVOLVED.
C                          = 2 THE OCCURRENCE OF ROUNDOFF ERROR IS
C                              DETECTED, WHICH PREVENTS THE REQUESTED
C                              TOLERANCE FROM BEING ACHIEVED.
C                          = 3 EXTREMELY BAD INTEGRAND BEHAVIOUR OCCURS
C                              AT SOME POINTS OF THE INTEGRATION
C                              INTERVAL.
C                          = 6 THE INPUT IS INVALID, BECAUSE
C                              EPSABS.LT.0 AND EPSREL.LT.0,
C                              RESULT, ABSERR, NEVAL ARE SET TO ZERO.
C
C 4.      SUBROUTINES OR FUNCTIONS NEEDED
C              - QAGE
```

```
C              - QK15, QK21, QK31, QK41, QK51 AND QK61
C              - QSORT
C              - F (USER-PROVIDED FUNCTION)
C              - QMACO
C              - FORTRAN ABS, AMAX1, AMIN1
C
C ..................................................................
C
      REAL A,ABSERR,ALIST,B,BLIST,ELIST,EPSABS,EPSREL,F,RESULT,RLIST
      INTEGER IER,IORD,KEY,LAST,LIMIT,NEVAL
C
      DIMENSION ALIST(500),BLIST(500),ELIST(500),IORD(500),RLIST(500)
C
      EXTERNAL F
C
C          LIMIT IS THE MAXIMUM NUMBER OF SUBINTERVALS ALLOWED IN
C          THE SUBDIVISION PROCESS OF QAGE. TAKE CARE THAT LIMIT.GE.1.
C
      DATA LIMIT/500/
C
C          KEY DETERMINES THE QUADRATURE FORMULAE TO BE APPLIED
C          BY DQAGE.
C
C
C***FIRST EXECUTABLE STATEMENT
      CALL QAGE(F,A,B,EPSABS,EPSREL,KEY,LIMIT,RESULT,ABSERR,NEVAL,IER,
     *  ALIST,BLIST,RLIST,ELIST,IORD,LAST)
C
      RETURN
      END
```

```
      SUBROUTINE QAGE(F,A,B,EPSABS,EPSREL,KEY,LIMIT,RESULT,ABSERR,NEVAL,
     *  IER,ALIST,BLIST,RLIST,ELIST,IORD,LAST)
C
C ......................................................................
C
C 1.     QAGE
C        COMPUTATION OF A DEFINITE INTEGRAL
C           STANDARD FORTRAN SUBROUTINE
C
C 2.     PURPOSE
C           THE ROUTINE CALCULATES AN APPROXIMATION  RESULT  TO A GIVEN
C           DEFINITE INTEGRAL   I = INTEGRAL OF  F  OVER (A,B),
C           HOPEFULLY SATISFYING FOLLOWING CLAIM FOR ACCURACY
C           ABS(I-RESULT).LE.MAX(EPSABS,EPSREL*ABS(I)).
C
C 3.     CALLING SEQUENCE
C           CALL QAGE(F,A,B,EPSABS,EPSREL,KEY,LIMIT,RESULT,ABSERR,NEVAL,
C                     IER,ALIST,BLIST,RLIST,ELIST,IORD,LAST)
C
C        PARAMETERS
C         ON ENTRY
C            F      - REAL
C                     FUNCTION SUBPROGRAM DEFINING THE INTEGRAND
C                     FUNCTION F(X). THE ACTUAL NAME FOR F NEEDS TO BE
C                     DECLARED E X T E R N A L IN THE DRIVER PROGRAM.
C
C            A      - REAL
C                     LOWER LIMIT OF INTEGRATION
C
C            B      - REAL
C                     UPPER LIMIT OF INTEGRATION
C
C            EPSABS - REAL
C                     ABSOLUTE ACCURACY REQUESTED
C            EPSREL - REAL
C                     RELATIVE ACCURACY REQUESTED
C                     IF  EPSABS.LT.0 AND EPSREL.LT.0,
C                     THE ROUTINE WILL END WITH IER = 6.
C
C            KEY    - INTEGER
C                     KEY FOR CHOICE OF LOCAL INTEGRATION RULE
C                     A GAUSS-KRONROD PAIR IS USED WITH
C                         7 - 15 POINTS IF KEY.LT.2,
C                        10 - 21 POINTS IF KEY = 2,
C                        15 - 31 POINTS IF KEY = 3,
C                        20 - 41 POINTS IF KEY = 4,
C                        25 - 51 POINTS IF KEY = 5,
C                        30 - 61 POINTS IF KEY.GT.5.
```

```
C
C            LIMIT  - INTEGER
C                     GIVES AN UPPERBOUND ON THE NUMBER OF SUBINTERVALS
C                     IN THE PARTITION OF (A,B), LIMIT.GE.1.
C
C         ON RETURN
C            RESULT - REAL
C                     APPROXIMATION TO THE INTEGRAL
C
C            ABSERR - REAL
C                     ESTIMATE OF THE MODULUS OF THE ABSOLUTE ERROR,
C                     WHICH SHOULD EQUAL OR EXCEED ABS(I-RESULT)
C
C            NEVAL  - INTEGER
C                     NUMBER OF INTEGRAND EVALUATIONS
C
C            IER    - INTEGER
C                     IER = 0 NORMAL AND RELIABLE TERMINATION OF THE
C                             ROUTINE. IT IS ASSUMED THAT THE REQUESTED
C                             ACCURACY HAS BEEN ACHIEVED.
C                     IER.GT.0 ABNORMAL TERMINATION OF THE ROUTINE
C                             THE ESTIMATES FOR RESULT AND ERROR ARE
C                             LESS RELIABLE. IT IS ASSUMED THAT THE
C                             REQUESTED ACCURACY HAS NOT BEEN ACHIEVED.
C                     IER = 1 MAXIMUM NUMBER OF SUBDIVISIONS ALLOWED
C                             HAS BEEN ACHIEVED. ONE CAN ALLOW MORE
C                             SUBDIVISIONS BY INCREASING THE VALUE OF
C                             LIMIT. HOWEVER, IF THIS YIELDS NO
C                             IMPROVEMENT IT IS ADVISED TO ANALYZE THE
C                             INTEGRAND IN ORDER TO DETERMINE THE
C                             INTEGRATION DIFFICULTIES. IF THE POSITION
C                             OF A LOCAL DIFFICULTY CAN BE DETERMINED
C                             (E.G. SINGULARITY, DISCONTINUITY WITHIN
C                             THE INTERVAL) ONE WILL PROBABLY GAIN FROM
C                             SPLITTING UP THE INTERVAL AT THIS POINT
C                             AND CALLING THE INTEGRATOR ON THE
C                             SUBRANGES. IF POSSIBLE, AN APPROPRIATE
C                             SPECIAL-PURPOSE INTEGRATOR SHOULD BE USED
C                             WHICH IS DESIGNED FOR HANDLING THE TYPE OF
C                             DIFFICULTY INVOLVED.
C                         = 2 THE OCCURRENCE OF ROUNDOFF ERROR IS
C                             DETECTED, WHICH PREVENTS THE REQUESTED
C                             TOLERANCE FROM BEING ACHIEVED.
C                         = 3 EXTREMELY BAD INTEGRAND BEHAVIOUR OCCURS
C                             AT SOME POINTS OF THE INTEGRATION
C                             INTERVAL.
C                         = 6 THE INPUT IS INVALID, BECAUSE
C                             EPSABS.LT.0 AND EPSREL.LT.0,
```

```
C                               RESULT, ABSERR, NEVAL, LAST, RLIST(1) ,
C                               ELIST(1) AND IORD(1) ARE SET TO ZERO.
C                               ALIST(1) AND BLIST(1) ARE SET TO A AND B
C                               RESPECTIVELY.
C
C            ALIST    - REAL
C                         VECTOR OF DIMENSION AT LEAST LIMIT, THE FIRST
C                          LAST  ELEMENTS OF WHICH ARE THE LEFT END POINTS
C                         OF THE SUBINTERVALS IN THE PARTITION OF THE GIVEN
C                         INTEGRATION RANGE (A,B)
C
C            BLIST    - REAL
C                         VECTOR OF DIMENSION AT LEAST LIMIT, THE FIRST
C                          LAST  ELEMENTS OF WHICH ARE THE RIGHT END POINTS
C                         OF THE SUBINTERVALS IN THE PARTITION OF THE GIVEN
C                         INTEGRATION RANGE (A,B)
C
C            RLIST    - REAL
C                         VECTOR OF DIMENSION AT LEAST LIMIT, THE FIRST
C                          LAST  ELEMENTS OF WHICH ARE THE INTEGRAL
C                         APPROXIMATIONS ON THE SUBINTERVALS
C
C            ELIST    - REAL
C                         VECTOR OF DIMENSION AT LEAST LIMIT, THE FIRST
C                          LAST  ELEMENTS OF WHICH ARE THE MODULI OF THE
C                         ABSOLUTE ERROR ESTIMATES ON THE SUBINTERVALS
C
C            IORD     - INTEGER
C                         VECTOR OF DIMENSION AT LEAST LIMIT, THE FIRST K
C                         ELEMENTS OF WHICH ARE POINTERS TO THE ERROR
C                         ESTIMATES OVER THE SUBINTERVALS, SUCH THAT
C                         ELIST(IORD(1)), ..., ELIST(IORD(K)) FORM A
C                         DECREASING SEQUENCE, WITH K = LAST
C                         IF LAST.LE.(LIMIT/2+2), AND K = LIMIT+1-LAST
C                         OTHERWISE
C
C            LAST     - INTEGER
C                         NUMBER OF SUBINTERVALS ACTUALLY PRODUCED IN THE
C                         SUBDIVISION PROCESS
C
C 4.      SUBROUTINES OR FUNCTIONS NEEDED
C               - QK15, QK21, QK31, QK41, QK51, QK61
C               - QSORT
C               - F (USER-PROVIDED FUNCTION)
C               - QMACO
C               - FORTRAN ABS, AMAX1, AMIN1
C
C ......................................................................
```

```
C
      REAL A,ABSERR,ALIST,AREA,AREA1,AREA12,AREA2,A1,A2,B,BLIST,B1,B2,C,
     *  DEFABS,DEFAB1,DEFAB2,ELIST,EPMACH,EPSABS,EPSREL,ERRBND,ERRMAX,
     *  ERROR1,ERROR2,ERRO12,ERRSUM,F,OFLOW,RESABS,RESULT,RLIST,UFLOW
      INTEGER IER,IORD,IROFF1,IROFF2,K,KEY,KEYF,LAST,LIMIT,MAXERR,NEVAL,
     *  NRMAX
C
      DIMENSION ALIST(LIMIT),BLIST(LIMIT),ELIST(LIMIT),IORD(LIMIT),
     *  RLIST(LIMIT)
C
      EXTERNAL F
C
C            LIST OF MAJOR VARIABLES
C            -----------------------
C
C           ALIST     - LIST OF LEFT END POINTS OF ALL SUBINTERVALS
C                       CONSIDERED UP TO NOW
C           BLIST     - LIST OF RIGHT END POINTS OF ALL SUBINTERVALS
C                       CONSIDERED UP TO NOW
C           RLIST(I)  - APPROXIMATION TO THE INTEGRAL OVER
C                       (ALIST(I),BLIST(I))
C           ELIST(I)  - ERROR ESTIMATE APPLYING TO RLIST(I)
C           MAXERR    - POINTER TO THE INTERVAL WITH LARGEST ERROR
C                       ESTIMATE
C           ERRMAX    - ELIST(MAXERR)
C           AREA      - SUM OF THE INTEGRALS OVER THE SUBINTERVALS
C           ERRSUM    - SUM OF THE ERRORS OVER THE SUBINTERVALS
C           ERRBND    - REQUESTED ACCURACY MAX(EPSABS,EPSREL*
C                       ABS(RESULT))
C           *****1    - VARIABLE FOR THE LEFT SUBINTERVAL
C           *****2    - VARIABLE FOR THE RIGHT SUBINTERVAL
C           LAST      - INDEX FOR SUBDIVISION
C
C
C           MACHINE DEPENDENT CONSTANTS
C           ---------------------------
C
C           EPMACH IS THE LARGEST RELATIVE SPACING.
C           UFLOW IS THE SMALLEST POSITIVE MAGNITUDE.
C           OFLOW IS THE LARGEST MAGNITUDE.
C
C***FIRST EXECUTABLE STATEMENT
      CALL QMACO(EPMACH,UFLOW,OFLOW)
C
C           TEST ON VALIDITY OF PARAMETERS
C           ------------------------------
C
      IER = 0
```

```
      NEVAL = 0
      LAST = 0
      RESULT = 0.0E+00
      ABSERR = 0.0E+00
      ALIST(1) = A
      BLIST(1) = B
      RLIST(1) = 0.0E+00
      ELIST(1) = 0.0E+00
      IORD(1) = 0
      IF(EPSABS.LT.0.0E+00.AND.EPSREL.LT.0.0E+00) IER = 6
      IF(IER.EQ.6) GO TO 999
C
C           FIRST APPROXIMATION TO THE INTEGRAL
C           -----------------------------------
C
      KEYF = KEY
      IF(KEY.LE.0) KEYF = 1
      IF(KEY.GE.7) KEYF = 6
      C = KEYF
      NEVAL = 0
      IF(KEYF.EQ.1) CALL QK15(F,A,B,RESULT,ABSERR,DEFABS,RESABS)
      IF(KEYF.EQ.2) CALL QK21(F,A,B,RESULT,ABSERR,DEFABS,RESABS)
      IF(KEYF.EQ.3) CALL QK31(F,A,B,RESULT,ABSERR,DEFABS,RESABS)
      IF(KEYF.EQ.4) CALL QK41(F,A,B,RESULT,ABSERR,DEFABS,RESABS)
      IF(KEYF.EQ.5) CALL QK51(F,A,B,RESULT,ABSERR,DEFABS,RESABS)
      IF(KEYF.EQ.6) CALL QK61(F,A,B,RESULT,ABSERR,DEFABS,RESABS)
      LAST = 1
      RLIST(1) = RESULT
      ELIST(1) = ABSERR
      IORD(1) = 1
C
C           TEST ON ACCURACY.
C
      ERRBND = AMAX1(EPSABS,EPSREL*ABS(RESULT))
      IF(ABSERR.LE.5.0E+01*EPMACH*DEFABS.AND.ABSERR.GT.ERRBND) IER = 2
      IF(LIMIT.EQ.1) IER = 1
      IF(IER.NE.0.OR.(ABSERR.LE.ERRBND.AND.ABSERR.NE.RESABS)
     *  .OR.ABSERR.EQ.0.0E+00) GO TO 60
C
C           INITIALIZATION
C           --------------
C
C
      ERRMAX = ABSERR
      MAXERR = 1
      AREA = RESULT
      ERRSUM = ABSERR
      NRMAX = 1
```

```
      IROFF1 = 0
      IROFF2 = 0
C
C           MAIN DO-LOOP
C           ------------
C
      DO 30 LAST = 2,LIMIT
C
C           BISECT THE SUBINTERVAL WITH THE LARGEST ERROR ESTIMATE.
C
        A1 = ALIST(MAXERR)
        B1 = 5.0E-01*(ALIST(MAXERR)+BLIST(MAXERR))
        A2 = B1
        B2 = BLIST(MAXERR)
        IF(KEYF.EQ.1) CALL QK15(F,A1,B1,AREA1,ERROR1,RESABS,DEFAB1)
        IF(KEYF.EQ.2) CALL QK21(F,A1,B1,AREA1,ERROR1,RESABS,DEFAB1)
        IF(KEYF.EQ.3) CALL QK31(F,A1,B1,AREA1,ERROR1,RESABS,DEFAB1)
        IF(KEYF.EQ.4) CALL QK41(F,A1,B1,AREA1,ERROR1,RESABS,DEFAB1)
        IF(KEYF.EQ.5) CALL QK51(F,A1,B1,AREA1,ERROR1,RESABS,DEFAB1)
        IF(KEYF.EQ.6) CALL QK61(F,A1,B1,AREA1,ERROR1,RESABS,DEFAB1)
        IF(KEYF.EQ.1) CALL QK15(F,A2,B2,AREA2,ERROR2,RESABS,DEFAB2)
        IF(KEYF.EQ.2) CALL QK21(F,A2,B2,AREA2,ERROR2,RESABS,DEFAB2)
        IF(KEYF.EQ.3) CALL QK31(F,A2,B2,AREA2,ERROR2,RESABS,DEFAB2)
        IF(KEYF.EQ.4) CALL QK41(F,A2,B2,AREA2,ERROR2,RESABS,DEFAB2)
        IF(KEYF.EQ.5) CALL QK51(F,A2,B2,AREA2,ERROR2,RESABS,DEFAB2)
        IF(KEYF.EQ.6) CALL QK61(F,A2,B2,AREA2,ERROR2,RESABS,DEFAB2)
C
C           IMPROVE PREVIOUS APPROXIMATIONS TO INTEGRAL AND ERROR AND
C           TEST FOR ACCURACY.
C
        NEVAL = NEVAL+1
        AREA12 = AREA1+AREA2
        ERRO12 = ERROR1+ERROR2
        ERRSUM = ERRSUM+ERRO12-ERRMAX
        AREA = AREA+AREA12-RLIST(MAXERR)
        IF(DEFAB1.EQ.ERROR1.OR.DEFAB2.EQ.ERROR2) GO TO 5
        IF(ABS(RLIST(MAXERR)-AREA12).LE.1.0E-05*ABS(AREA12)
     *  .AND.ERRO12.GE.9.9E-01*ERRMAX) IROFF1 = IROFF1+1
        IF(LAST.GT.10.AND.ERRO12.GT.ERRMAX) IROFF2 = IROFF2+1
    5   RLIST(MAXERR) = AREA1
        RLIST(LAST) = AREA2
        ERRBND = AMAX1(EPSABS,EPSREL*ABS(AREA))
        IF(ERRSUM.LE.ERRBND) GO TO 8
C
C           TEST FOR ROUNDOFF ERROR AND EVENTUALLY SET ERROR FLAG
C
        IF(IROFF1.GE.6.OR.IROFF2.GE.20) IER = 2
C
```

```
C           SET ERROR FLAG IN THE CASE THAT THE NUMBER OF SUBINTERVALS
C           EQUALS LIMIT.
C
        IF(LAST.EQ.LIMIT) IER = 1
C
C           SET ERROR FLAG IN THE CASE OF BAD INTEGRAND BEHAVIOUR
C           AT A POINT OF THE INTEGRATION RANGE.
C
        IF(AMAX1(ABS(A1),ABS(B2)).LE.(1.0E+00+C*1.0E+03*
     *  EPMACH)*(ABS(A2)+1.0E+04*UFLOW)) IER = 3
C
C           APPEND THE NEWLY-CREATED INTERVALS TO THE LIST.
C
    8   IF(ERROR2.GT.ERROR1) GO TO 10
        ALIST(LAST) = A2
        BLIST(MAXERR) = B1
        BLIST(LAST) = B2
        ELIST(MAXERR) = ERROR1
        ELIST(LAST) = ERROR2
        GO TO 20
   10   ALIST(MAXERR) = A2
        ALIST(LAST) = A1
        BLIST(LAST) = B1
        RLIST(MAXERR) = AREA2
        RLIST(LAST) = AREA1
        ELIST(MAXERR) = ERROR2
        ELIST(LAST) = ERROR1
C
C           CALL SUBROUTINE QSORT TO MAINTAIN THE DESCENDING ORDERING
C           IN THE LIST OF ERROR ESTIMATES AND SELECT THE SUBINTERVAL
C           WITH THE LARGEST ERROR ESTIMATE (TO BE BISECTED NEXT).
C
   20   CALL QSORT(LIMIT,LAST,MAXERR,ERRMAX,ELIST,IORD,NRMAX)
C***JUMP OUT OF DO-LOOP
        IF(IER.NE.0.OR.ERRSUM.LE.ERRBND) GO TO 40
   30 CONTINUE
C
C           COMPUTE FINAL RESULT.
C           ---------------------
C
   40 RESULT = 0.0E+00
      DO 50 K=1,LAST
        RESULT = RESULT+RLIST(K)
   50 CONTINUE
      ABSERR = ERRSUM
   60 IF(KEYF.NE.1) NEVAL = (10*KEYF+1)*(2*NEVAL+1)
      IF(KEYF.EQ.1) NEVAL = 30*NEVAL+15
  999 RETURN
      END
```

```
      SUBROUTINE QAGS(F,A,B,EPSABS,EPSREL,RESULT,ABSERR,NEVAL,IER)
C
C ......................................................................
C
C 1.     QAGS
C        COMPUTATION OF A DEFINITE INTEGRAL
C           STANDARD FORTRAN SUBROUTINE
C
C 2.     PURPOSE
C           THE ROUTINE CALCULATES AN APPROXIMATION  RESULT  TO A GIVEN
C           DEFINITE INTEGRAL   I = INTEGRAL OF  F  OVER (A,B),
C           HOPEFULLY SATISFYING FOLLOWING CLAIM FOR ACCURACY
C           ABS(I-RESULT).LE.MAX(EPSABS,EPSREL*ABS(I)).
C
C 3.     CALLING SEQUENCE
C           CALL QAGS(F,A,B,EPSABS,EPSREL,RESULT,ABSERR,NEVAL,IER)
C
C        PARAMETERS
C         ON ENTRY
C            F      - REAL
C                     FUNCTION SUBPROGRAM DEFINING THE INTEGRAND
C                     FUNCTION F(X). THE ACTUAL NAME FOR F NEEDS TO BE
C                     DECLARED E X T E R N A L IN THE DRIVER PROGRAM.
C
C            A      - REAL
C                     LOWER LIMIT OF INTEGRATION
C
C            B      - REAL
C                     UPPER LIMIT OF INTEGRATION
C
C            EPSABS - REAL
C                     ABSOLUTE ACCURACY REQUESTED
C            EPSREL - REAL
C                     RELATIVE ACCURACY REQUESTED
C                     IF  EPSABS.LT.0 AND EPSREL.LT.0,
C                     THE ROUTINE WILL END WITH IER = 6.
C
C         ON RETURN
C            RESULT - REAL
C                     APPROXIMATION TO THE INTEGRAL
C
C            ABSERR - REAL
C                     ESTIMATE OF THE MODULUS OF THE ABSOLUTE ERROR,
C                     WHICH SHOULD EQUAL OR EXCEED ABS(I-RESULT)
C
C            NEVAL  - INTEGER
C                     NUMBER OF INTEGRAND EVALUATIONS
C
```

```
C            IER    - INTEGER
C                     IER = 0 NORMAL AND RELIABLE TERMINATION OF THE
C                             ROUTINE. IT IS ASSUMED THAT THE REQUESTED
C                             ACCURACY HAS BEEN ACHIEVED.
C                     IER.GT.0 ABNORMAL TERMINATION OF THE ROUTINE
C                             THE ESTIMATES FOR INTEGRAL AND ERROR ARE
C                             LESS RELIABLE. IT IS ASSUMED THAT THE
C                             REQUESTED ACCURACY HAS NOT BEEN ACHIEVED.
C                         = 1 MAXIMUM NUMBER OF SUBDIVISIONS ALLOWED
C                             HAS BEEN ACHIEVED. ONE CAN ALLOW MORE SUB-
C                             DIVISIONS BY INCREASING THE DATA VALUE OF
C                             LIMIT IN QAGS (AND TAKING THE ACCORDING
C                             DIMENSION ADJUSTMENTS INTO ACCOUNT).
C                             HOWEVER, IF THIS YIELDS NO IMPROVEMENT
C                             IT IS ADVISED TO ANALYZE THE INTEGRAND
C                             IN ORDER TO DETERMINE THE INTEGRATION
C                             DIFFICULTIES. IF THE POSITION OF A
C                             LOCAL DIFFICULTY CAN BE DETERMINED (E.G.
C                             SINGULARITY, DISCONTINUITY WITHIN THE
C                             INTERVAL) ONE WILL PROBABLY GAIN FROM
C                             SPLITTING UP THE INTERVAL AT THIS POINT
C                             AND CALLING THE INTEGRATOR ON THE SUB-
C                             RANGES. IF POSSIBLE, AN APPROPRIATE
C                             SPECIAL-PURPOSE INTEGRATOR SHOULD BE USED,
C                             WHICH IS DESIGNED FOR HANDLING THE TYPE
C                             OF DIFFICULTY INVOLVED.
C                         = 2 THE OCCURRENCE OF ROUNDOFF ERROR IS DETEC-
C                             TED, WHICH PREVENTS THE REQUESTED
C                             TOLERANCE FROM BEING ACHIEVED.
C                             THE ERROR MAY BE UNDER-ESTIMATED.
C                         = 3 EXTREMELY BAD INTEGRAND BEHAVIOUR OCCURS
C                             AT SOME  POINTS OF THE INTEGRATION
C                             INTERVAL.
C                         = 4 THE ALGORITHM DOES NOT CONVERGE. ROUNDOFF
C                             ERROR IS DETECTED IN THE EXTRAPOLATION
C                             TABLE. IT IS PRESUMED THAT THE REQUESTED
C                             TOLERANCE CANNOT BE ACHIEVED, AND THAT THE
C                             RETURNED RESULT IS THE BEST WHICH CAN BE
C                             OBTAINED.
C                         = 5 THE INTEGRAL IS PROBABLY DIVERGENT, OR
C                             SLOWLY CONVERGENT. IT MUST BE NOTED THAT
C                             DIVERGENCE CAN OCCUR WITH ANY OTHER VALUE
C                             OF IER.
C                         = 6 THE INPUT IS INVALID, BECAUSE
C                             EPSABS.LT.0 AND EPSREL.LT.0,
C                             RESULT, ABSERR AND NEVAL ARE SET TO ZERO.
C
C 4.      SUBROUTINES OR FUNCTIONS NEEDED
```

```
C               - QK21
C               - QSORT
C               - QEXT
C               - F (USER-PROVIDED FUNCTION)
C               - QMACO
C               - FORTRAN ABS, MAX1, AMIN1
C
C ...................................................................
C
      REAL A,ABSEPS,ABSERR,ALIST,AREA,AREA1,AREA12,AREA2,A1,A2,B,BLIST,
     *  B1,B2,CORREC,DEFABS,DEFAB1,DEFAB2,DRES,ELIST,EPMACH,EPSABS,
     *  EPSREL,ERLARG,ERLAST,ERRBND,ERRMAX,ERROR1,ERROR2,ERRO12,ERRSUM,
     *  ERTEST,F,OFLOW,RESABS,RESEPS,RESULT,RES3LA,RLIST,RLIST2,SMALL,
     *  UFLOW
      INTEGER ID,IER,IERRO,IORD,IROFF1,IROFF2,IROFF3,JUPBND,K,KSGN,
     *  KTMIN,LAST,LIMIT,MAXERR,NEVAL,NRES,NRMAX,NUMRL2
      LOGICAL EXTRAP,NOEXT
C
      DIMENSION ALIST(500),BLIST(500),ELIST(500),IORD(500),RES3LA(3),
     *  RLIST(500),RLIST2(52)
C
C             THE DIMENSION OF RLIST2 IS DETERMINED BY THE VALUE OF
C             LIMEXP IN SUBROUTINE QEXT (RLIST2 SHOULD BE OF DIMENSION
C             (LIMEXP+2) AT LEAST).
C
      EXTERNAL F
C
C           LIMIT IS THE MAXIMUM NUMBER OF SUBINTERVALS ALLOWED IN THE
C           SUBDIVISION PROCESS. TAKE CARE THAT LIMIT.GE.1.
C
      DATA LIMIT/500/
C
C             LIST OF MAJOR VARIABLES
C             -----------------------
C
C            ALIST     - LIST OF LEFT END POINTS OF ALL SUBINTERVALS
C                        CONSIDERED UP TO NOW
C            BLIST     - LIST OF RIGHT END POINTS OF ALL SUBINTERVALS
C                        CONSIDERED UP TO NOW
C            RLIST(I)  - APPROXIMATION TO THE INTEGRAL OVER
C                        (ALIST(I),BLIST(I))
C            RLIST2    - ARRAY OF DIMENSION AT LEAST LIMEXP+2 CONTAINING
C                        THE PART OF THE EPSILON TABLE WHICH IS STILL
C                        NEEDED FOR FURTHER COMPUTATIONS
C            ELIST(I)  - ERROR ESTIMATE APPLYING TO RLIST(I)
C            MAXERR    - POINTER TO THE INTERVAL WITH LARGEST ERROR
C                        ESTIMATE
C            ERRMAX    - ELIST(MAXERR)
```

```
C           ERLAST    - ERROR ON THE INTERVAL CURRENTLY SUBDIVIDED
C                       (BEFORE THAT SUBDIVISION HAS TAKEN PLACE)
C           AREA      - SUM OF THE INTEGRALS OVER THE SUBINTERVALS
C           ERRSUM    - SUM OF THE ERRORS OVER THE SUBINTERVALS
C           ERRBND    - REQUESTED ACCURACY MAX(EPSABS,EPSREL*
C                       ABS(RESULT))
C           *****1    - VARIABLE FOR THE LEFT INTERVAL
C           *****2    - VARIABLE FOR THE RIGHT INTERVAL
C           LAST      - INDEX FOR SUBDIVISION
C           NRES      - NUMBER OF CALLS TO THE EXTRAPOLATION ROUTINE
C           NUMRL2    - NUMBER OF ELEMENTS CURRENTLY IN RLIST2. IF AN
C                       APPROPRIATE APPROXIMATION TO THE COMPOUNDED
C                       INTEGRAL HAS BEEN OBTAINED IT IS PUT IN
C                       RLIST2(NUMRL2) AFTER NUMRL2 HAS BEEN INCREASED
C                       BY ONE.
C           SMALL     - LENGTH OF THE SMALLEST INTERVAL CONSIDERED
C                       UP TO NOW, MULTIPLIED BY 1.5
C           ERLARG    - SUM OF THE ERRORS OVER THE INTERVALS LARGER
C                       THAN THE SMALLEST INTERVAL CONSIDERED UP TO NOW
C           EXTRAP    - LOGICAL VARIABLE DENOTING THAT THE ROUTINE IS
C                       ATTEMPTING TO PERFORM EXTRAPOLATION I.E. BEFORE
C                       SUBDIVIDING THE SMALLEST INTERVAL WE TRY TO
C                       DECREASE THE VALUE OF ERLARG.
C           NOEXT     - LOGICAL VARIABLE DENOTING THAT EXTRAPOLATION
C                       IS NO LONGER ALLOWED (TRUE VALUE)
C
C            MACHINE DEPENDENT CONSTANTS
C            ---------------------------
C
C           EPMACH IS THE LARGEST RELATIVE SPACING.
C           UFLOW IS THE SMALLEST POSITIVE MAGNITUDE.
C           OFLOW IS THE LARGEST POSITIVE MAGNITUDE.
C
C***FIRST EXECUTABLE STATEMENT
      CALL QMACO(EPMACH,UFLOW,OFLOW)
C
C            TEST ON VALIDITY OF PARAMETERS
C            ------------------------------
      IER = 0
      NEVAL = 0
      LAST = 0
      RESULT = 0.0E+00
      ABSERR = 0.0E+00
      ALIST(1) = A
      BLIST(1) = B
      RLIST(1) = 0.0E+00
      ELIST(1) = 0.0E+00
      IF(EPSABS.LT.0.0E+00.AND.EPSREL.LT.0.0E+00) IER = 6
```

```
      IF(IER.EQ.6) GO TO 999
C
C           FIRST APPROXIMATION TO THE INTEGRAL
C           -----------------------------------
C
      IERRO = 0
      CALL QK21(F,A,B,RESULT,ABSERR,DEFABS,RESABS)
C
C           TEST ON ACCURACY.
C
      DRES = ABS(RESULT)
      ERRBND = AMAX1(EPSABS,EPSREL*DRES)
      LAST = 1
      RLIST(1) = RESULT
      ELIST(1) = ABSERR
      IORD(1) = 1
      IF(ABSERR.LE.1.0E+02*EPMACH*DEFABS.AND.ABSERR.GT.ERRBND) IER = 2
      IF(LIMIT.EQ.1) IER = 1
      IF(IER.NE.0.OR.(ABSERR.LE.ERRBND.AND.ABSERR.NE.RESABS).OR.
     *  ABSERR.EQ.0.0E+00) GO TO 140
C
C           INITIALIZATION
C           --------------
C
      RLIST2(1) = RESULT
      ERRMAX = ABSERR
      MAXERR = 1
      AREA = RESULT
      ERRSUM = ABSERR
      ABSERR = OFLOW
      NRMAX = 1
      NRES = 0
      NUMRL2 = 2
      KTMIN = 0
      EXTRAP = .FALSE.
      NOEXT = .FALSE.
      IROFF1 = 0
      IROFF2 = 0
      IROFF3 = 0
      KSGN = -1
      IF(DRES.GE.(1.0E+00-5.0E+01*EPMACH)*DEFABS) KSGN = 1
C
C           MAIN DO-LOOP
C           ------------
C
      DO 90 LAST = 2,LIMIT
C
C           BISECT THE SUBINTERVAL WITH THE NRMAX-TH LARGEST ERROR
```

```
C           ESTIMATE.
C
        A1 = ALIST(MAXERR)
        B1 = 5.0E-01*(ALIST(MAXERR)+BLIST(MAXERR))
        A2 = B1
        B2 = BLIST(MAXERR)
        ERLAST = ERRMAX
        CALL QK21(F,A1,B1,AREA1,ERROR1,RESABS,DEFAB1)
        CALL QK21(F,A2,B2,AREA2,ERROR2,RESABS,DEFAB2)
C
C           IMPROVE PREVIOUS APPROXIMATIONS TO INTEGRAL AND ERROR
C           AND TEST FOR ACCURACY.
C
        AREA12 = AREA1+AREA2
        ERRO12 = ERROR1+ERROR2
        ERRSUM = ERRSUM+ERRO12-ERRMAX
        AREA = AREA+AREA12-RLIST(MAXERR)
        IF(DEFAB1.EQ.ERROR1.OR.DEFAB2.EQ.ERROR2) GO TO 15
        IF(ABS(RLIST(MAXERR)-AREA12).GT.1.0E-05*ABS(AREA12)
     * .OR.ERRO12.LT.9.9E-01*ERRMAX) GO TO 10
        IF(EXTRAP) IROFF2 = IROFF2+1
        IF(.NOT.EXTRAP) IROFF1 = IROFF1+1
   10   IF(LAST.GT.10.AND.ERRO12.GT.ERRMAX) IROFF3 = IROFF3+1
   15   RLIST(MAXERR) = AREA1
        RLIST(LAST) = AREA2
        ERRBND = AMAX1(EPSABS,EPSREL*ABS(AREA))
C
C           TEST FOR ROUNDOFF ERROR AND EVENTUALLY SET ERROR FLAG.
C
        IF(IROFF1+IROFF2.GE.10.OR.IROFF3.GE.20) IER = 2
        IF(IROFF2.GE.5) IERRO = 3
C
C           SET ERROR FLAG IN THE CASE THAT THE NUMBER OF SUBINTERVALS
C           EQUALS LIMIT.
C
        IF(LAST.EQ.LIMIT) IER = 1
C
C           SET ERROR FLAG IN THE CASE OF BAD INTEGRAND BEHAVIOUR
C           AT A POINT OF THE INTEGRATION RANGE.
C
        IF(AMAX1(ABS(A1),ABS(B2)).LE.(1.0E+00+1.0E+03*EPMACH)*
     * (ABS(A2)+1.0E+03*UFLOW)) IER = 4
C
C           APPEND THE NEWLY-CREATED INTERVALS TO THE LIST.
C
        IF(ERROR2.GT.ERROR1) GO TO 20
        ALIST(LAST) = A2
        BLIST(MAXERR) = B1
```

```
      BLIST(LAST) = B2
      ELIST(MAXERR) = ERROR1
      ELIST(LAST) = ERROR2
      GO TO 30
   20 ALIST(MAXERR) = A2
      ALIST(LAST) = A1
      BLIST(LAST) = B1
      RLIST(MAXERR) = AREA2
      RLIST(LAST) = AREA1
      ELIST(MAXERR) = ERROR2
      ELIST(LAST) = ERROR1
C
C           CALL SUBROUTINE QSORT TO MAINTAIN THE DESCENDING ORDERING
C           IN THE LIST OF ERROR ESTIMATES AND SELECT THE SUBINTERVAL
C           WITH NRMAX-TH LARGEST ERROR ESTIMATE (TO BE BISECTED NEXT).
C
   30 CALL QSORT(LIMIT,LAST,MAXERR,ERRMAX,ELIST,IORD,NRMAX)
C***JUMP OUT OF DO-LOOP
      IF(ERRSUM.LE.ERRBND) GO TO 115
C***JUMP OUT OF DO-LOOP
      IF(IER.NE.0) GO TO 100
      IF(LAST.EQ.2) GO TO 80
      IF(NOEXT) GO TO 90
      ERLARG = ERLARG-ERLAST
      IF(ABS(B1-A1).GT.SMALL) ERLARG = ERLARG+ERRO12
      IF(EXTRAP) GO TO 40
C
C           TEST WHETHER THE INTERVAL TO BE BISECTED NEXT IS THE
C           SMALLEST INTERVAL.
C
      IF(ABS(BLIST(MAXERR)-ALIST(MAXERR)).GT.SMALL) GO TO 90
      EXTRAP = .TRUE.
      NRMAX = 2
   40 IF(IERRO.EQ.3.OR.ERLARG.LE.ERTEST) GO TO 60
C
C           THE SMALLEST INTERVAL HAS THE LARGEST ERROR.
C           BEFORE BISECTING DECREASE THE SUM OF THE ERRORS OVER THE
C           LARGER INTERVALS (ERLARG) AND PERFORM EXTRAPOLATION.
C
      ID = NRMAX
      JUPBND = LAST
      IF(LAST.GT.(2+LIMIT/2)) JUPBND = LIMIT+3-LAST
      DO 50 K = ID,JUPBND
        MAXERR = IORD(NRMAX)
        ERRMAX = ELIST(MAXERR)
C***JUMP OUT OF DO-LOOP
        IF(ABS(BLIST(MAXERR)-ALIST(MAXERR)).GT.SMALL) GO TO 90
        NRMAX = NRMAX+1
```

```
   50   CONTINUE
C
C           PERFORM EXTRAPOLATION.
C
   60   NUMRL2 = NUMRL2+1
        RLIST2(NUMRL2) = AREA
        CALL QEXT(NUMRL2,RLIST2,RESEPS,ABSEPS,RES3LA,NRES)
        KTMIN = KTMIN+1
        IF(KTMIN.GT.5.AND.ABSERR.LT.1.0E-03*ERRSUM) IER = 5
        IF(ABSEPS.GE.ABSERR) GO TO 70
        KTMIN = 0
        ABSERR = ABSEPS
        RESULT = RESEPS
        CORREC = ERLARG
        ERTEST = AMAX1(EPSABS,EPSREL*ABS(RESEPS))
C***JUMP OUT OF DO-LOOP
        IF(ABSERR.LE.ERTEST) GO TO 100
C
C           PREPARE BISECTION OF THE SMALLEST INTERVAL.
C
   70   IF(NUMRL2.EQ.1) NOEXT = .TRUE.
        IF(IER.EQ.5) GO TO 100
        MAXERR = IORD(1)
        ERRMAX = ELIST(MAXERR)
        NRMAX = 1
        EXTRAP = .FALSE.
        SMALL = SMALL*5.0E-01
        ERLARG = ERRSUM
        GO TO 90
   80   SMALL = ABS(B-A)*3.75E-01
        ERLARG = ERRSUM
        ERTEST = ERRBND
        RLIST2(2) = AREA
   90 CONTINUE
C
C           SET FINAL RESULT AND ERROR ESTIMATE.
C           ------------------------------------
C
  100 IF(ABSERR.EQ.OFLOW) GO TO 115
      IF(IER+IERRO.EQ.0) GO TO 110
      IF(IERRO.EQ.3) ABSERR = ABSERR+CORREC
      IF(IER.EQ.0) IER = 3
      IF(RESULT.NE.0.0E+00.AND.AREA.NE.0.0E+00) GO TO 105
      IF(ABSERR.GT.ERRSUM) GO TO 115
      IF(AREA.EQ.0.0E+00) GO TO 130
      GO TO 110
  105 IF(ABSERR/ABS(RESULT).GT.ERRSUM/ABS(AREA)) GO TO 115
C
```

```
C           TEST ON DIVERGENCE.
C
  110 IF(KSGN.EQ.(-1).AND.AMAX1(ABS(RESULT),ABS(AREA)).LE.
     * DEFABS*1.0E-02) GO TO 130
      IF(1.0E-02.GT.(RESULT/AREA).OR.(RESULT/AREA).GT.1.0E+02
     * .OR.ERRSUM.GT.ABS(AREA)) IER = 6
      GO TO 130
C
C           COMPUTE GLOBAL INTEGRAL SUM.
C
  115 RESULT = 0.0E+00
      DO 120 K = 1,LAST
         RESULT = RESULT+RLIST(K)
  120 CONTINUE
      ABSERR = ERRSUM
  130 IF(IER.GT.2) IER = IER-1
  140 NEVAL = 42*LAST-21
  999 RETURN
      END
```

```
      SUBROUTINE QAGP(F,A,B,NPTS2,POINTS,EPSABS,EPSREL,RESULT,ABSERR,
     *  NEVAL,IER)
C
C ..........................................................................
C
C 1.     QAGP
C        COMPUTATION OF A DEFINITE INTEGRAL
C          STANDARD FORTRAN SUBROUTINE
C
C 2.     PURPOSE
C           THE ROUTINE CALCULATES AN APPROXIMATION  RESULT  TO A GIVEN
C           DEFINITE INTEGRAL   I = INTEGRAL OF  F  OVER (A,B),
C           HOPEFULLY SATISFYING FOLLOWING CLAIM FOR ACCURACY
C           ABS(I-RESULT).LE.MAX(EPSABS,EPSREL*ABS(I)).
C           BREAK POINTS  OF THE INTEGRATION INTERVAL, WHERE LOCAL
C           DIFFICULTIES OF THE INTEGRAND MAY OCCUR (E.G. SINGULARITIES,
C           DISCONTINUITIES), ARE PROVIDED BY THE USER.
C
C 3.     CALLING SEQUENCE
C           CALL QAGP(F,A,B,NPTS2,POINTS,EPSABS,EPSREL,RESULT,ABSERR,
C                     NEVAL,IER)
C
C        PARAMETERS
C         ON ENTRY
C            F      - REAL
C                     FUNCTION SUBPROGRAM DEFINING THE INTEGRAND
C                     FUNCTION F(X). THE ACTUAL NAME FOR F NEEDS TO BE
C                     DECLARED E X T E R N A L IN THE DRIVER PROGRAM.
C
C            A      - REAL
C                     LOWER LIMIT OF INTEGRATION
C
C            B      - REAL
C                     UPPER LIMIT OF INTEGRATION
C
C            NPTS2  - INTEGER
C                     NUMBER EQUAL TO TWO MORE THAN THE NUMBER OF
C                     USER-SUPPLIED INTERIOR BREAK POINTS WITHIN THE
C                     INTEGRATION RANGE, NPTS2.GE.2.
C                     IF NPTS2.LT.2, THE ROUTINE WILL END WITH IER = 6.
C
C            POINTS - REAL
C                     VECTOR OF DIMENSION NPTS2, THE FIRST (NPTS2-2)
C                     ELEMENTS OF WHICH ARE THE USER PROVIDED INTERIOR
C                     BREAK POINTS. IF THESE POINTS DO NOT CONSTITUTE
C                     AN ASCENDING SEQUENCE THERE WILL BE AN AUTOMATIC
C                     SORTING.
C
```

```
C            EPSABS - REAL
C                     ABSOLUTE ACCURACY REQUESTED
C            EPSREL - REAL
C                     RELATIVE ACCURACY REQUESTED
C                     IF  EPSABS.LT.0 AND EPSREL.LT.0,
C                     THE ROUTINE WILL END WITH IER = 6.
C
C         ON RETURN
C            RESULT - REAL
C                     APPROXIMATION TO THE INTEGRAL
C
C            ABSERR - REAL
C                     ESTIMATE OF THE MODULUS OF THE ABSOLUTE ERROR,
C                     WHICH SHOULD EQUAL OR EXCEED ABS(I-RESULT)
C
C            NEVAL  - INTEGER
C                     NUMBER OF INTEGRAND EVALUATIONS
C
C            IER    - INTEGER
C                     IER = 0 NORMAL AND RELIABLE TERMINATION OF THE
C                             ROUTINE. IT IS ASSUMED THAT THE REQUESTED
C                             ACCURACY HAS BEEN ACHIEVED.
C                     IER.GT.0 ABNORMAL TERMINATION OF THE ROUTINE.
C                             THE ESTIMATES FOR INTEGRAL AND ERROR ARE
C                             LESS RELIABLE. IT IS ASSUMED THAT THE
C                             REQUESTED ACCURACY HAS NOT BEEN ACHIEVED.
C                     IER = 1 MAXIMUM NUMBER OF SUBDIVISIONS ALLOWED
C                             HAS BEEN ACHIEVED. ONE CAN ALLOW MORE
C                             SUBDIVISIONS BY INCREASING THE DATA VALUE
C                             OF LIMIT IN QAGP(AND TAKING THE ACCORDING
C                             DIMENSION ADJUSTMENTS INTO ACCOUNT).
C                             HOWEVER, IF THIS YIELDS NO IMPROVEMENT
C                             IT IS ADVISED TO ANALYZE THE INTEGRAND
C                             IN ORDER TO DETERMINE THE INTEGRATION
C                             DIFFICULTIES. IF THE POSITION OF A LOCAL
C                             DIFFICULTY CAN BE DETERMINED (I.E.
C                             SINGULARITY, DISCONTINUITY WITHIN THE
C                             INTERVAL), IT SHOULD BE SUPPLIED TO THE
C                             ROUTINE AS AN ELEMENT OF THE VECTOR
C                             POINTS. IF NECESSARY, AN APPROPRIATE
C                             SPECIAL-PURPOSE INTEGRATOR MUST BE USED,
C                             WHICH IS DESIGNED FOR HANDLING THE TYPE
C                             OF DIFFICULTY INVOLVED.
C                         = 2 THE OCCURRENCE OF ROUNDOFF ERROR IS
C                             DETECTED, WHICH PREVENTS THE REQUESTED
C                             TOLERANCE FROM BEING ACHIEVED.
C                             THE ERROR MAY BE UNDER-ESTIMATED.
C                         = 3 EXTREMELY BAD INTEGRAND BEHAVIOUR OCCURS
```

```
C                                  AT SOME POINTS OF THE INTEGRATION
C                                  INTERVAL.
C                              = 4 THE ALGORITHM DOES NOT CONVERGE. ROUNDOFF
C                                  ERROR IS DETECTED IN THE EXTRAPOLATION
C                                  TABLE. IT IS PRESUMED THAT THE REQUESTED
C                                  TOLERANCE CANNOT BE ACHIEVED, AND THAT
C                                  THE RETURNED RESULT IS THE BEST WHICH
C                                  CAN BE OBTAINED.
C                              = 5 THE INTEGRAL IS PROBABLY DIVERGENT, OR
C                                  SLOWLY CONVERGENT. IT MUST BE NOTED THAT
C                                  DIVERGENCE CAN OCCUR WITH ANY OTHER VALUE
C                                  OF IER.GT.0.
C                              = 6 THE INPUT IS INVALID BECAUSE
C                                  NPTS2.LT.2 OR
C                                  BREAK POINTS ARE SPECIFIED OUTSIDE
C                                  THE INTEGRATION RANGE OR
C                                  EPSABS.LT.0 AND EPSREL.LT.0,
C                                  OR LIMIT.LT.NPTS2.
C                                  RESULT, ABSERR, NEVAL ARE SET TO ZERO.
C
C
C 4.      SUBROUTINES OR FUNCTIONS NEEDED
C               - QK21
C               - QSORT
C               - QEXT
C               - F (USER-PROVIDED FUNCTION)
C               - QMACO
C               - FORTRAN ABS, AMAX1, AMIN1
C
C .....................................................................
      REAL A,ABSEPS,ABSERR,ALIST,AREA,AREA1,AREA12,AREA2,A1,A2,B,BLIST,
     * B1,B2,CORREC,DEFABS,DEFAB1,DEFAB2,DRES,ELIST,EPMACH,EPSABS,
     * EPSREL,ERLARG,ERLAST,ERRBND,ERRMAX,ERROR1,ERRO12,ERROR2,ERRSUM,
     * ERTEST,F,OFLOW,POINTS,PTS,RESA,RESABS,RESEPS,RESULT,RES3LA,
     * RLIST,RLIST2,SIGN,TEMP,UFLOW
      INTEGER I,ID,IER,IERRO,IND1,IND2,IORD,IP1,IROFF1,IROFF2,IROFF3,J,
     * JLOW,JUPBND,K,KSGN,KTMIN,LAST,LEVCUR,LEVEL,LEVMAX,LIMIT,MAXERR,
     * NDIN,NEVAL,NINT,NINTP1,NPTS,NPTS2,NRES,NRMAX,NUMRL2
      LOGICAL EXTRAP,NOEXT
C
      DIMENSION ALIST(500),BLIST(500),ELIST(500),IORD(500),LEVEL(500),
     * NDIN(40),POINTS(40),PTS(40),RES3LA(3),RLIST(500),RLIST2(52)
C
C         THE DIMENSION OF RLIST2 IS DETERMINED BY THE VALUE OF
C         LIMEXP IN SUBROUTINE EPSALG (RLIST2 SHOULD BE OF DIMENSION
C         (LIMEXP+2) AT LEAST).
C
      EXTERNAL F
```

```
C
C          LIMIT IS THE MAXIMUM NUMBER OF SUBINTERVALS ALLOWED IN THE
C          SUBDIVISION PROCESS OF QAGP. TAKE CARE THAT LIMIT.GE.2.
C
      DATA LIMIT/500/
C
C          LIST OF MAJOR VARIABLES
C          -----------------------
C
C         ALIST     - LIST OF LEFT END POINTS OF ALL SUBINTERVALS
C                     CONSIDERED UP TO NOW
C         BLIST     - LIST OF RIGHT END POINTS OF ALL SUBINTERVALS
C                     CONSIDERED UP TO NOW
C         RLIST(I)  - APPROXIMATION TO THE INTEGRAL OVER
C                     (ALIST(I),BLIST(I))
C         RLIST2    - ARRAY OF DIMENSION AT LEAST LIMEXP+2
C                     CONTAINING THE PART OF THE EPSILON TABLE WHICH
C                     IS STILL NEEDED FOR FURTHER COMPUTATIONS
C         ELIST(I)  - ERROR ESTIMATE APPLYING TO RLIST(I)
C         MAXERR    - POINTER TO THE INTERVAL WITH LARGEST ERROR
C                     ESTIMATE
C         ERRMAX    - ELIST(MAXERR)
C         ERLAST    - ERROR ON THE INTERVAL CURRENTLY SUBDIVIDED
C                     (BEFORE THAT SUBDIVISION HAS TAKEN PLACE)
C         AREA      - SUM OF THE INTEGRALS OVER THE SUBINTERVALS
C         ERRSUM    - SUM OF THE ERRORS OVER THE SUBINTERVALS
C         ERRBND    - REQUESTED ACCURACY MAX(EPSABS,EPSREL*
C                     ABS(RESULT))
C         *****1    - VARIABLE FOR THE LEFT SUBINTERVAL
C         *****2    - VARIABLE FOR THE RIGHT SUBINTERVAL
C         LAST      - INDEX FOR SUBDIVISION
C         NRES      - NUMBER OF CALLS TO THE EXTRAPOLATION ROUTINE
C         NUMRL2    - NUMBER OF ELEMENTS IN RLIST2. IF AN APPROPRIATE
C                     APPROXIMATION TO THE COMPOUNDED INTEGRAL HAS
C                     OBTAINED, IT IS PUT IN RLIST2(NUMRL2) AFTER
C                     NUMRL2 HAS BEEN INCREASED BY ONE.
C         ERLARG    - SUM OF THE ERRORS OVER THE INTERVALS LARGER
C                     THAN THE SMALLEST INTERVAL CONSIDERED UP TO NOW
C         EXTRAP    - LOGICAL VARIABLE DENOTING THAT THE ROUTINE
C                     IS ATTEMPTING TO PERFORM EXTRAPOLATION. I.E.
C                     BEFORE SUBDIVIDING THE SMALLEST INTERVAL WE
C                     TRY TO DECREASE THE VALUE OF ERLARG.
C         NOEXT     - LOGICAL VARIABLE DENOTING THAT EXTRAPOLATION IS
C                     NO LONGER ALLOWED (TRUE-VALUE)
C
C          MACHINE DEPENDENT CONSTANTS
C          ---------------------------
C
```

```
C          EPMACH IS THE LARGEST RELATIVE SPACING.
C          UFLOW IS THE SMALLEST POSITIVE MAGNITUDE.
C          OFLOW IS THE LARGEST POSITIVE MAGNITUDE.
C
C***FIRST EXECUTABLE STATEMENT
      CALL QMACO(EPMACH,UFLOW,OFLOW)
C
C           TEST ON VALIDITY OF PARAMETERS
C           -----------------------------
C
      IER = 0
      NEVAL = 0
      LAST = 0
      RESULT = 0.0E+00
      ABSERR = 0.0E+00
      ALIST(1) = A
      BLIST(1) = B
      RLIST(1) = 0.0E+00
      ELIST(1) = 0.0E+00
      IORD(1) = 0
      LEVEL(1) = 0
      NPTS = NPTS2-2
      IF(NPTS2.LT.2.OR.LIMIT.LE.NPTS.OR.(EPSABS.LT.0.0E+00.AND.
     *  EPSREL.LT.0.0E+00)) IER = 6
      IF(IER.EQ.6) GO TO 999
C
C           IF ANY BREAK POINTS ARE PROVIDED, SORT THEM INTO AN
C           ASCENDING SEQUENCE.
C
      SIGN = 1.0E+00
      IF(A.GT.B) SIGN = -1.0E+00
      PTS(1) = AMIN1(A,B)
      IF(NPTS.EQ.0) GO TO 15
      DO 10 I = 1,NPTS
        PTS(I+1) = POINTS(I)
   10 CONTINUE
   15 PTS(NPTS+2) = AMAX1(A,B)
      NINT = NPTS+1
      A1 = PTS(1)
      IF(NPTS.EQ.0) GO TO 40
      NINTP1 = NINT+1
      DO 20 I = 1,NINT
        IP1 = I+1
        DO 20 J = IP1,NINTP1
          IF(PTS(I).LE.PTS(J)) GO TO 20
          TEMP = PTS(I)
          PTS(I) = PTS(J)
          PTS(J) = TEMP
```

```
   20 CONTINUE
      IF(PTS(1).NE.AMIN1(A,B).OR.PTS(NINTP1).NE.AMAX1(A,B)) IER = 6
      IF(IER.EQ.6) GO TO 999
C
C            COMPUTE FIRST INTEGRAL AND ERROR APPROXIMATIONS.
C            ------------------------------------------------
C
   40 RESABS = 0.0E+00
      DO 50 I = 1,NINT
        B1 = PTS(I+1)
        CALL QK21(F,A1,B1,AREA1,ERROR1,DEFABS,RESA)
        ABSERR = ABSERR+ERROR1
        RESULT = RESULT+AREA1
        NDIN(I) = 0
        IF(ERROR1.EQ.RESA.AND.ERROR1.NE.0.0E+00) NDIN(I) = 1
        RESABS = RESABS+DEFABS
        LEVEL(I) = 0
        ELIST(I) = ERROR1
        ALIST(I) = A1
        BLIST(I) = B1
        RLIST(I) = AREA1
        IORD(I) = I
        A1 = B1
   50 CONTINUE
      ERRSUM = 0.0E+00
      DO 55 I = 1,NINT
        IF(NDIN(I).EQ.1) ELIST(I) = ABSERR
        ERRSUM = ERRSUM+ELIST(I)
   55 CONTINUE
C
C           TEST ON ACCURACY.
C
      LAST = NINT
      NEVAL = 21*NINT
      DRES = ABS(RESULT)
      ERRBND = AMAX1(EPSABS,EPSREL*DRES)
      IF(ABSERR.LE.1.0E+02*EPMACH*RESABS.AND.ABSERR.GT.ERRBND) IER = 2
      IF(NINT.EQ.1) GO TO 80
      DO 70 I = 1,NPTS
        JLOW = I+1
        IND1 = IORD(I)
        DO 60 J = JLOW,NINT
          IND2 = IORD(J)
          IF(ELIST(IND1).GT.ELIST(IND2)) GO TO 60
          IND1 = IND2
          K = J
   60   CONTINUE
        IF(IND1.EQ.IORD(I)) GO TO 70
```

```
        IORD(K) = IORD(I)
        IORD(I) = IND1
   70 CONTINUE
      IF(LIMIT.LT.NPTS2) IER = 1
   80 IF(IER.NE.0.OR.ABSERR.LE.ERRBND) GO TO 999
C
C           INITIALIZATION
C           --------------
C
      RLIST2(1) = RESULT
      MAXERR = IORD(1)
      ERRMAX = ELIST(MAXERR)
      AREA = RESULT
      NRMAX = 1
      NRES = 0
      NUMRL2 = 1
      KTMIN = 0
      EXTRAP = .FALSE.
      NOEXT = .FALSE.
      ERLARG = ERRSUM
      ERTEST = ERRBND
      LEVMAX = 1
      IROFF1 = 0
      IROFF2 = 0
      IROFF3 = 0
      IERRO = 0
      ABSERR = OFLOW
      KSGN = -1
      IF(DRES.GE.(1.0E+00-5.0E+01*EPMACH)*RESABS) KSGN = 1
C
C           MAIN DO-LOOP
C           ------------
C
      DO 160 LAST = NPTS2,LIMIT
C
C           BISECT THE SUBINTERVAL WITH THE NRMAX-TH LARGEST ERROR
C           ESTIMATE.
C
        LEVCUR = LEVEL(MAXERR)+1
        A1 = ALIST(MAXERR)
        B1 = 5.0E-01*(ALIST(MAXERR)+BLIST(MAXERR))
        A2 = B1
        B2 = BLIST(MAXERR)
        ERLAST = ERRMAX
        CALL QK21(F,A1,B1,AREA1,ERROR1,RESA,DEFAB1)
        CALL QK21(F,A2,B2,AREA2,ERROR2,RESA,DEFAB2)
C
C           IMPROVE PREVIOUS APPROXIMATIONS TO INTEGRAL AND ERROR
```

```
C           AND TEST FOR ACCURACY.
C
      NEVAL = NEVAL+42
      AREA12 = AREA1+AREA2
      ERRO12 = ERROR1+ERROR2
      ERRSUM = ERRSUM+ERRO12-ERRMAX
      AREA = AREA+AREA12-RLIST(MAXERR)
      IF(DEFAB1.EQ.ERROR1.OR.DEFAB2.EQ.ERROR2) GO TO 95
      IF(ABS(RLIST(MAXERR)-AREA12).GT.1.0E-05*ABS(AREA12)
     * .OR.ERRO12.LT.9.9E-01*ERRMAX) GO TO 90
      IF(EXTRAP) IROFF2 = IROFF2+1
      IF(.NOT.EXTRAP) IROFF1 = IROFF1+1
   90 IF(LAST.GT.10.AND.ERRO12.GT.ERRMAX) IROFF3 = IROFF3+1
   95 LEVEL(MAXERR) = LEVCUR
      LEVEL(LAST) = LEVCUR
      RLIST(MAXERR) = AREA1
      RLIST(LAST) = AREA2
      ERRBND = AMAX1(EPSABS,EPSREL*ABS(AREA))
C
C           TEST FOR ROUNDOFF ERROR AND EVENTUALLY SET ERROR FLAG.
C
      IF(IROFF1+IROFF2.GE.10.OR.IROFF3.GE.20) IER = 2
      IF(IROFF2.GE.5) IERRO = 3
C
C           SET ERROR FLAG IN THE CASE THAT THE NUMBER OF SUBINTERVALS
C           EQUALS LIMIT.
C
      IF(LAST.EQ.LIMIT) IER = 1
C
C           SET ERROR FLAG IN THE CASE OF BAD INTEGRAND BEHAVIOUR
C           AT A POINT OF THE INTEGRATION RANGE
C
      IF(AMAX1(ABS(A1),ABS(B2)).LE.(1.0E+00+1.0E+03*EPMACH)*
     * (ABS(A2)+1.0E+03*UFLOW)) IER = 4
C
C           APPEND THE NEWLY-CREATED INTERVALS TO THE LIST.
C
      IF(ERROR2.GT.ERROR1) GO TO 100
      ALIST(LAST) = A2
      BLIST(MAXERR) = B1
      BLIST(LAST) = B2
      ELIST(MAXERR) = ERROR1
      ELIST(LAST) = ERROR2
      GO TO 110
  100 ALIST(MAXERR) = A2
      ALIST(LAST) = A1
      BLIST(LAST) = B1
      RLIST(MAXERR) = AREA2
```

```
      RLIST(LAST) = AREA1
      ELIST(MAXERR) = ERROR2
      ELIST(LAST) = ERROR1
C
C           CALL SUBROUTINE QSORT TO MAINTAIN THE DESCENDING ORDERING
C           IN THE LIST OF ERROR ESTIMATES AND SELECT THE SUBINTERVAL
C           WITH NRMAX-TH LARGEST ERROR ESTIMATE (TO BE BISECTED NEXT).
C
  110 CALL QSORT(LIMIT,LAST,MAXERR,ERRMAX,ELIST,IORD,NRMAX)
C***JUMP OUT OF DO-LOOP
      IF(ERRSUM.LE.ERRBND) GO TO 190
C***JUMP OUT OF DO-LOOP
      IF(IER.NE.0) GO TO 170
      IF(NOEXT) GO TO 160
      ERLARG = ERLARG-ERLAST
      IF(LEVCUR+1.LE.LEVMAX) ERLARG = ERLARG+ERRO12
      IF(EXTRAP) GO TO 120
C
C           TEST WHETHER THE INTERVAL TO BE BISECTED NEXT IS THE
C           SMALLEST INTERVAL.
C
      IF(LEVEL(MAXERR)+1.LE.LEVMAX) GO TO 160
      EXTRAP = .TRUE.
      NRMAX = 2
  120 IF(IERRO.EQ.3.OR.ERLARG.LE.ERTEST) GO TO 140
C
C           THE SMALLEST INTERVAL HAS THE LARGEST ERROR.
C           BEFORE BISECTING DECREASE THE SUM OF THE ERRORS OVER THE
C           LARGER INTERVALS (ERLARG) AND PERFORM EXTRAPOLATION.
C
      ID = NRMAX
      JUPBND = LAST
      IF(LAST.GT.(2+LIMIT/2)) JUPBND = LIMIT+3-LAST
      DO 130 K = ID,JUPBND
        MAXERR = IORD(NRMAX)
        ERRMAX = ELIST(MAXERR)
C***JUMP OUT OF DO-LOOP
        IF(LEVEL(MAXERR)+1.LE.LEVMAX) GO TO 160
        NRMAX = NRMAX+1
  130 CONTINUE
C
C           PERFORM EXTRAPOLATION.
C
  140 NUMRL2 = NUMRL2+1
      RLIST2(NUMRL2) = AREA
      IF(NUMRL2.LE.2) GO TO 155
      CALL QEXT(NUMRL2,RLIST2,RESEPS,ABSEPS,RES3LA,NRES)
      KTMIN = KTMIN+1
```

```
        IF(KTMIN.GT.5.AND.ABSERR.LT.1.0E-03*ERRSUM) IER = 5
        IF(ABSEPS.GE.ABSERR) GO TO 150
        KTMIN = 0
        ABSERR = ABSEPS
        RESULT = RESEPS
        CORREC = ERLARG
        ERTEST = AMAX1(EPSABS,EPSREL*ABS(RESEPS))
C***JUMP OUT OF DO-LOOP
        IF(ABSERR.LT.ERTEST) GO TO 170
C
C           PREPARE BISECTION OF THE SMALLEST INTERVAL.
C
  150   IF(NUMRL2.EQ.1) NOEXT = .TRUE.
        IF(IER.GE.5) GO TO 170
  155   MAXERR = IORD(1)
        ERRMAX = ELIST(MAXERR)
        NRMAX = 1
        EXTRAP = .FALSE.
        LEVMAX = LEVMAX+1
        ERLARG = ERRSUM
  160 CONTINUE
C
C           SET THE FINAL RESULT.
C           ---------------------
C
C
  170 IF(ABSERR.EQ.OFLOW) GO TO 190
      IF((IER+IERRO).EQ.0) GO TO 180
      IF(IERRO.EQ.3) ABSERR = ABSERR+CORREC
      IF(IER.EQ.0) IER = 3
      IF(RESULT.NE.0.0E+00.AND.AREA.NE.0.0E+00)GO TO 175
      IF(ABSERR.GT.ERRSUM)GO TO 190
      IF(AREA.EQ.0.0E+00) GO TO 210
      GO TO 180
  175 IF(ABSERR/ABS(RESULT).GT.ERRSUM/ABS(AREA))GO TO 190
C
C           TEST ON DIVERGENCE.
C
  180 IF(KSGN.EQ.(-1).AND.AMAX1(ABS(RESULT),ABS(AREA)).LE.
     *  DEFABS*1.0E-02) GO TO 210
      IF(1.0E-02.GT.(RESULT/AREA).OR.(RESULT/AREA).GT.1.0E+02.OR.
     *  ERRSUM.GT.ABS(AREA)) IER = 6
      GO TO 210
C
C           COMPUTE GLOBAL INTEGRAL SUM.
C
  190 RESULT = 0.0E+00
      DO 200 K = 1,LAST
```

```
        RESULT = RESULT+RLIST(K)
  200 CONTINUE
      ABSERR = ERRSUM
  210 IF(IER.GT.2) IER = IER - 1
      RESULT = RESULT*SIGN
 999  RETURN
      END
```

```
      SUBROUTINE QAGI(F,BOUND,INF,EPSABS,EPSREL,RESULT,ABSERR,NEVAL,IER)
C
C ..........................................................................
C
C 1.     QAGI
C        INTEGRATION OVER INFINITE INTERVALS
C           STANDARD FORTRAN SUBROUTINE
C
C 2.     PURPOSE
C           THE ROUTINE CALCULATES AN APPROXIMATION  RESULT  TO A GIVEN
C           INTEGRAL    I = INTEGRAL OF  F  OVER (BOUND,+INFINITY)
C                    OR I = INTEGRAL OF  F  OVER (-INFINITY,BOUND)
C                    OR I = INTEGRAL OF  F  OVER (-INFINITY,+INFINITY),
C           HOPEFULLY SATISFYING FOLLOWING CLAIM FOR ACCURACY
C           ABS(I-RESULT).LE.MAX(EPSABS,EPSREL*ABS(I)).
C
C 3.     CALLING SEQUENCE
C           CALL QAGI(F,BOUND,INF,EPSABS,EPSREL,RESULT,ABSERR,NEVAL,IER)
C
C        PARAMETERS
C         ON ENTRY
C            F      - REAL
C                     FUNCTION SUBPROGRAM DEFINING THE INTEGRAND
C                     FUNCTION F(X). THE ACTUAL NAME FOR F NEEDS TO BE
C                     DECLARED E X T E R N A L IN THE DRIVER PROGRAM.
C
C            BOUND  - REAL
C                     FINITE BOUND OF INTEGRATION RANGE
C                     (HAS NO MEANING IF INTERVAL IS DOUBLY-INFINITE)
C
C            INF    - REAL
C                     INDICATING THE KIND OF INTEGRATION RANGE INVOLVED
C                     INF = 1 CORRESPONDS TO  (BOUND,+INFINITY),
C                     INF = -1            TO  (-INFINITY,BOUND),
C                     INF = 2             TO (-INFINITY,+INFINITY).
C
C            EPSABS - REAL
C                     ABSOLUTE ACCURACY REQUESTED
C            EPSREL - REAL
C                     RELATIVE ACCURACY REQUESTED
C                     IF  EPSABS.LT.0 AND EPSREL.LT.0,
C                     THE ROUTINE WILL END WITH IER = 6.
C
C         ON RETURN
C            RESULT - REAL
C                     APPROXIMATION TO THE INTEGRAL
C
C            ABSERR - REAL
```

```
C                    ESTIMATE OF THE MODULUS OF THE ABSOLUTE ERROR,
C                    WHICH SHOULD EQUAL OR EXCEED ABS(I-RESULT)
C
C            NEVAL  - INTEGER
C                    NUMBER OF INTEGRAND EVALUATIONS
C
C            IER    - INTEGER
C                    IER = 0 NORMAL AND RELIABLE TERMINATION OF THE
C                            ROUTINE. IT IS ASSUMED THAT THE REQUESTED
C                            ACCURACY HAS BEEN ACHIEVED.
C                  - IER.GT.0 ABNORMAL TERMINATION OF THE ROUTINE. THE
C                            ESTIMATES FOR RESULT AND ERROR ARE LESS
C                            RELIABLE. IT IS ASSUMED THAT THE REQUESTED
C                            ACCURACY HAS NOT BEEN ACHIEVED.
C                    IER = 1 MAXIMUM NUMBER OF SUBDIVISIONS ALLOWED
C                            HAS BEEN ACHIEVED. ONE CAN ALLOW MORE
C                            SUBDIVISIONS BY INCREASING THE DATA VALUE
C                            OF LIMIT IN QAGI (AND TAKING THE ACCORDING
C                            DIMENSION ADJUSTMENTS INTO ACCOUNT).
C                            HOWEVER, IF THIS YIELDS NO IMPROVEMENT
C                            IT IS ADVISED TO ANALYZE THE INTEGRAND
C                            IN ORDER TO DETERMINE THE INTEGRATION
C                            DIFFICULTIES. IF THE POSITION OF A LOCAL
C                            DIFFICULTY CAN BE DETERMINED (E.G.
C                            SINGULARITY, DISCONTINUITY WITHIN THE
C                            INTERVAL) ONE WILL PROBABLY GAIN FROM
C                            SPLITTING UP THE INTERVAL AT THIS POINT
C                            AND CALLING THE INTEGRATOR ON THE
C                            SUBRANGES. IF POSSIBLE, AN APPROPRIATE
C                            SPECIAL-PURPOSE INTEGRATOR SHOULD BE USED,
C                            WHICH IS DESIGNED FOR HANDLING THE TYPE
C                            OF DIFFICULTY INVOLVED.
C                        = 2 THE OCCURRENCE OF ROUNDOFF ERROR IS
C                            DETECTED, WHICH PREVENTS THE REQUESTED
C                            TOLERANCE FROM BEING ACHIEVED.
C                            THE ERROR MAY BE UNDER-ESTIMATED.
C                        = 3 EXTREMELY BAD INTEGRAND BEHAVIOUR OCCURS
C                            AT SOME POINTS OF THE INTEGRATION
C                            INTERVAL.
C                        = 4 THE ALGORITHM DOES NOT CONVERGE.
C                            ROUNDOFF ERROR IS DETECTED IN THE
C                            EXTRAPOLATION TABLE.
C                            IT IS ASSUMED THAT THE REQUESTED TOLERANCE
C                            CANNOT BE ACHIEVED, AND THAT THE RETURNED
C                            RESULT IS THE BEST WHICH CAN BE OBTAINED.
C                        = 5 THE INTEGRAL IS PROBABLY DIVERGENT, OR
C                            SLOWLY CONVERGENT. IT MUST BE NOTED THAT
C                            DIVERGENCE CAN OCCUR WITH ANY OTHER VALUE
```

```
C                              OF IER.
C                          = 6 THE INPUT IS INVALID, BECAUSE
C                              INF.NE.1 AND INF.NE.-1 AND INF.NE.2, OR
C                              EPSABS.LT.0 AND EPSREL.LT.0,
C                              RESULT, ABSERR, NEVAL ARE SET TO ZERO.
C
C 4.      SUBROUTINES OR FUNCTIONS NEEDED
C               - QK15I
C               - QSORT
C               - QEXT
C               - F (USER-PROVIDED FUNCTION)
C               - QMACO
C               - FORTRAN ABS, AMAX1
C
C ...................................................................
C
      REAL ABSEPS,ABSERR,ALIST,AREA,AREA1,AREA12,AREA2,A1,A2,BLIST,BOUN,
     *  BOUND,B1,B2,CORREC,DEFABS,DEFAB1,DEFAB2,DRES,ELIST,EPMACH,
     *  EPSABS,EPSREL,ERLARG,ERLAST,ERRBND,ERRMAX,ERROR1,ERROR2,ERRO12,
     *  ERRSUM,ERTEST,F,OFLOW,RESABS,RESEPS,RESULT,RES3LA,RLIST,RLIST2,
     *  SMALL,UFLOW
      INTEGER ID,IER,IERRO,INF,IORD,IROFF1,IROFF2,IROFF3,JUPBND,K,KSGN,
     *  KTMIN,LAST,LIMIT,MAXERR,NEVAL,NRES,NRMAX,NUMRL2
      LOGICAL EXTRAP,NOEXT
C
      DIMENSION ALIST(500),BLIST(500),ELIST(500),IORD(500),RES3LA(3),
     *  RLIST(500),RLIST2(52)
C
C           THE DIMENSION OF RLIST2 IS DETERMINED BY THE VALUE OF
C           LIMEXP IN SUBROUTINE QEXT.
C
      EXTERNAL F
C
C           LIMIT IS THE MAXIMUM NUMBER IF SUBINTERVALS ALLOWED IN THE
C           SUBDIVISION PROCESS. TAKE CARE THAT LIMIT.GE.1.
C
      DATA LIMIT/500/
C
C           LIST OF MAJOR VARIABLES
C           -----------------------
C
C          ALIST     - LIST OF LEFT END POINTS OF ALL SUBINTERVALS
C                      CONSIDERED UP TO NOW
C          BLIST     - LIST OF RIGHT END POINTS OF ALL SUBINTERVALS
C                      CONSIDERED UP TO NOW
C          RLIST(I)  - APPROXIMATION TO THE INTEGRAL OVER
C                      (ALIST(I),BLIST(I))
C          RLIST2    - ARRAY OF DIMENSION AT LEAST (LIMEXP+2),
```

```
C                        CONTAINING THE PART OF THE EPSILON TABLE
C                        WHICH IS STILL NEEDED FOR FURTHER COMPUTATIONS
C           ELIST(I)   - ERROR ESTIMATE APPLYING TO RLIST(I)
C           MAXERR     - POINTER TO THE INTERVAL WITH LARGEST ERROR
C                        ESTIMATE
C           ERRMAX     - ELIST(MAXERR)
C           ERLAST     - ERROR ON THE INTERVAL CURRENTLY SUBDIVIDED
C                        (BEFORE THAT SUBDIVISION HAS TAKEN PLACE)
C           AREA       - SUM OF THE INTEGRALS OVER THE SUBINTERVALS
C           ERRSUM     - SUM OF THE ERRORS OVER THE SUBINTERVALS
C           ERRBND     - REQUESTED ACCURACY MAX(EPSABS,EPSREL*
C                        ABS(RESULT))
C           *****1     - VARIABLE FOR THE LEFT SUBINTERVAL
C           *****2     - VARIABLE FOR THE RIGHT SUBINTERVAL
C           LAST       - INDEX FOR SUBDIVISION
C           NRES       - NUMBER OF CALLS TO THE EXTRAPOLATION ROUTINE
C           NUMRL2     - NUMBER OF ELEMENTS CURRENTLY IN RLIST2. IF AN
C                        APPROPRIATE APPROXIMATION TO THE COMPOUNDED
C                        INTEGRAL HAS BEEN OBTAINED, IT IS PUT IN
C                        RLIST2(NUMRL2) AFTER NUMRL2 HAS BEEN INCREASED
C                        BY ONE.
C           SMALL      - LENGTH OF THE SMALLEST INTERVAL CONSIDERED UP
C                        TO NOW, MULTIPLIED BY 1.5
C           ERLARG     - SUM OF THE ERRORS OVER THE INTERVALS LARGER
C                        THAN THE SMALLEST INTERVAL CONSIDERED UP TO NOW
C           EXTRAP     - LOGICAL VARIABLE DENOTING THAT THE ROUTINE
C                        IS ATTEMPTING TO PERFORM EXTRAPOLATION. I.E.
C                        BEFORE SUBDIVIDING THE SMALLEST INTERVAL WE
C                        TRY TO DECREASE THE VALUE OF ERLARG.
C           NOEXT      - LOGICAL VARIABLE DENOTING THAT EXTRAPOLATION
C                        IS NO LONGER ALLOWED (TRUE-VALUE)
C
C            MACHINE DEPENDENT CONSTANTS
C            ---------------------------
C
C           EPMACH IS THE LARGEST RELATIVE SPACING.
C           UFLOW IS THE SMALLEST POSITIVE MAGNITUDE.
C           OFLOW IS THE LARGEST POSITIVE MAGNITUDE.
C
C***FIRST EXECUTABLE STATEMENT
      CALL QMACO(EPMACH,UFLOW,OFLOW)
C
C           TEST ON VALIDITY OF PARAMETERS
C           -----------------------------
C
      IER = 0
      NEVAL = 0
      LAST = 0
```

```
      RESULT = 0.0E+00
      ABSERR = 0.0E+00
      ALIST(1) = 0.0E+00
      BLIST(1) = 1.0E+00
      RLIST(1) = 0.0E+00
      ELIST(1) = 0.0E+00
      IORD(1) = 0
      IF(EPSABS.LT.0.0E+00.AND.EPSREL.LT.0.0E+00) IER = 6
      IF(IER.EQ.6) GO TO 999
C
C
C           FIRST APPROXIMATION TO THE INTEGRAL
C           -----------------------------------
C
C           DETERMINE THE INTERVAL TO BE MAPPED ONTO (0,1).
C           IF INF = 2 THE INTEGRAL IS COMPUTED AS I = I1+I2, WHERE
C           I1 = INTEGRAL OF F OVER (-INFINITY,0),
C           I2 = INTEGRAL OF F OVER (0,+INFINITY).
C
      BOUN = BOUND
      IF(INF.EQ.2) BOUN = 0.0E+00
      CALL QK15I(F,BOUN,INF,0.0E+00,1.0E+00,RESULT,ABSERR,DEFABS,RESABS)
C
C           TEST ON ACCURACY
C
      LAST = 1
      RLIST(1) = RESULT
      ELIST(1) = ABSERR
      IORD(1) = 1
      DRES = ABS(RESULT)
      ERRBND = AMAX1(EPSABS,EPSREL*DRES)
      IF(ABSERR.LE.1.0E+02*EPMACH*DEFABS.AND.ABSERR.GT.ERRBND) IER = 2
      IF(LIMIT.EQ.1) IER = 1
      IF(IER.NE.0.OR.(ABSERR.LE.ERRBND.AND.ABSERR.NE.RESABS).OR.
     *  ABSERR.EQ.0.0E+00) GO TO 130
C
C           INITIALIZATION
C           --------------
C
      RLIST2(1) = RESULT
      ERRMAX = ABSERR
      MAXERR = 1
      AREA = RESULT
      ERRSUM = ABSERR
      ABSERR = OFLOW
      NRMAX = 1
      NRES = 0
      KTMIN = 0
```

```
      NUMRL2 = 2
      EXTRAP = .FALSE.
      NOEXT = .FALSE.
      IERRO = 0
      IROFF1 = 0
      IROFF2 = 0
      IROFF3 = 0
      KSGN = -1
      IF(DRES.GE.(1.0E+00-5.0E+01*EPMACH)*DEFABS) KSGN = 1
C
C           MAIN DO-LOOP
C           ------------
C
      DO 90 LAST = 2,LIMIT
C
C           BISECT THE SUBINTERVAL WITH NRMAX-TH LARGEST ERROR ESTIMATE.
C
        A1 = ALIST(MAXERR)
        B1 = 5.0E-01*(ALIST(MAXERR)+BLIST(MAXERR))
        A2 = B1
        B2 = BLIST(MAXERR)
        ERLAST = ERRMAX
        CALL QK15I(F,BOUN,INF,A1,B1,AREA1,ERROR1,RESABS,DEFAB1)
        CALL QK15I(F,BOUN,INF,A2,B2,AREA2,ERROR2,RESABS,DEFAB2)
C
C           IMPROVE PREVIOUS APPROXIMATIONS TO INTEGRAL AND ERROR
C           AND TEST FOR ACCURACY.
C
        AREA12 = AREA1+AREA2
        ERRO12 = ERROR1+ERROR2
        ERRSUM = ERRSUM+ERRO12-ERRMAX
        AREA = AREA+AREA12-RLIST(MAXERR)
        IF(DEFAB1.EQ.ERROR1.OR.DEFAB2.EQ.ERROR2)GO TO 15
        IF(ABS(RLIST(MAXERR)-AREA12).GT.1.0E-05*ABS(AREA12)
     * .OR.ERRO12.LT.9.9E-01*ERRMAX) GO TO 10
        IF(EXTRAP) IROFF2 = IROFF2+1
        IF(.NOT.EXTRAP) IROFF1 = IROFF1+1
   10   IF(LAST.GT.10.AND.ERRO12.GT.ERRMAX) IROFF3 = IROFF3+1
   15   RLIST(MAXERR) = AREA1
        RLIST(LAST) = AREA2
        ERRBND = AMAX1(EPSABS,EPSREL*ABS(AREA))
C
C           TEST FOR ROUNDOFF ERROR AND EVENTUALLY SET ERROR FLAG.
C
        IF(IROFF1+IROFF2.GE.10.OR.IROFF3.GE.20) IER = 2
        IF(IROFF2.GE.5) IERRO = 3
C
C           SET ERROR FLAG IN THE CASE THAT THE NUMBER OF SUBINTERVALS
```

```
C           EQUALS LIMIT.
C
      IF(LAST.EQ.LIMIT) IER = 1
C
C           SET ERROR FLAG IN THE CASE OF BAD INTEGRAND BEHAVIOUR
C           AT SOME POINTS OF THE INTEGRATION RANGE.
C
      IF(AMAX1(ABS(A1),ABS(B2)).LE.(1.0E+00+1.0E+03*EPMACH)*
     * (ABS(A2)+1.0E+03*UFLOW)) IER = 4
C
C           APPEND THE NEWLY-CREATED INTERVALS TO THE LIST.
C
      IF(ERROR2.GT.ERROR1) GO TO 20
      ALIST(LAST) = A2
      BLIST(MAXERR) = B1
      BLIST(LAST) = B2
      ELIST(MAXERR) = ERROR1
      ELIST(LAST) = ERROR2
      GO TO 30
   20 ALIST(MAXERR) = A2
      ALIST(LAST) = A1
      BLIST(LAST) = B1
      RLIST(MAXERR) = AREA2
      RLIST(LAST) = AREA1
      ELIST(MAXERR) = ERROR2
      ELIST(LAST) = ERROR1
C
C           CALL SUBROUTINE QSORT TO MAINTAIN THE DESCENDING ORDERING
C           IN THE LIST OF ERROR ESTIMATES AND SELECT THE SUBINTERVAL
C           WITH NRMAX-TH LARGEST ERROR ESTIMATE (TO BE BISECTED NEXT).
C
   30 CALL QSORT(LIMIT,LAST,MAXERR,ERRMAX,ELIST,IORD,NRMAX)
      IF(ERRSUM.LE.ERRBND) GO TO 115
      IF(IER.NE.0) GO TO 100
      IF(LAST.EQ.2) GO TO 80
      IF(NOEXT) GO TO 90
      ERLARG = ERLARG-ERLAST
      IF(ABS(B1-A1).GT.SMALL) ERLARG = ERLARG+ERRO12
      IF(EXTRAP) GO TO 40
C
C           TEST WHETHER THE INTERVAL TO BE BISECTED NEXT IS THE
C           SMALLEST INTERVAL.
C
      IF(ABS(BLIST(MAXERR)-ALIST(MAXERR)).GT.SMALL) GO TO 90
      EXTRAP = .TRUE.
      NRMAX = 2
   40 IF(IERRO.EQ.3.OR.ERLARG.LE.ERTEST) GO TO 60
C
```

```
C           THE SMALLEST INTERVAL HAS THE LARGEST ERROR.
C           BEFORE BISECTING DECREASE THE SUM OF THE ERRORS OVER THE
C           LARGER INTERVALS (ERLARG) AND PERFORM EXTRAPOLATION.
C
        ID = NRMAX
        JUPBND = LAST
        IF(LAST.GT.(2+LIMIT/2)) JUPBND = LIMIT+3-LAST
        DO 50 K = ID,JUPBND
          MAXERR = IORD(NRMAX)
          ERRMAX = ELIST(MAXERR)
          IF(ABS(BLIST(MAXERR)-ALIST(MAXERR)).GT.SMALL) GO TO 90
          NRMAX = NRMAX+1
   50   CONTINUE
C
C           PERFORM EXTRAPOLATION.
C
   60   NUMRL2 = NUMRL2+1
        RLIST2(NUMRL2) = AREA
        CALL QEXT(NUMRL2,RLIST2,RESEPS,ABSEPS,RES3LA,NRES)
        KTMIN = KTMIN+1
        IF(KTMIN.GT.5.AND.ABSERR.LT.1.0E-03*ERRSUM) IER = 5
        IF(ABSEPS.GE.ABSERR) GO TO 70
        KTMIN = 0
        ABSERR = ABSEPS
        RESULT = RESEPS
        CORREC = ERLARG
        ERTEST = AMAX1(EPSABS,EPSREL*ABS(RESEPS))
        IF(ABSERR.LE.ERTEST) GO TO 100
C
C            PREPARE BISECTION OF THE SMALLEST INTERVAL.
C
   70   IF(NUMRL2.EQ.1) NOEXT = .TRUE.
        IF(IER.EQ.5) GO TO 100
        MAXERR = IORD(1)
        ERRMAX = ELIST(MAXERR)
        NRMAX = 1
        EXTRAP = .FALSE.
        SMALL = SMALL*5.0E-01
        ERLARG = ERRSUM
        GO TO 90
   80   SMALL = 3.75E-01
        ERLARG = ERRSUM
        ERTEST = ERRBND
        RLIST2(2) = AREA
   90 CONTINUE
C
C           SET FINAL RESULT AND ERROR ESTIMATE.
C           ------------------------------------
```

```
C
  100 IF(ABSERR.EQ.OFLOW) GO TO 115
      IF((IER+IERRO).EQ.0) GO TO 110
      IF(IERRO.EQ.3) ABSERR = ABSERR+CORREC
      IF(IER.EQ.0) IER = 3
      IF(RESULT.NE.0.0E+00.AND.AREA.NE.0.0E+00)GO TO 105
      IF(ABSERR.GT.ERRSUM)GO TO 115
      IF(AREA.EQ.0.0E+00) GO TO 130
      GO TO 110
  105 IF(ABSERR/ABS(RESULT).GT.ERRSUM/ABS(AREA))GO TO 115
C
C           TEST ON DIVERGENCE
C
  110 IF(KSGN.EQ.(-1).AND.AMAX1(ABS(RESULT),ABS(AREA)).LE.
     * DEFABS*1.0E-02) GO TO 130
      IF(1.0E-02.GT.(RESULT/AREA).OR.(RESULT/AREA).GT.1.0E+02.
     *OR.ERRSUM.GT.ABS(AREA)) IER = 6
      GO TO 130
C
C           COMPUTE GLOBAL INTEGRAL SUM.
C
  115 RESULT = 0.0E+00
      DO 120 K = 1,LAST
        RESULT = RESULT+RLIST(K)
  120 CONTINUE
      ABSERR = ERRSUM
  130 NEVAL = 30*LAST-15
      IF(INF.EQ.2) NEVAL = 2*NEVAL
      IF(IER.GT.2) IER=IER-1
  999 RETURN
      END
```

```
      SUBROUTINE QAWO(F,A,B,OMEGA,INTEGR,EPSABS,EPSREL,RESULT,ABSERR,
     *  NEVAL,IER)
C
C ..........................................................................
C
C 1.     QAWO
C        COMPUTATION OF OSCILLATORY INTEGRALS
C           STANDARD FORTRAN SUBROUTINE
C
C 2.     PURPOSE
C           THE ROUTINE CALCULATES AN APPROXIMATION RESULT TO A GIVEN
C           DEFINITE INTEGRAL
C                I = INTEGRAL OF F(X)*W(X) OVER (A,B)
C                    WHERE W(X) = COS(OMEGA*X)
C                       OR W(X) = SIN(OMEGA*X),
C           HOPEFULLY SATISFYING FOLLOWING CLAIM FOR ACCURACY
C           ABS(I-RESULT).LE.MAX(EPSABS,EPSREL*ABS(I)).
C
C 3.     CALLING SEQUENCE
C           CALL QAWO(F,A,B,OMEGA,INTEGR,EPSABS,EPSREL,RESULT,ABSERR,
C                     NEVAL,IER)
C
C        PARAMETERS
C         ON ENTRY
C            F      - REAL
C                     FUNCTION SUBPROGRAM DEFINING THE
C                     FUNCTION F(X) THE ACTUAL NAME FOR F NEEDS TO BE
C                     DECLARED E X T E R N A L IN THE DRIVER PROGRAM.
C
C            A      - REAL
C                     LOWER LIMIT OF INTEGRATION
C
C            B      - REAL
C                     UPPER LIMIT OF INTEGRATION
C
C            OMEGA  - REAL
C                     PARAMETER IN THE INTEGRAND WEIGHT FUNCTION
C
C            INTEGR - INTEGER
C                     INDICATES WHICH OF THE WEIGHT FUNCTIONS IS USED
C                     INTEGR = 1      W(X) = COS(OMEGA*X)
C                     INTEGR = 2      W(X) = SIN(OMEGA*X)
C                     IF INTEGR.NE.1.AND.INTEGR.NE.2, THE ROUTINE
C                     WILL END WITH IER = 6.
C
C            EPSABS - REAL
C                     ABSOLUTE ACCURACY REQUESTED
C            EPSREL - REAL
```

```
C                    RELATIVE ACCURACY REQUESTED
C                    IF EPSABS.LT.0 AND EPSREL.LT.0,
C                    THE ROUTINE WILL END WITH IER = 6.
C
C          ON RETURN
C             RESULT - REAL
C                    APPROXIMATION TO THE INTEGRAL
C
C             ABSERR - REAL
C                    ESTIMATE OF THE MODULUS OF THE ABSOLUTE ERROR,
C                    WHICH SHOULD EQUAL OR EXCEED ABS(I-RESULT)
C
C             NEVAL  - INTEGER
C                    NUMBER OF INTEGRAND EVALUATIONS
C
C             IER    - INTEGER
C                    IER = 0 NORMAL AND RELIABLE TERMINATION OF THE
C                            ROUTINE. IT IS ASSUMED THAT THE
C                            REQUESTED ACCURACY HAS BEEN ACHIEVED.
C                  - IER.GT.0 ABNORMAL TERMINATION OF THE ROUTINE.
C                            THE ESTIMATES FOR INTEGRAL AND ERROR ARE
C                            LESS RELIABLE. IT IS ASSUMED THAT THE
C                            REQUESTED ACCURACY HAS NOT BEEN ACHIEVED.
C                    IER = 1 MAXIMUM NUMBER OF SUBDIVISIONS ALLOWED
C                            (= LIMIT) HAS BEEN ACHIEVED. ONE CAN
C                            ALLOW MORE SUBDIVISIONS BY INCREASING THE
C                            VALUE OF LIMIT (AND TAKING THE ACCORDING
C                            DIMENSION ADJUSTMENTS INTO ACCOUNT).
C                            HOWEVER, IF THIS YIELDS NO IMPROVEMENT IT
C                            IS ADVISED TO ANALYZE THE INTEGRAND IN
C                            ORDER TO DETERMINE THE INTEGRATION
C                            DIFFICULTIES. IF THE POSITION OF A LOCAL
C                            DIFFICULTY CAN BE DETERMINED (E.G.
C                            SINGULARITY, DISCONTINUITY WITHIN THE
C                            INTERVAL) ONE WILL PROBABLY GAIN FROM
C                            SPLITTING UP THE INTERVAL AT THIS POINT
C                            AND CALLING THE INTEGRATOR ON THE
C                            SUBRANGES. IF POSSIBLE, AN APPROPRIATE
C                            SPECIAL-PURPOSE INTEGRATOR SHOULD
C                            BE USED WHICH IS DESIGNED FOR HANDLING
C                            THE TYPE OF DIFFICULTY INVOLVED.
C                        = 2 THE OCCURRENCE OF ROUNDOFF ERROR IS
C                            DETECTED, WHICH PREVENTS THE REQUESTED
C                            TOLERANCE FROM BEING ACHIEVED.
C                            THE ERROR MAY BE UNDER-ESTIMATED.
C                        = 3 EXTREMELY BAD INTEGRAND BEHAVIOUR OCCURS
C                            AT SOME INTERIOR POINTS OF THE INTEGRATION
C                            INTERVAL.
```

```
C                             = 4 THE ALGORITHM DOES NOT CONVERGE. ROUNDOFF
C                                 ERROR IS DETECTED IN THE EXTRAPOLATION
C                                 TABLE. IT IS PRESUMED THAT THE REQUESTED
C                                 TOLERANCE CANNOT BE ACHIEVED DUE TO
C                                 ROUNDOFF IN THE EXTRAPOLATION TABLE,
C                                 AND THAT THE RETURNED RESULT IS THE BEST
C                                 WHICH CAN BE OBTAINED.
C                             = 5 THE INTEGRAL IS PROBABLY DIVERGENT, OR
C                                 SLOWLY CONVERGENT. IT MUST BE NOTED THAT
C                                 DIVERGENCE CAN OCCUR WITH ANY OTHER VALUE
C                                 OF IER.
C                             = 6 THE INPUT IS INVALID, BECAUSE
C                                 EPSABS.LT.0 AND EPSREL.LT.0,
C                                 RESULT, ABSERR, NEVAL ARE SET TO ZERO.
C
C 4.      SUBROUTINES OR FUNCTIONS NEEDED
C               - QFOUR
C               - QEXT
C               - QSORT
C               - QC25O
C               - QK15W
C               - QCHEB
C               - QWGTO
C               - QMACO
C               - FORTRAN ABS, AMAX1, AMIN1
C               - F (USER-PROVIDED FUNCTION)
C
C ......................................................................
C
      REAL A,ABSERR,ALIST,B,BLIST,CHEBMO,ELIST,EPSABS,EPSREL,F,OMEGA,
     * RESULT,RLIST
      INTEGER IER,INTEGR,IORD,LIMIT,MAXP1,MOMCOM,NEVAL,NNLOG
C
      DIMENSION ALIST(500),BLIST(500),CHEBMO(21,25),ELIST(500),
     * IORD(500),NNLOG(500),RLIST(500)
C
      EXTERNAL F
C
C        LIMIT IS THE MAXIMUM NUMBER OF SUBINTERVALS ALLOWED IN THE
C        SUBDIVISION PROCESS OF QFOUR. TAKE CARE THAT LIMIT.GE.1.
C
      DATA LIMIT/500/
C
C        MAXP1 GIVES AN UPPER BOUND ON THE NUMBER OF CHEBYSHEV MOMENTS
C        WHICH CAN BE STORED, I.E. FOR THE INTERVALS OF LENGTHS
C        ABS(B-A)*2**(-L), L = 0, 1, ... , MAXP1-2. TAKE CARE THAT
C        MAXP1.GE.1.
C
```

```
      DATA MAXP1/21/
C
C***FIRST EXECUTABLE STATEMENT
      CALL QFOUR(F,A,B,OMEGA,INTEGR,EPSABS,EPSREL,LIMIT,1,MAXP1,RESULT,
     *   ABSERR,NEVAL,IER,ALIST,BLIST,RLIST,ELIST,IORD,NNLOG,MOMCOM,
     *   CHEBMO)
C
      RETURN
      END
```

```
      SUBROUTINE QAWF(F,A,OMEGA,INTEGR,EPSABS,RESULT,ABSERR,
     *  NEVAL,IER)
C
C ......................................................................
C
C 1.     QAWF
C        COMPUTATION OF FOURIER INTEGRALS
C           STANDARD FORTRAN SUBROUTINE
C
C 2.     PURPOSE
C           THE ROUTINE CALCULATES AN APPROXIMATION  RESULT  TO A GIVEN
C           FOURIER INTEGRAL
C                I = INTEGRAL OF F(X)*W(X) OVER (A,INFINITY)
C                    WHERE W(X) = COS(OMEGA*X)
C                       OR W(X) = SIN(OMEGA*X),
C           HOPEFULLY SATISFYING FOLLOWING CLAIM FOR ACCURACY
C           ABS(I-RESULT).LE.EPSABS.
C
C 3.     CALLING SEQUENCE
C           CALL QAWF(F,A,OMEGA,INTEGR,EPSABS,RESULT,ABSERR,NEVAL,IER)
C
C        PARAMETERS
C         ON ENTRY
C            F      - REAL
C                     FUNCTION SUBPROGRAM DEFINING THE INTEGRAND
C                     FUNCTION F(X). THE ACTUAL NAME FOR F NEEDS TO BE
C                     DECLARED E X T E R N A L IN THE DRIVER PROGRAM.
C
C            A      - REAL
C                     LOWER LIMIT OF INTEGRATION
C
C            OMEGA  - REAL
C                     PARAMETER IN THE INTEGRAND WEIGHT FUNCTION
C
C            INTEGR - INTEGER
C                     INDICATES WHICH OF THE WEIGHT FUNCTIONS IS USED
C                     INTEGR = 1      W(X) = COS(OMEGA*X)
C                     INTEGR = 2      W(X) = SIN(OMEGA*X)
C                     IF INTEGR.NE.1.AND.INTEGR.NE.2, THE ROUTINE
C                     WILL END WITH IER = 6.
C
C            EPSABS - REAL
C                     ABSOLUTE ACCURACY REQUESTED, EPSABS.GT.0.
C                     IF EPSABS.LE.0, THE ROUTINE WILL END WITH IER = 6.
C
C         ON RETURN
C            RESULT - REAL
C                     APPROXIMATION TO THE INTEGRAL
```

```
C
C            ABSERR - REAL
C                     ESTIMATE OF THE MODULUS OF THE ABSOLUTE ERROR,
C                     WHICH SHOULD EQUAL OR EXCEED ABS(I-RESULT)
C
C            NEVAL  - INTEGER
C                     NUMBER OF INTEGRAND EVALUATIONS
C
C            IER    - INTEGER
C                     IER = 0 NORMAL AND RELIABLE TERMINATION OF THE
C                             ROUTINE. IT IS ASSUMED THAT THE
C                             REQUESTED ACCURACY HAS BEEN ACHIEVED.
C                     IER.GT.0 ABNORMAL TERMINATION OF THE ROUTINE.
C                             THE ESTIMATES FOR INTEGRAL AND ERROR ARE
C                             LESS RELIABLE. IT IS ASSUMED THAT THE
C                             REQUESTED ACCURACY HAS NOT BEEN ACHIEVED.
C                    IF OMEGA.NE.0
C                     IER = 6 THE INPUT IS INVALID BECAUSE
C                             (INTEGR.NE.1 AND INTEGR.NE.2) OR
C                              EPSABS.LE.0
C                              RESULT, ABSERR, NEVAL, LST ARE SET TO
C                              ZERO.
C                         = 7 ABNORMAL TERMINATION OF THE COMPUTATION
C                             OF ONE OR MORE SUBINTEGRALS
C                         = 8 MAXIMUM NUMBER OF CYCLES ALLOWED
C                             HAS BEEN ACHIEVED, I.E. OF SUBINTERVALS
C                             (A+(K-1)C,A+KC) WHERE
C                             C = (2*INT(ABS(OMEGA))+1)*PI/ABS(OMEGA),
C                             FOR K = 1, 2, ...
C                         = 9 THE EXTRAPOLATION TABLE CONSTRUCTED FOR
C                             CONVERGENCE ACCELERATION OF THE SERIES
C                             FORMED BY THE INTERGRAL CONTRIBUTIONS
C                             OVER THE CYCLES, DOES NOT CONVERGE TO
C                             WITHIN THE REQUESTED ACCURACY.
C
C                    IF OMEGA = 0 AND INTEGR = 1,
C                    THE INTEGRAL IS CALCULATED BY MEANS OF QAGI
C                    AND IER HAS THE MEANING DESCRIBED IN THE
C                    COMMENTS OF QAGI
C
C
C 4.      SUBROUTINES OR FUNCTIONS NEEDED
C              - QAWFE
C              - QAGI
C              - QFOUR
C              - QEXT
C              - QK15I
C              - QSORT
```

```
C                 - QC25O
C                 - QK15W
C                 - QCHEB
C                 - QWGTO
C                 - QMACO
C                 - FORTRAN ABS, AMAX1, AMIN1
C                 - F (USER-PROVIDED FUNCTION)
C
C ......................................................................
C
      REAL A,ABSERR,EPSABS,F,OMEGA,RESULT,ALIST,BLIST,ELIST,RLIST,
     *  RSLST,ERLST,CHEBMO
      INTEGER IER,INTEGR,LAST,LIMIT,LIMLST,IORD,IERLST,NNLOG,
     *  LST,MAXP1,NEVAL
C
      DIMENSION ALIST(500),BLIST(500),RLIST(500),ELIST(500),
     *    CHEBMO(21,25),ERLST(50),RSLST(50),IERLST(50),
     *  IORD(500),NNLOG(500)
C
      EXTERNAL F
C
C***FIRST EXECUTABLE STATEMENT
      IER = 6
      NEVAL = 0
      LAST = 0
      RESULT = 0.0E+00
      ABSERR = 0.0E+00
C          DIMENSIONING PARAMETERS
C             LIMLST - INTEGER
C                      LIMLST GIVES AN UPPER BOUND ON
C                      THE NUMBER OF CYCLES, LIMLST.GE.3.
C                      IF LIMLST.LT.3, THE ROUTINE WILL END WITH IER = 6.
C
C
C             MAXP1  - INTEGER
C                      MAXP1 GIVES AN UPPER BOUND ON THE
C                      NUMBER OF CHEBYSHEV MOMENTS WHICH CAN BE
C                      STORED, I.E. FOR THE INTERVALS OF LENGTHS
C                      ABS(B-A)*2**(-L), L = 0,1, ..., MAXP1-2,
C                      MAXP1.GE.1.
C                      IF MAXP1.LT.1, THE ROUTINE WILL END WITH IER = 6.
      MAXP1 = 21
      LIMLST = 50
      LIMIT = 500
C
C          CHECK VALIDITY OF LIMLST AND MAXP1
C
      IF(LIMLST.LT.3.OR.MAXP1.LT.1) GO TO 10
```

```
C
C          PREPARE CALL FOR QAWFE
C
      CALL QAWFE(F,A,OMEGA,INTEGR,EPSABS,LIMLST,LIMIT,MAXP1,RESULT,
     *  ABSERR,NEVAL,IER,RSLST,ERLST,IERLST,LST,ALIST,
     *  BLIST,RLIST,ELIST,IORD,NNLOG,CHEBMO)
10    RETURN
      END
```

```
      SUBROUTINE QAWFE(F,A,OMEGA,INTEGR,EPSABS,LIMLST,LIMIT,MAXP1,
     *  RESULT,ABSERR,NEVAL,IER,RSLST,ERLST,IERLST,LST,ALIST,BLIST,
     *  RLIST,ELIST,IORD,NNLOG,CHEBMO)
C
C ......................................................................
C
C 1.     QAWFE
C        COMPUTATION OF FOURIER INTEGRALS
C           STANDARD FORTRAN SUBROUTINE
C
C 2.     PURPOSE
C           THE ROUTINE CALCULATES AN APPROXIMATION  RESULT  TO A
C           GIVEN FOURIER INTEGRAL
C                 I = INTEGRAL OF F(X)*W(X) OVER (A,INFINITY)
C                     WHERE W(X) = COS(OMEGA*X)
C                        OR W(X) = SIN(OMEGA*X),
C           HOPEFULLY SATISFYING FOLLOWING CLAIM FOR ACCURACY
C           ABS(I-RESULT).LE.EPSABS.
C
C 3.     CALLING SEQUENCE
C           CALL QAWFE(F,A,OMEGA,INTEGR,EPSABS,LIMLST,LIMIT,MAXP1,
C                RESULT,ABSERR,NEVAL,IER,RSLST,ERLST,IERLST,LST,ALIST,
C                BLIST,RLIST,ELIST,IORD,NNLOG,CHEBMO)
C
C        PARAMETERS
C         ON ENTRY
C            F      - REAL
C                     FUNCTION SUBPROGRAM DEFINING THE INTEGRAND
C                     FUNCTION F(X). THE ACTUAL NAME FOR F NEEDS TO BE
C                     DECLARED E X T E R N A L IN THE DRIVER PROGRAM.
C
C            A      - REAL
C                     LOWER LIMIT OF INTEGRATION
C
C            OMEGA  - REAL
C                     PARAMETER IN THE WEIGHT FUNCTION
C
C            INTEGR - INTEGER
C                     INDICATES WHICH WEIGHT FUNCTION IS USED
C                     INTEGR = 1      W(X) = COS(OMEGA*X)
C                     INTEGR = 2      W(X) = SIN(OMEGA*X)
C                     IF INTEGR.NE.1.AND.INTEGR.NE.2, THE ROUTINE WILL
C                     END WITH IER = 6.
C
C            EPSABS - REAL
C                     ABSOLUTE ACCURACY REQUESTED, EPSABS.GT.0
C                     IF EPSABS.LE.0, THE ROUTINE WILL END WITH IER = 6.
C
```

```
C           LIMLST - INTEGER
C                    LIMLST GIVES AN UPPER BOUND ON THE
C                    NUMBER OF CYCLES, LIMLST.GE.1.
C                    IF LIMLST.LT.3, THE ROUTINE WILL END WITH IER = 6.
C
C           LIMIT  - INTEGER
C                    GIVES AN UPPER BOUND ON THE NUMBER OF
C                    SUBINTERVALS ALLOWED IN THE PARTITION OF
C                    EACH CYCLE, LIMIT.GE.1.
C
C           MAXP1  - INTEGER
C                    GIVES AN UPPER BOUND ON THE NUMBER OF
C                    CHEBYSHEV MOMENTS WHICH CAN BE STORED, I.E.
C                    FOR THE INTERVALS OF LENGTHS ABS(B-A)*2**(-L),
C                    L=0,1, ..., MAXP1-2, MAXP1.GE.1
C
C         ON RETURN
C           RESULT - REAL
C                    APPROXIMATION TO THE INTEGRAL
C
C           ABSERR - REAL
C                    ESTIMATE OF THE MODULUS OF THE ABSOLUTE ERROR,
C                    WHICH SHOULD EQUAL OR EXCEED ABS(I-RESULT)
C
C           NEVAL  - INTEGER
C                    NUMBER OF INTEGRAND EVALUATIONS
C
C           IER    - IER = 0 NORMAL AND RELIABLE TERMINATION OF
C                            THE ROUTINE. IT IS ASSUMED THAT THE
C                            REQUESTED ACCURACY HAS BEEN ACHIEVED.
C                    IER.GT.0 ABNORMAL TERMINATION OF THE ROUTINE
C                            THE ESTIMATES FOR INTEGRAL AND ERROR
C                            ARE LESS RELIABLE. IT IS ASSUMED THAT
C                            THE REQUESTED ACCURACY HAS NOT BEEN
C                            ACHIEVED.
C                   IF OMEGA.NE.0
C                    IER = 6 THE INPUT IS INVALID BECAUSE
C                            (INTEGR.NE.1 AND INTEGR.NE.2) OR
C                             EPSABS.LE.0 OR LIMLST.LT.3.
C                             RESULT, ABSERR, NEVAL, LST ARE SET
C                             TO ZERO.
C                        = 7 BAD INTEGRAND BEHAVIOUR OCCURS WITHIN
C                            ONE OR MORE OF THE CYCLES. LOCATION
C                            AND TYPE OF THE DIFFICULTY INVOLVED
C                            CAN BE DETERMINED FROM THE VECTOR IERLST.
C                            HERE LST IS THE NUMBER OF CYCLES ACTUALLY
C                            NEEDED (SEE BELOW).
C                        = 8 MAXIMUM NUMBER OF  CYCLES  ALLOWED
```

```
C                                   HAS BEEN ACHIEVED., I.E. OF SUBINTERVALS
C                                   (A+(K-1)C,A+KC) WHERE
C                                   C = (2*INT(ABS(OMEGA))+1)*PI/ABS(OMEGA),
C                                   FOR K = 1, 2, ..., LST.
C                                   ONE CAN ALLOW MORE CYCLES BY INCREASING
C                                   THE VALUE OF LIMLST (AND TAKING THE
C                                   ACCORDING DIMENSION ADJUSTMENTS INTO
C                                   ACCOUNT).
C                                   EXAMINE THE ARRAY IERLST WHICH CONTAINS
C                                   THE ERROR FLAGS ON THE CYCLES, IN ORDER
C                                   TO EVENTUAL LOOK FOR LOCAL INTEGRATION
C                                   DIFFICULTIES.
C                                   IF THE POSITION OF A LOCAL DIFFICULTY CAN
C                                   BE DETERMINED (E.G. SINGULARITY,
C                                   DISCONTINUITY WITHIN THE INTERVAL)
C                                   ONE WILL PROBABLY GAIN FROM SPLITTING
C                                   UP THE INTERVAL AT THIS POINT AND
C                                   CALLING APPROPRIATE INTEGRATORS ON
C                                   THE SUBRANGES.
C                               = 9 THE EXTRAPOLATION TABLE CONSTRUCTED FOR
C                                   CONVERGENCE ACCELERATION OF THE SERIES
C                                   FORMED BY THE INTEGRAL CONTRIBUTIONS
C                                   OVER THE CYCLES, DOES NOT CONVERGE TO
C                                   WITHIN THE REQUESTED ACCURACY.
C                                   AS IN THE CASE OF IER = 8, IT IS ADVISED
C                                   TO EXAMINE THE ARRAY IERLST WHICH CONTAINS
C                                   THE ERROR FLAGS ON THE CYCLES.
C                                   IERLST(K) = 1 THE MAXIMUM NUMBER OF
C                                                 SUBDIVISIONS (= LIMIT)
C                                                 HAS BEEN ACHIEVED ON THE
C                                                 K TH CYCLE.
C                                             = 2 OCCURENCE OF ROUNDOFF
C                                                 ERROR IS DETECTED AND
C                                                 PREVENTS THE TOLERANCE
C                                                 IMPOSED ON THE K TH CYCLE
C                                                 FROM BEING ACHIEVED.
C                                             = 3 EXTREMELY BAD INTEGRAND
C                                                 BEHAVIOUR OCCURS AT SOME
C                                                 POINTS OF THE K TH CYCLE.
C                                             = 4 THE INTEGRATION PROCEDURE
C                                                 OVER THE K TH CYCLE DOES
C                                                 NOT CONVERGE (TO WITHIN THE
C                                                 REQUIRED ACCURACY) DUE TO
C                                                 ROUNDOFF IN THE
C                                                 EXTRAPOLATION PROCEDURE
C                                                 INVOKED ON THIS CYCLE. IT
C                                                 IS ASSUMED THAT THE RESULT
C                                                 ON THIS INTERVAL IS THE
```

```
C                                          BEST WHICH CAN BE OBTAINED.
C                                      = 5 THE INTEGRAL OVER THE K TH
C                                          CYCLE IS PROBABLY DIVERGENT
C                                          OR SLOWLY CONVERGENT. IT
C                                          MUST BE NOTED THAT
C                                          DIVERGENCE CAN OCCUR WITH
C                                          ANY OTHER VALUE OF
C                                          IERLST(K).
C                    IF OMEGA = 0 AND INTEGR = 1,
C                    THE INTEGRAL IS CALCULATED BY MEANS OF QAGI
C                    AND IER HAS THE MEANING DESCRIBED IN THE
C                    COMMENTS OF QAGI
C
C             RSLST  - REAL
C                      VECTOR OF DIMENSION AT LEAST LIMLST
C                      RSLST(K) CONTAINS THE INTEGRAL CONTRIBUTION
C                      OVER THE INTERVAL (A+(K-1)C,A+KC) WHERE
C                      C = (2*INT(ABS(OMEGA))+1)*PI/ABS(OMEGA),
C                      K = 1, 2, ..., LST.
C                      NOTE THAT, IF OMEGA = 0, RSLST(1) CONTAINS
C                      THE VALUE OF THE INTEGRAL OVER (A,INFINITY).
C
C             ERLST  - REAL
C                      VECTOR OF DIMENSION AT LEAST LIMLST
C                      ERLST(K) CONTAINS THE ERROR ESTIMATE
C                      CORRESPONDING WITH RSLST(K).
C
C             IERLST - INTEGER
C                      VECTOR OF DIMENSION AT LEAST LIMLST
C                      IERLST(K) CONTAINS THE ERROR FLAG CORRESPONDING
C                      WITH RSLST(K). FOR THE MEANING OF THE LOCAL ERROR
C                      FLAGS SEE DESCRIPTION OF OUTPUT PARAMETER IER.
C
C             LST    - INTEGER
C                      NUMBER OF SUBINTEGRALS NEEDED FOR THE INTEGRATION
C                      IF OMEGA = 0 THEN LST IS SET TO 1.
C
C             ALIST, BLIST, RLIST, ELIST - REAL
C                      WORKVECTORS OF DIMENSION AT LEAST LIMIT.
C
C             IORD, NNLOG - INTEGER
C                      WORKVECTORS OF DIMENSION AT LEAST LIMIT, PROVIDING
C                      SPACE FOR THE QUANTITIES NEEDED IN THE
C                      SUBDIVISION PROCESS OF EACH CYCLE
C
C             CHEBMO - REAL
C                      WORKARRAY OF DIMENSION AT LEAST (MAXP1,25),
C                      PROVIDING SPACE FOR THE CHEBYSHEV MOMENTS
```

```
C                        NEEDED WITHIN THE CYCLES
C
C 4.      SUBROUTINES OR FUNCTIONS NEEDED
C                - QFOUR
C                - QAGI
C                - QEXT
C                - QK15I
C                - QSORT
C                - QC25O
C                - QK15W
C                - QCHEB
C                - F (USER-PROVIDED FUNCTION)
C                - QWGTO
C                - QMACO
C                - FORTRAN ABS, AMAX1, AMIN1
C
C .............................................................
C
      REAL A,ABSEPS,ABSERR,ALIST,BLIST,CHEBMO,CORREC,CYCLE,C1,C2,DL,DLA,
     *  DRL,ELIST,EP,EPS,EPSA,EPSABS,ERLST,ERRSUM,F,FACT,OMEGA,P,PI,P1,
     *  PSUM,RESEPS,RESULT,RES3LA,RLIST,RSLST,UFLOW,EPMACH,OFLOW
      INTEGER IER,IERLST,INTEGR,IORD,KTMIN,L,LST,LIMIT,LIMLST,LL,
     *     MAXP1,MOMCOM,NEV,NEVAL,NNLOG,NRES,NUMRL2
C
      DIMENSION ALIST(LIMIT),BLIST(LIMIT),CHEBMO(MAXP1,25),ELIST(LIMIT),
     *  ERLST(LIMLST),IERLST(LIMLST),IORD(LIMIT),NNLOG(LIMIT),PSUM(52),
     *  RES3LA(3),RLIST(LIMIT),RSLST(LIMLST)
C
      EXTERNAL F
C
C
C             THE DIMENSION OF  PSUM  IS DETERMINED BY THE VALUE OF
C             LIMEXP IN SUBROUTINE QEXT (PSUM MUST BE
C             OF DIMENSION (LIMEXP+2) AT LEAST).
C
C            LIST OF MAJOR VARIABLES
C            -----------------------
C
C            C1, C2     - END POINTS OF SUBINTERVAL (OF LENGTH
C                           CYCLE)
C            CYCLE      - (2*INT(ABS(OMEGA))+1)*PI/ABS(OMEGA)
C            PSUM       - VECTOR OF DIMENSION AT LEAST (LIMEXP+2)
C                           (SEE ROUTINE QEXT)
C                           PSUM CONTAINS THE PART OF THE EPSILON TABLE
C                           WHICH IS STILL NEEDED FOR FURTHER COMPUTATIONS.
C                           EACH ELEMENT OF PSUM IS A PARTIAL SUM OF
C                           THE SERIES WHICH SHOULD SUM TO THE VALUE OF
C                           THE INTEGRAL.
```

```
C            ERRSUM     - SUM OF ERROR ESTIMATES OVER THE
C                         SUBINTERVALS, CALCULATED CUMULATIVELY
C            EPSA       - ABSOLUTE TOLERANCE REQUESTED OVER CURRENT
C                         SUBINTERVAL
C            CHEBMO     - ARRAY CONTAINING THE MODIFIED CHEBYSHEV
C                         MOMENTS (SEE ALSO ROUTINE QC25F)
C
      DATA P/9.0E-01/,PI/3.1415926535897932E+00/
C
C            MACHINE DEPENDENT CONSTANTS
C            ---------------------------
C            EPMACH IS THE LARGEST RELATIVE SPACING.
C            UFLOW IS THE SMALLEST POSITIVE MAGNITUDE.
C            OFLOW IS THE LARGEST MAGNITUDE.
C
C***FIRST EXECUTABLE STATEMENT
      CALL QMACO(EPMACH,UFLOW,OFLOW)
C
C            TEST ON VALIDITY OF PARAMETERS
C            ------------------------------
C
      RESULT = 0.0E+00
      ABSERR = 0.0E+00
      NEVAL = 0
      LST = 0
      IER = 0
      IF((INTEGR.NE.1.AND.INTEGR.NE.2).OR.EPSABS.LE.0.0E+00.OR.
     *  LIMLST.LT.3) IER = 6
      IF(IER.EQ.6) GO TO 999
      IF(OMEGA.NE.0.0E+00) GO TO 10
C
C            INTEGRATION BY QAGI IF OMEGA IS ZERO
C            ------------------------------------
C
      IF(INTEGR.EQ.1) CALL QAGI(F,0.0E+00,1,EPSABS,0.0E+00,
     *  RESULT,ABSERR,NEVAL,IER)
      RSLST(1) = RESULT
      ERLST(1) = ABSERR
      IERLST(1) = IER
      LST = 1
      GO TO 999
C
C            INITIALIZATIONS
C            ---------------
C
   10 L = ABS(OMEGA)
      DL = 2*L+1
      CYCLE = DL*PI/ABS(OMEGA)
```

```
-----------------------------------------------------------------------
      IER = 0
      KTMIN = 0
      NEVAL = 0
      NUMRL2 = 0
      NRES = 0
      C1 = A
      C2 = CYCLE+A
      P1 = 1.0E+00-P
      EPS = EPSABS
      IF(EPSABS.GT.UFLOW/P1) EPS = EPSABS*P1
      EP = EPS
      FACT = 1.0E+00
      CORREC = 0.0E+00
      ABSERR = 0.0E+00
      ERRSUM = 0.0E+00
C
C           MAIN DO-LOOP
C           ------------
C
      DO 50 LST = 1,LIMLST
C
C           INTEGRATE OVER CURRENT SUBINTERVAL.
C
        DLA = LST
        EPSA = EPS*FACT
        CALL QFOUR(F,C1,C2,OMEGA,INTEGR,EPSA,0.0E+00,LIMIT,LST,MAXP1,
     *  RSLST(LST),ERLST(LST),NEV,IERLST(LST),ALIST,BLIST,RLIST,ELIST,
     *  IORD,NNLOG,MOMCOM,CHEBMO)
        NEVAL = NEVAL+NEV
        FACT = FACT*P
        ERRSUM = ERRSUM+ERLST(LST)
        DRL = 5.0E+01*ABS(RSLST(LST))
C
C           TEST ON ACCURACY WITH PARTIAL SUM
C
        IF((ERRSUM+DRL).LE.EPSABS.AND.LST.GE.6) GO TO 80
        CORREC = AMAX1(CORREC,ERLST(LST))
        IF(IERLST(LST).NE.0) EPS = AMAX1(EP,CORREC*P1)
        IF(IERLST(LST).NE.0) IER = 7
        IF(IER.EQ.7.AND.(ERRSUM+DRL).LE.CORREC*1.0E+01.AND.
     *  LST.GT.5) GO TO 80
        NUMRL2 = NUMRL2+1
        IF(LST.GT.1) GO TO 20
        PSUM(1) = RSLST(1)
        GO TO 40
   20   PSUM(NUMRL2) = PSUM(LL)+RSLST(LST)
        IF(LST.EQ.2) GO TO 40
C
```

```
C            TEST ON MAXIMUM NUMBER OF SUBINTERVALS
C
        IF(LST.EQ.LIMLST) IER = 8
C
C            PERFORM NEW EXTRAPOLATION
C
        CALL QEXT(NUMRL2,PSUM,RESEPS,ABSEPS,RES3LA,NRES)
C
C            TEST WHETHER EXTRAPOLATED RESULT IS INFLUENCED BY ROUNDOFF
C
        KTMIN = KTMIN+1
        IF(KTMIN.GE.15.AND.ABSERR.LE.1.0E-03*(ERRSUM+DRL)) IER = 9
        IF(ABSEPS.GT.ABSERR.AND.LST.NE.3) GO TO 30
        ABSERR = ABSEPS
        RESULT = RESEPS
        KTMIN = 0
C
C            IF IER .NE. 0, CHECK WHETHER DIRECT RESULT (PARTIAL
C            SUM) OR EXTRAPOLATED RESULT YIELDS THE BEST INTEGRAL
C            APPROXIMATION
C
        IF((ABSERR+1.0E+01*CORREC).LE.EPSABS.OR.
     *  (ABSERR.LE.EPSABS.AND.1.0E+01*CORREC.GE.EPSABS)) GO TO 60
   30   IF(IER.NE.0.AND.IER.NE.7) GO TO 60
   40   LL = NUMRL2
        C1 = C2
        C2 = C2+CYCLE
   50 CONTINUE
C
C          SET FINAL RESULT AND ERROR ESTIMATE
C          ------------------------------------
C
   60 ABSERR = ABSERR+1.0E+01*CORREC
      IF(IER.EQ.0) GO TO 999
      IF(RESULT.NE.0.0E+00.AND.PSUM(NUMRL2).NE.0.0E+00) GO TO 70
      IF(ABSERR.GT.ERRSUM) GO TO 80
      IF(PSUM(NUMRL2).EQ.0.0E+00) GO TO 999
   70 IF(ABSERR/ABS(RESULT).GT.(ERRSUM+DRL)/ABS(PSUM(NUMRL2))) GO TO 80
      IF(IER.GE.1.AND.IER.NE.7) ABSERR = ABSERR+DRL
      GO TO 999
   80 RESULT = PSUM(NUMRL2)
      ABSERR = ERRSUM+DRL
  999 RETURN
      END
```

```
      SUBROUTINE QAWS(F,A,B,ALFA,BETA,INTEGR,EPSABS,EPSREL,RESULT,
     *  ABSERR,NEVAL,IER)
C
C ......................................................................
C
C 1.     QAWS
C        INTEGRATION OF FUNCTIONS HAVING ALGEBRAICO-LOGARITHMIC
C        END POINT SINGULARITIES
C           STANDARD FORTRAN SUBROUTINE
C
C 2.     PURPOSE
C           THE ROUTINE CALCULATES AN APPROXIMATION  RESULT  TO A GIVEN
C           DEFINITE INTEGRAL   I = INTEGRAL OF F*W OVER (A,B) (WHERE W
C           SHOWS A SINGULAR BEHAVIOUR AT THE END POINTS, SEE PARAMETER
C           INTEGR), HOPEFULLY SATISFYING FOLLOWING CLAIM FOR ACCURACY
C           ABS(I-RESULT).LE.MAX(EPSABS,EPSREL*ABS(I)).
C
C 3.     CALLING SEQUENCE
C           CALL QAWS(F,A,B,ALFA,BETA,INTEGR,EPSABS,EPSREL,RESULT,
C                     ABSERR,NEVAL,IER)
C
C        PARAMETERS
C         ON ENTRY
C           F      - REAL
C                    FUNCTION SUBPROGRAM DEFINING THE INTEGRAND
C                    FUNCTION F(X). THE ACTUAL NAME FOR F NEEDS TO BE
C                    DECLARED E X T E R N A L IN THE DRIVER PROGRAM.
C
C           A      - REAL
C                    LOWER LIMIT OF INTEGRATION
C
C           B      - REAL
C                    UPPER LIMIT OF INTEGRATION, B.GT.A
C                    IF B.LE.A, THE ROUTINE WILL END WITH IER = 6.
C
C           ALFA   - REAL
C                    PARAMETER IN THE INTEGRAND FUNCTION, ALFA.GT.(-1)
C                    IF ALFA.LE.(-1), THE ROUTINE WILL END WITH
C                    IER = 6.
C
C           BETA   - REAL
C                    PARAMETER IN THE INTEGRAND FUNCTION, BETA.GT.(-1)
C                    IF BETA.LE.(-1), THE ROUTINE WILL END WITH
C                    IER = 6.
C
C           INTEGR - INTEGER
C                    INDICATES WHICH WEIGHT FUNCTION IS TO BE USED
C                    = 1  (X-A)**ALFA*(B-X)**BETA
```

```
C                     = 2  (X-A)**ALFA*(B-X)**BETA*LOG(X-A)
C                     = 3  (X-A)**ALFA*(B-X)**BETA*LOG(B-X)
C                     = 4  (X-A)**ALFA*(B-X)**BETA*LOG(X-A)*LOG(B-X)
C                     IF INTEGR.LT.1 OR INTEGR.GT.4, THE ROUTINE WILL
C                     END WITH IER = 6.
C
C           EPSABS - REAL
C                     ABSOLUTE ACCURACY REQUESTED
C           EPSREL - REAL
C                     RELATIVE ACCURACY REQUESTED
C                     IF  EPSABS.LT.0 AND EPSREL.LT.0,
C                     THE ROUTINE WILL END WITH IER = 6.
C
C         ON RETURN
C           RESULT - REAL
C                     APPROXIMATION TO THE INTEGRAL
C
C           ABSERR - REAL
C                     ESTIMATE OF THE MODULUS OF THE ABSOLUTE ERROR,
C                     WHICH SHOULD EQUAL OR EXCEED ABS(I-RESULT)
C
C           NEVAL  - INTEGER
C                     NUMBER OF INTEGRAND EVALUATIONS
C
C           IER    - INTEGER
C                     IER = 0 NORMAL AND RELIABLE TERMINATION OF THE
C                             ROUTINE. IT IS ASSUMED THAT THE REQUESTED
C                             ACCURACY HAS BEEN ACHIEVED.
C                     IER.GT.0 ABNORMAL TERMINATION OF THE ROUTINE
C                             THE ESTIMATES FOR THE INTEGRAL AND ERROR
C                             ARE LESS RELIABLE. IT IS ASSUMED THAT THE
C                             REQUESTED ACCURACY HAS NOT BEEN ACHIEVED.
C                     IER = 1 MAXIMUM NUMBER OF SUBDIVISIONS ALLOWED
C                             HAS BEEN ACHIEVED. ONE CAN ALLOW MORE
C                             SUBDIVISIONS BY INCREASING THE DATA VALUE
C                             OF LIMIT IN QAWS (AND TAKING THE ACCORDING
C                             DIMENSION ADJUSTMENTS INTO ACCOUNT).
C                             HOWEVER, IF THIS YIELDS NO IMPROVEMENT IT
C                             IS ADVISED TO ANALYZE THE INTEGRAND, IN
C                             ORDER TO DETERMINE THE INTEGRATION
C                             DIFFICULTIES WHICH PREVENT THE REQUESTED
C                             TOLERANCE FROM BEING ACHIEVED. IN CASE OF
C                             A JUMP DISCONTINUITY OR A LOCAL
C                             SINGULARITY OF ALGEBRAICO-LOGARITHMIC TYPE
C                             AT ONE OR MORE INTERIOR POINTS OF THE
C                             INTEGRATION RANGE, ONE SHOULD PROCEED BY
C                             SPLITTING UP THE INTERVAL AT THESE POINTS
C                             AND CALLING THE INTEGRATOR ON THE
```

```
C                                   SUBRANGES.
C                               = 2 THE OCCURRENCE OF ROUNDOFF ERROR IS
C                                   DETECTED, WHICH PREVENTS THE REQUESTED
C                                   TOLERANCE FROM BEING ACHIEVED.
C                               = 3 EXTREMELY BAD INTEGRAND BEHAVIOUR OCCURS
C                                   AT SOME POINTS OF THE INTEGRATION
C                                   INTERVAL.
C                               = 6 THE INPUT IS INVALID, BECAUSE
C                                   B.LE.A OR ALFA.LE.(-1) OR BETA.LE.(-1) OR
C                                   OR INTEGR.LT.1 OR INTEGR.GT.4 OR
C                                   EPSABS.LT.0 AND EPSREL.LT.0,
C                                   RESULT, ABSERR, NEVAL ARE SET TO ZERO.
C
C 4.      SUBROUTINES OR FUNCTIONS NEEDED
C               - QAWSE
C               - QMOMO
C               - QC25S
C               - QSORT
C               - QK15W
C               - QWGTS
C               - QCHEB
C               - F (USER-PROVIDED FUNCTION)
C               - QMACO
C               - FORTRAN ABS, AMAX1, AMIN1
C
C ......................................................................
C
      REAL A,ABSERR,ALFA,ALIST,B,BLIST,BETA,ELIST,EPSABS,EPSREL,F,
     *  RESULT,RLIST
      INTEGER IER,INTEGR,IORD,LAST,LIMIT,NEVAL
C
      DIMENSION ALIST(500),BLIST(500),ELIST(500),IORD(500),RLIST(500)
C
      EXTERNAL F
C
C          LIMIT IS THE MAXIMUM NUMBER OF SUBINTERVALS ALLOWED IN THE
C          SUBDIVISION PROCESS OF QAWSE. TAKE CARE THAT LIMIT.GE.2.
C
      DATA LIMIT/500/
C
C***FIRST EXECUTABLE STATEMENT
C
      CALL QAWSE(F,A,B,ALFA,BETA,INTEGR,EPSABS,EPSREL,LIMIT,RESULT,
     *  ABSERR,NEVAL,IER,ALIST,BLIST,RLIST,ELIST,IORD,LAST)
      RETURN
      END
```

```
      SUBROUTINE QAWSE(F,A,B,ALFA,BETA,INTEGR,EPSABS,EPSREL,LIMIT,
     *  RESULT,ABSERR,NEVAL,IER,ALIST,BLIST,RLIST,ELIST,IORD,LAST)
C
C ......................................................................
C
C 1.     QAWSE
C        INTEGRATION OF FUNCTIONS HAVING ALGEBRAICO-LOGARITHMIC
C        END POINT SINGULARITIES
C           STANDARD FORTRAN SUBROUTINE
C
C 2.     PURPOSE
C           THE ROUTINE CALCULATES AN APPROXIMATION  RESULT  TO A GIVEN
C           DEFINITE INTEGRAL   I = INTEGRAL OF F*W OVER (A,B), (WHERE W
C           SHOWS A SINGULAR BEHAVIOUR AT THE END POINTS, SEE PARAMETER
C           INTEGR), HOPEFULLY SATISFYING FOLLOWING CLAIM FOR ACCURACY
C           ABS(I-RESULT).LE.MAX(EPSABS,EPSREL*ABS(I)).
C
C 3.     CALLING SEQUENCE
C           CALL QAWSE(F,A,B,ALFA,BETA,INTEGR,EPSABS,EPSREL,LIMIT,
C                      RESULT,ABSERR,NEVAL,IER,ALIST,BLIST,RLIST,ELIST,
C                      IORD,LAST)
C
C        PARAMETERS
C         ON ENTRY
C            F      - REAL
C                     FUNCTION SUBPROGRAM DEFINING THE INTEGRAND
C                     FUNCTION F(X). THE ACTUAL NAME FOR F NEEDS TO BE
C                     DECLARED E X T E R N A L IN THE DRIVER PROGRAM.
C
C            A      - REAL
C                     LOWER LIMIT OF INTEGRATION
C
C            B      - REAL
C                     UPPER LIMIT OF INTEGRATION, B.GT.A
C                     IF B.LE.A, THE ROUTINE WILL END WITH IER = 6.
C
C            ALFA   - REAL
C                     PARAMETER IN THE WEIGHT FUNCTION, ALFA.GT.(-1)
C                     IF ALFA.LE.(-1), THE ROUTINE WILL END WITH
C                     IER = 6.
C
C            BETA   - REAL
C                     PARAMETER IN THE WEIGHT FUNCTION, BETA.GT.(-1)
C                     IF BETA.LE.(-1), THE ROUTINE WILL END WITH
C                     IER = 6.
C
C            INTEGR - INTEGER
C                     INDICATES WHICH WEIGHT FUNCTION IS TO BE USED
```

```
C                          = 1   (X-A)**ALFA*(B-X)**BETA
C                          = 2   (X-A)**ALFA*(B-X)**BETA*LOG(X-A)
C                          = 3   (X-A)**ALFA*(B-X)**BETA*LOG(B-X)
C                          = 4   (X-A)**ALFA*(B-X)**BETA*LOG(X-A)*LOG(B-X)
C                          IF INTEGR.LT.1 OR INTEGR.GT.4, THE ROUTINE WILL
C                          END WITH IER = 6.
C
C              EPSABS  - REAL
C                          ABSOLUTE ACCURACY REQUESTED
C              EPSREL  - REAL
C                          RELATIVE ACCURACY REQUESTED
C                          IF  EPSABS.LT.0 AND EPSREL.LT.0,
C                          THE ROUTINE WILL END WITH IER = 6.
C
C              LIMIT   - INTEGER
C                          GIVES AN UPPER BOUND ON THE NUMBER OF SUBINTERVALS
C                          IN THE PARTITION OF (A,B), LIMIT.GE.2
C                          IF LIMIT.LT.2, THE ROUTINE WILL END WITH IER = 6.
C
C           ON RETURN
C              RESULT  - REAL
C                          APPROXIMATION TO THE INTEGRAL
C
C              ABSERR  - REAL
C                          ESTIMATE OF THE MODULUS OF THE ABSOLUTE ERROR,
C                          WHICH SHOULD EQUAL OR EXCEED ABS(I-RESULT)
C
C              NEVAL   - INTEGER
C                          NUMBER OF INTEGRAND EVALUATIONS
C
C              IER     - INTEGER
C                          IER = 0 NORMAL AND RELIABLE TERMINATION OF THE
C                                  ROUTINE. IT IS ASSUMED THAT THE REQUESTED
C                                  ACCURACY HAS BEEN ACHIEVED.
C                          IER.GT.0 ABNORMAL TERMINATION OF THE ROUTINE
C                                  THE ESTIMATES FOR THE INTEGRAL AND ERROR
C                                  ARE LESS RELIABLE. IT IS ASSUMED THAT THE
C                                  REQUESTED ACCURACY HAS NOT BEEN ACHIEVED.
C                              = 1 MAXIMUM NUMBER OF SUBDIVISIONS ALLOWED
C                                  HAS BEEN ACHIEVED. ONE CAN ALLOW MORE
C                                  SUBDIVISIONS BY INCREASING THE VALUE OF
C                                  LIMIT. HOWEVER, IF THIS YIELDS NO
C                                  IMPROVEMENT IT IS ADVISED TO ANALYZE THE
C                                  INTEGRAND, IN ORDER TO DETERMINE THE
C                                  INTERGRATION DIFFICULTIES WHICH PREVENT
C                                  THE REQUESTED TOLERANCE FROM BEING
C                                  ACHIEVED. IN CASE OF A JUMP DISCONTINUITY
C                                  OR A LOCAL SINGULARITY OF ALGEBRAICO-
```

```
C                             LOGARITHMIC TYPE AT ONE OR MORE INTERIOR
C                             POINTS OF THE INTEGRATION RANGE, ONE
C                             SHOULD PROCEED BY SPLITTING UP THE
C                             INTERVAL AT THESE POINTS AND CALLING THE
C                             INTEGRATOR ON THE SUBRANGES.
C                         = 2 THE OCCURRENCE OF ROUNDOFF ERROR IS
C                             DETECTED, WHICH PREVENTS THE REQUESTED
C                             TOLERANCE FROM BEING ACHIEVED.
C                         = 3 EXTREMELY BAD INTEGRAND BEHAVIOUR OCCURS
C                             AT SOME POINTS OF THE INTEGRATION
C                             INTERVAL.
C                         = 6 THE INPUT IS INVALID, BECAUSE
C                             B.LE.A OR ALFA.LE.(-1) OR BETA.LE.(-1) OR
C                             INTEGR.LT.1 OR INTEGR.GT.4, OR
C                             EPSABS.LT.0 AND EPSREL.LT.0,
C                             OR LIMIT.LT.2.
C                             RESULT, ABSERR, NEVAL, RLIST(1), ELIST(1),
C                             IORD(1) AND LAST ARE SET TO ZERO.
C                             ALIST(1) AND BLIST(1) ARE SET TO A AND B
C                             RESPECTIVELY.
C
C            ALIST  - REAL
C                     VECTOR OF DIMENSION AT LEAST LIMIT, THE FIRST
C                      LAST  ELEMENTS OF WHICH ARE THE LEFT END POINTS
C                     OF THE SUBINTERVALS IN THE PARTITION OF THE GIVEN
C                     INTEGRATION RANGE (A,B)
C
C            BLIST  - REAL
C                     VECTOR OF DIMENSION AT LEAST LIMIT, THE FIRST
C                      LAST  ELEMENTS OF WHICH ARE THE RIGHT END POINTS
C                     OF THE SUBINTERVALS IN THE PARTITION OF THE GIVEN
C                     INTEGRATION RANGE (A,B)
C
C            RLIST  - REAL
C                     VECTOR OF DIMENSION AT LEAST LIMIT,THE FIRST
C                      LAST  ELEMENTS OF WHICH ARE THE INTEGRAL
C                     APPROXIMATIONS ON THE SUBINTERVALS
C
C            ELIST  - REAL
C                     VECTOR OF DIMENSION AT LEAST LIMIT, THE FIRST
C                      LAST  ELEMENTS OF WHICH ARE THE MODULI OF THE
C                     ABSOLUTE ERROR ESTIMATES ON THE SUBINTERVALS
C
C            IORD   - INTEGER
C                     VECTOR OF DIMENSION AT LEAST LIMIT, THE FIRST K
C                     ELEMENTS OF WHICH ARE POINTERS TO THE ERROR
C                     ESTIMATES OVER THE SUBINTERVALS, SO THAT
C                     ELIST(IORD(1)), ..., ELIST(IORD(K)) WITH K = LAST
```

```
C                       IF LAST.LE.(LIMIT/2+2), AND K = LIMIT+1-LAST
C                       OTHERWISE, FORM A DECREASING SEQUENCE.
C
C            LAST    - INTEGER
C                       NUMBER OF SUBINTERVALS ACTUALLY PRODUCED IN THE
C                       SUBDIVISION PROCESS
C
C 4.      SUBROUTINES OR FUNCTIONS NEEDED
C                 - QMOMO
C                 - QC25S
C                 - QSORT
C                 - QK15W
C                 - QCHEB
C                 - QMACO
C                 - QWGTS
C                 - F (USER-PROVIDED FUNCTION)
C                 - FORTRAN ABS, AMAX1, AMIN1
C
C .................................................................
C
      REAL A,ABSERR,ALFA,ALIST,AREA,AREA1,AREA12,AREA2,A1,A2,B,
     *  BETA,BLIST,B1,B2,CENTRE,ELIST,EPMACH,EPSABS,EPSREL,ERRBND,
     *  ERRMAX,ERROR1,ERRO12,ERROR2,ERRSUM,F,OFLOW,RESAS1,RESAS2,RESULT,
     *  RG,RH,RI,RJ,RLIST,UFLOW
      INTEGER IER,INTEGR,IORD,IROFF1,IROFF2,K,LAST,LIMIT,MAXERR,NEV,
     *  NEVAL,NRMAX
C
      EXTERNAL F
C
      DIMENSION ALIST(LIMIT),BLIST(LIMIT),RLIST(LIMIT),ELIST(LIMIT),
     *  IORD(LIMIT),RI(25),RJ(25),RH(25),RG(25)
C
C            LIST OF MAJOR VARIABLES
C            -----------------------
C
C           ALIST      - LIST OF LEFT END POINTS OF ALL SUBINTERVALS
C                        CONSIDERED UP TO NOW
C           BLIST      - LIST OF RIGHT END POINTS OF ALL SUBINTERVALS
C                        CONSIDERED UP TO NOW
C           RLIST(I)   - APPROXIMATION TO THE INTEGRAL OVER
C                        (ALIST(I),BLIST(I))
C           ELIST(I)   - ERROR ESTIMATE APPLYING TO RLIST(I)
C           MAXERR     - POINTER TO THE INTERVAL WITH LARGEST ERROR
C                        ESTIMATE
C           ERRMAX     - ELIST(MAXERR)
C           AREA       - SUM OF THE INTEGRALS OVER THE SUBINTERVALS
C           ERRSUM     - SUM OF THE ERRORS OVER THE SUBINTERVALS
C           ERRBND     - REQUESTED ACCURACY MAX(EPSABS,EPSREL*
```

```
C                         ABS(RESULT))
C           *****1     - VARIABLE FOR THE LEFT SUBINTERVAL
C           *****2     - VARIABLE FOR THE RIGHT SUBINTERVAL
C           LAST       - INDEX FOR SUBDIVISION
C
C
C            MACHINE DEPENDENT CONSTANTS
C            ---------------------------
C
C           EPMACH IS THE LARGEST RELATIVE SPACING.
C           UFLOW IS THE SMALLEST POSITIVE MAGNITUDE.
C           OFLOW IS THE LARGEST MAGNITUDE.
C
C***FIRST EXECUTABLE STATEMENT
      CALL QMACO(EPMACH,UFLOW,OFLOW)
C
C           TEST ON VALIDITY OF PARAMETERS
C           ------------------------------
C
      IER = 6
      NEVAL = 0
      LAST = 0
      RLIST(1) = 0.0E+00
      ELIST(1) = 0.0E+00
      IORD(1) = 0
      RESULT = 0.0E+00
      ABSERR = 0.0E+00
      IF(B.LE.A.OR.(EPSABS.LT.0.0E+00.AND.EPSREL.LT.0.0E+00).OR.
     *  ALFA.LE.(-1.0E+00).OR.BETA.LE.(-1.0E+00).OR.INTEGR.LT.1.
     *  OR.INTEGR.GT.4.OR.LIMIT.LT.2) GO TO 999
      IER = 0
C
C           COMPUTE THE MODIFIED CHEBYSHEV MOMENTS.
C
      CALL QMOMO(ALFA,BETA,RI,RJ,RG,RH,INTEGR)
C
C           INTEGRATE OVER THE INTERVALS (A,(A+B)/2) AND ((A+B)/2,B).
C
      CENTRE = 5.0E-01*(B+A)
      CALL QC25S(F,A,B,A,CENTRE,ALFA,BETA,RI,RJ,RG,RH,AREA1,ERROR1,
     *  RESAS1,INTEGR,NEV)
      NEVAL = NEV
      CALL QC25S(F,A,B,CENTRE,B,ALFA,BETA,RI,RJ,RG,RH,AREA2,ERROR2,
     *  RESAS2,INTEGR,NEV)
      LAST = 2
      NEVAL = NEVAL+NEV
      RESULT = AREA1+AREA2
      ABSERR = ERROR1+ERROR2
```

```
C
C           TEST ON ACCURACY.
C
      ERRBND = AMAX1(EPSABS,EPSREL*ABS(RESULT))
C
C           INITIALIZATION
C           --------------
C
      IF(ERROR2.GT.ERROR1) GO TO 10
      ALIST(1) = A
      ALIST(2) = CENTRE
      BLIST(1) = CENTRE
      BLIST(2) = B
      RLIST(1) = AREA1
      RLIST(2) = AREA2
      ELIST(1) = ERROR1
      ELIST(2) = ERROR2
      GO TO 20
   10 ALIST(1) = CENTRE
      ALIST(2) = A
      BLIST(1) = B
      BLIST(2) = CENTRE
      RLIST(1) = AREA2
      RLIST(2) = AREA1
      ELIST(1) = ERROR2
      ELIST(2) = ERROR1
   20 IORD(1) = 1
      IORD(2) = 2
      IF(LIMIT.EQ.2) IER = 1
      IF(ABSERR.LE.ERRBND.OR.IER.EQ.1) GO TO 999
      ERRMAX = ELIST(1)
      MAXERR = 1
      NRMAX = 1
      AREA = RESULT
      ERRSUM = ABSERR
      IROFF1 = 0
      IROFF2 = 0
C
C            MAIN DO-LOOP
C            ------------
C
      DO 60 LAST = 3,LIMIT
C
C           BISECT THE SUBINTERVAL WITH LARGEST ERROR ESTIMATE.
C
        A1 = ALIST(MAXERR)
        B1 = 5.0E-01*(ALIST(MAXERR)+BLIST(MAXERR))
        A2 = B1
```

```
      B2 = BLIST(MAXERR)
C
      CALL QC25S(F,A,B,A1,B1,ALFA,BETA,RI,RJ,RG,RH,AREA1,
     * ERROR1,RESAS1,INTEGR,NEV)
      NEVAL = NEVAL+NEV
      CALL QC25S(F,A,B,A2,B2,ALFA,BETA,RI,RJ,RG,RH,AREA2,
     * ERROR2,RESAS2,INTEGR,NEV)
      NEVAL = NEVAL+NEV
C
C           IMPROVE PREVIOUS APPROXIMATIONS INTEGRAL AND ERROR AND
C           TEST FOR ACCURACY.
C
      AREA12 = AREA1+AREA2
      ERRO12 = ERROR1+ERROR2
      ERRSUM = ERRSUM+ERRO12-ERRMAX
      AREA = AREA+AREA12-RLIST(MAXERR)
      IF(A.EQ.A1.OR.B.EQ.B2) GO TO 30
      IF(RESAS1.EQ.ERROR1.OR.RESAS2.EQ.ERROR2) GO TO 30
C
C           TEST FOR ROUNDOFF ERROR.
C
      IF(ABS(RLIST(MAXERR)-AREA12).LT.1.0E-05*ABS(AREA12)
     * .AND.ERRO12.GE.9.9E-01*ERRMAX) IROFF1 = IROFF1+1
      IF(LAST.GT.10.AND.ERRO12.GT.ERRMAX) IROFF2 = IROFF2+1
   30 RLIST(MAXERR) = AREA1
      RLIST(LAST) = AREA2
C
C           TEST ON ACCURACY.
C
      ERRBND = AMAX1(EPSABS,EPSREL*ABS(AREA))
      IF(ERRSUM.LE.ERRBND) GO TO 35
C
C           SET ERROR FLAG IN THE CASE THAT THE NUMBER OF INTERVAL
C           BISECTIONS EXCEEDS LIMIT.
C
      IF(LAST.EQ.LIMIT) IER = 1
C
C
C           SET ERROR FLAG IN THE CASE OF ROUNDOFF ERROR.
C
      IF(IROFF1.GE.6.OR.IROFF2.GE.20) IER = 2
C
C           SET ERROR FLAG IN THE CASE OF BAD INTEGRAND BEHAVIOUR
C           AT INTERIOR POINTS OF INTEGRATION RANGE.
C
      IF(AMAX1(ABS(A1),ABS(B2)).LE.(1.0E+00+1.0E+03*EPMACH)*
     * (ABS(A2)+1.0E+03*UFLOW)) IER = 3
C
```

```
C           APPEND THE NEWLY-CREATED INTERVALS TO THE LIST.
C
   35   IF(ERROR2.GT.ERROR1) GO TO 40
        ALIST(LAST) = A2
        BLIST(MAXERR) = B1
        BLIST(LAST) = B2
        ELIST(MAXERR) = ERROR1
        ELIST(LAST) = ERROR2
        GO TO 50
   40   ALIST(MAXERR) = A2
        ALIST(LAST) = A1
        BLIST(LAST) = B1
        RLIST(MAXERR) = AREA2
        RLIST(LAST) = AREA1
        ELIST(MAXERR) = ERROR2
        ELIST(LAST) = ERROR1
C
C           CALL SUBROUTINE QSORT TO MAINTAIN THE DESCENDING ORDERING
C           IN THE LIST OF ERROR ESTIMATES AND SELECT THE SUBINTERVAL
C           WITH LARGEST ERROR ESTIMATE (TO BE BISECTED NEXT).
C
   50   CALL QSORT(LIMIT,LAST,MAXERR,ERRMAX,ELIST,IORD,NRMAX)
C***JUMP OUT OF DO-LOOP
        IF (IER.NE.0.OR.ERRSUM.LE.ERRBND) GO TO 70
   60 CONTINUE
C
C           COMPUTE FINAL RESULT.
C           ---------------------
C
   70 RESULT = 0.0E+00
      DO 80 K=1,LAST
        RESULT = RESULT+RLIST(K)
   80 CONTINUE
      ABSERR = ERRSUM
  999 RETURN
      END
```

```
      SUBROUTINE QAWC(F,A,B,C,EPSABS,EPSREL,RESULT,ABSERR,NEVAL,IER)
C
C.......................................................................
C
C 1.     QAWC
C        COMPUTATION OF A CAUCHY PRINCIPAL VALUE
C           STANDARD FORTRAN SUBROUTINE
C
C 2.     PURPOSE
C           THE ROUTINE CALCULATES AN APPROXIMATION  RESULT  TO A
C           CAUCHY PRINCIPAL VALUE I = INTEGRAL OF F*W OVER (A,B)
C           (W(X) = 1/(X-C), C.NE.A, C.NE.B), HOPEFULLY SATISFYING
C           FOLLOWING CLAIM FOR ACCURACY
C           ABS(I-RESULT).LE.MAX(EPSABS,EPSREL*ABS(I)).
C
C 3.     CALLING SEQUENCE
C           CALL QAWC(F,A,B,C,EPSABS,EPSREL,RESULT,ABSERR,NEVAL,IER)
C
C        PARAMETERS
C         ON ENTRY
C            F      - REAL
C                     FUNCTION SUBPROGRAM DEFINING THE INTEGRAND
C                     FUNCTION F(X). THE ACTUAL NAME FOR F NEEDS TO BE
C                     DECLARED E X T E R N A L IN THE DRIVER PROGRAM.
C
C            A      - REAL
C                     LOWER LIMIT OF INTEGRATION
C
C            B      - REAL
C                     UPPER LIMIT OF INTEGRATION
C
C            C      - PARAMETER IN THE WEIGHT FUNCTION, C.NE.A, C.NE.B
C                     IF C = A OR C = B, THE ROUTINE WILL END WITH
C                     IER = 6 .
C
C            EPSABS - REAL
C                     ABSOLUTE ACCURACY REQUESTED
C            EPSREL - REAL
C                     RELATIVE ACCURACY REQUESTED
C                     IF  EPSABS.LT.0 AND EPSREL.LT.0,
C                     THE ROUTINE WILL END WITH IER = 6.
C
C         ON RETURN
C            RESULT - REAL
C                     APPROXIMATION TO THE INTEGRAL
C
C            ABSERR - REAL
C                     ESTIMATE OR THE MODULUS OF THE ABSOLUTE ERROR,
```

```
C                     WHICH SHOULD EQUAL OR EXCEED ABS(I-RESULT)
C
C            NEVAL  - INTEGER
C                     NUMBER OF INTEGRAND EVALUATIONS
C
C            IER    - INTEGER
C                     IER = 0 NORMAL AND RELIABLE TERMINATION OF THE
C                             ROUTINE. IT IS ASSUMED THAT THE REQUESTED
C                             ACCURACY HAS BEEN ACHIEVED.
C                     IER.GT.0 ABNORMAL TERMINATION OF THE ROUTINE
C                             THE ESTIMATES FOR INTEGRAL AND ERROR ARE
C                             LESS RELIABLE. IT IS ASSUMED THAT THE
C                             REQUESTED ACCURACY HAS NOT BEEN ACHIEVED.
C                     IER = 1 MAXIMUM NUMBER OF SUBDIVISIONS ALLOWED
C                             HAS BEEN ACHIEVED. ONE CAN ALLOW MORE SUB-
C                             DIVISIONS BY INCREASING THE DATA VALUE OF
C                             LIMIT IN QAWC (AND TAKING THE ACCORDING
C                             DIMENSION ADJUSTMENTS INTO ACCOUNT).
C                             HOWEVER, IF THIS YIELDS NO IMPROVEMENT IT
C                             IS ADVISED TO ANALYZE THE INTEGRAND IN
C                             ORDER TO DETERMINE THE INTEGRATION
C                             DIFFICULTIES. IF THE POSITION OF A LOCAL
C                             DIFFICULTY CAN BE DETERMINED (E.G.
C                             SINGULARITY, DISCONTINUITY WITHIN THE
C                             INTERVAL ONE WILL PROBABLY GAIN FROM
C                             SPLITTING UP THE INTERVAL AT THIS POINT
C                             AND CALLING APPROPRIATE INTEGRATORS ON THE
C                             SUBRANGES.
C                         = 2 THE OCCURRENCE OF ROUNDOFF ERROR IS DETEC-
C                             TED, WHICH PREVENTS THE REQUESTED
C                             TOLERANCE FROM BEING ACHIEVED.
C                         = 3 EXTREMELY BAD INTEGRAND BEHAVIOUR OCCURS
C                             AT SOME POINTS OF THE INTEGRATION
C                             INTERVAL.
C                         = 6 THE INPUT IS INVALID, BECAUSE
C                             C = A OR C = B OR
C                             EPSABS.LT.0 AND EPSREL.LT.0,
C                             RESULT, ABSERR, NEVAL ARE SET TO ZERO.
C
C 4.      SUBROUTINES OR FUNCTIONS NEEDED
C               - QAWCE
C               - QC25C
C               - QSORT
C               - QK15W
C               - QCHEB
C               - F (USER-PROVIDED FUNCTION)
C               - QWGTC
C               - QMACO
```

```
C              - FORTRAN ABS, AMAX1, AMIN1
C
C ......................................................................
C
      REAL A,ABSERR,ALIST,B,BLIST,ELIST,C,EPSABS,EPSREL,F,RESULT,RLIST
      INTEGER IER,IORD,LAST,LIMIT,NEVAL
C
      DIMENSION ALIST(500),BLIST(500),ELIST(500),IORD(500),RLIST(500)
C
      EXTERNAL F
C
C         LIMIT IS THE MAXIMUM NUMBER OF SUBINTERVALS ALLOWED IN THE
C         SUBDIVISION PROCESS OF QAWCE. TAKE CARE THAT LIMIT.GE.1.
C
      DATA LIMIT/500/
C
C***FIRST EXECUTABLE STATEMENT
      CALL QAWCE(F,A,B,C,EPSABS,EPSREL,LIMIT,RESULT,ABSERR,NEVAL,IER,
     *  ALIST,BLIST,RLIST,ELIST,IORD,LAST)
      RETURN
      END
```

```
      SUBROUTINE QAWCE(F,A,B,C,EPSABS,EPSREL,LIMIT,RESULT,ABSERR,NEVAL,
     *  IER,ALIST,BLIST,RLIST,ELIST,IORD,LAST)
C
C.......................................................................
C
C 1.     QAWCE
C        COMPUTATION OF A CAUCHY PRINCIPAL VALUE
C           STANDARD FORTRAN SUBROUTINE
C
C 2.     PURPOSE
C           THE ROUTINE CALCULATES AN APPROXIMATION  RESULT  TO A
C           CAUCHY PRINCIPAL VALUE I = INTEGRAL OF F*W OVER (A,B)
C           (W(X) = 1/(X-C), (C.NE.A, C.NE.B), HOPEFULLY SATISFYING
C           FOLLOWING CLAIM FOR ACCURACY
C           ABS(I-RESULT).LE.MAX(EPSABS,EPSREL*ABS(I)).
C
C 3.     CALLING SEQUENCE
C           CALL QAWCE(F,A,B,C,EPSABS,EPSREL,LIMIT,RESULT,ABSERR,NEVAL,
C                      IER,ALIST,BLIST,RLIST,ELIST,IORD,LAST)
C
C        PARAMETERS
C         ON ENTRY
C            F      - REAL
C                     FUNCTION SUBPROGRAM DEFINING THE INTEGRAND
C                     FUNCTION F(X). THE ACTUAL NAME FOR F NEEDS TO BE
C                     DECLARED E X T E R N A L IN THE DRIVER PROGRAM.
C
C            A      - REAL
C                     LOWER LIMIT OF  INTEGRATION
C
C            B      - REAL
C                     UPPER LIMIT OF  INTEGRATION
C
C            C      - REAL
C                     PARAMETER  IN THE WEIGHT FUNCTION, C.NE.A, C.NE.B
C                     IF C = A OR C = B, THE ROUTINE WILL END WITH
C                     IER = 6.
C
C            EPSABS - REAL
C                     ABSOLUTE ACCURACY REQUESTED
C            EPSREL - REAL
C                     RELATIVE ACCURACY REQUESTED
C                     IF  EPSABS.LT.0 AND EPSREL.LT.0,
C                     THE ROUTINE WILL END WITH IER = 6.
C
C            LIMIT  - INTEGER
C                     GIVES AN UPPER BOUND ON THE NUMBER OF SUBINTERVALS
C                     IN THE PARTITION OF (A,B), LIMIT.GE.1
```

```
C
C        ON RETURN
C           RESULT - REAL
C                    APPROXIMATION TO THE INTEGRAL
C
C           ABSERR - REAL
C                    ESTIMATE OF THE MODULUS OF THE ABSOLUTE ERROR,
C                    WHICH SHOULD EQUAL OR EXCEED ABS(I-RESULT)
C
C           NEVAL  - INTEGER
C                    NUMBER OF INTEGRAND EVALUATIONS
C
C           IER    - INTEGER
C                    IER = 0 NORMAL AND RELIABLE TERMINATION OF THE
C                            ROUTINE. IT IS ASSUMED THAT THE REQUESTED
C                            ACCURACY HAS BEEN ACHIEVED.
C                    IER.GT.0 ABNORMAL TERMINATION OF THE ROUTINE
C                            THE ESTIMATES FOR INTEGRAL AND ERROR ARE
C                            LESS RELIABLE. IT IS ASSUMED THAT THE
C                            REQUESTED ACCURACY HAS NOT BEEN ACHIEVED.
C                    IER = 1 MAXIMUM NUMBER OF SUBDIVISIONS ALLOWED
C                            HAS BEEN ACHIEVED. ONE CAN ALLOW MORE SUB-
C                            DIVISIONS BY INCREASING THE VALUE OF
C                            LIMIT. HOWEVER, IF THIS YIELDS NO
C                            IMPROVEMENT IT IS ADVISED TO ANALYZE THE
C                            INTEGRAND, IN ORDER TO DETERMINE THE
C                            INTEGRATION DIFFICULTIES.  IF THE POSITION
C                            OF A LOCAL DIFFICULTY CAN BE DETERMINED
C                            (E.G. SINGULARITY, DISCONTINUITY WITHIN
C                            THE INTERVAL) ONE WILL PROBABLY GAIN
C                            FROM SPLITTING UP THE INTERVAL AT THIS
C                            POINT AND CALLING APPROPRIATE INTEGRATORS
C                            ON THE SUBRANGES.
C                          = 2 THE OCCURRENCE OF ROUNDOFF ERROR IS DETEC-
C                            TED, WHICH PREVENTS THE REQUESTED
C                            TOLERANCE FROM BEING ACHIEVED.
C                          = 3 EXTREMELY BAD INTEGRAND BEHAVIOUR OCCURS
C                            AT SOME INTERIOR POINTS OF THE INTEGRATION
C                            INTERVAL.
C                          = 6 THE INPUT IS INVALID, BECAUSE
C                            C = A OR C = B OR
C                            EPSABS.LT.0 AND EPSREL.LT.0,
C                            OR LIMIT.LT.1.
C                            RESULT, ABSERR, NEVAL, RLIST(1), ELIST(1),
C                            IORD(1) AND LAST ARE SET TO ZERO.
C                            ALIST(1) AND BLIST(1) ARE SET TO A AND B
C                            RESPECTIVELY.
C
```

```
---------------------------------------------------------------------
C           ALIST    - REAL
C                      VECTOR OF DIMENSION AT LEAST LIMIT, THE FIRST
C                       LAST  ELEMENTS OF WHICH ARE THE LEFT END POINTS
C                      OF THE SUBINTERVALS IN THE PARTITION OF THE GIVEN
C                      INTEGRATION RANGE (A,B)
C
C           BLIST    - REAL
C                      VECTOR OF DIMENSION AT LEAST LIMIT, THE FIRST
C                       LAST  ELEMENTS OF WHICH ARE THE RIGHT END POINTS
C                      OF THE SUBINTERVALS IN THE PARTITION OF THE GIVEN
C                      INTEGRATION RANGE (A,B)
C
C           RLIST    - REAL
C                      VECTOR OF DIMENSION AT LEAST LIMIT, THE FIRST
C                       LAST  ELEMENTS OF WHICH ARE THE INTEGRAL
C                      APPROXIMATIONS ON THE SUBINTERVALS
C
C           ELIST    - REAL
C                      VECTOR OF DIMENSION LIMIT, THE FIRST  LAST
C                      ELEMENTS OF WHICH ARE THE MODULI OF THE ABSOLUTE
C                      ERROR ESTIMATES ON THE SUBINTERVALS
C
C           IORD     - INTEGER
C                      VECTOR OF DIMENSION AT LEAST LIMIT, THE FIRST K
C                      ELEMENTS OF WHICH ARE POINTERS TO THE ERROR
C                      ESTIMATES OVER THE SUBINTERVALS, SO THAT
C                      ELIST(IORD(1)), ...,  ELIST(IORD(K)) WITH
C                      K = LAST IF LAST.LE.(LIMIT/2+2), AND
C                      K = LIMIT+1-LAST OTHERWISE, FORM A DECREASING
C                      SEQUENCE.
C
C           LAST     - INTEGER
C                      NUMBER OF SUBINTERVALS ACTUALLY PRODUCED IN
C                      THE SUBDIVISION PROCESS
C
C 4.     SUBROUTINES OR FUNCTIONS NEEDED
C              - QC25C
C              - QSORT
C              - QK15W
C              - QCHEB
C              - QWGTC
C              - QMACO
C              - F (USER-PROVIDED FUNCTION)
C              - FORTRAN ABS, AMAX1, AMIN1
C
C .....................................................................
C
      REAL A,AA,ABSERR,ALIST,AREA,AREA1,AREA12,AREA2,A1,A2,B,BB,BLIST,
```

```
     *  B1,B2,C,ELIST,EPMACH,EPSABS,EPSREL,ERRBND,ERRMAX,ERROR1,ERROR2,
     *  ERRO12,ERRSUM,F,OFLOW,RESULT,RLIST,UFLOW
      INTEGER IER,IORD,IROFF1,IROFF2,K,KRULE,LAST,LIMIT,MAXERR,NEV,
     *  NEVAL,NRMAX
C
      DIMENSION ALIST(LIMIT),BLIST(LIMIT),RLIST(LIMIT),ELIST(LIMIT),
     *  IORD(LIMIT)
C
      EXTERNAL F
C
C            LIST OF MAJOR VARIABLES
C            -----------------------
C
C           ALIST     - LIST OF LEFT END POINTS OF ALL SUBINTERVALS
C                       CONSIDERED UP TO NOW
C           BLIST     - LIST OF RIGHT END POINTS OF ALL SUBINTERVALS
C                       CONSIDERED UP TO NOW
C           RLIST(I)  - APPROXIMATION TO THE INTEGRAL OVER
C                       (ALIST(I),BLIST(I))
C           ELIST(I)  - ERROR ESTIMATE APPLYING TO RLIST(I)
C           MAXERR    - POINTER TO THE INTERVAL WITH LARGEST ERROR
C                       ESTIMATE
C           ERRMAX    - ELIST(MAXERR)
C           AREA      - SUM OF THE INTEGRALS OVER THE SUBINTERVALS
C           ERRSUM    - SUM OF THE ERRORS OVER THE SUBINTERVALS
C           ERRBND    - REQUESTED ACCURACY MAX(EPSABS,EPSREL*
C                       ABS(RESULT))
C           *****1    - VARIABLE FOR THE LEFT SUBINTERVAL
C           *****2    - VARIABLE FOR THE RIGHT SUBINTERVAL
C           LAST      - INDEX FOR SUBDIVISION
C
C
C            MACHINE DEPENDENT CONSTANTS
C            ---------------------------
C
C           EPMACH IS THE LARGEST RELATIVE SPACING.
C           UFLOW IS THE SMALLEST POSITIVE MAGNITUDE.
C           OFLOW IS THE LARGEST MAGNITUDE.
C
C***FIRST EXECUTABLE STATEMENT
      CALL QMACO(EPMACH,UFLOW,OFLOW)
C
C
C           TEST ON VALIDITY OF PARAMETERS
C           ------------------------------
C
      IER = 6
      NEVAL = 0
```

```
      LAST = 0
      ALIST(1) = A
      BLIST(1) = B
      RLIST(1) = 0.0E+00
      ELIST(1) = 0.0E+00
      IORD(1) = 0
      RESULT = 0.0E+00
      ABSERR = 0.0E+00
      IF(C.EQ.A.OR.C.EQ.B.OR.(EPSABS.LT.0.0E+00.AND.EPSREL.LT.0.0E+00))
     *  GO TO 999
C
C           FIRST APPROXIMATION TO THE INTEGRAL
C           -----------------------------------
C
      AA=A
      BB=B
      IF (A.LE.B) GO TO 10
      AA=B
      BB=A
10    IER=0
      KRULE = 1
      CALL QC25C(F,AA,BB,C,RESULT,ABSERR,KRULE,NEVAL)
      LAST = 1
      RLIST(1) = RESULT
      ELIST(1) = ABSERR
      IORD(1) = 1
      ALIST(1) = A
      BLIST(1) = B
C
C           TEST ON ACCURACY
C
      ERRBND = AMAX1(EPSABS,EPSREL*ABS(RESULT))
      IF(LIMIT.EQ.1) IER = 1
      IF(ABSERR.LT.AMIN1(1.0E-02*ABS(RESULT),ERRBND).OR.IER.EQ.1)
     *  GO TO 70
C
C           INITIALIZATION
C           --------------
C
      ALIST(1) = AA
      BLIST(1) = BB
      RLIST(1) = RESULT
      ERRMAX = ABSERR
      MAXERR = 1
      AREA = RESULT
      ERRSUM = ABSERR
      NRMAX = 1
      IROFF1 = 0
```

```
      IROFF2 = 0
C
C           MAIN DO-LOOP
C           ------------
C
      DO 40 LAST = 2,LIMIT
C
C           BISECT THE SUBINTERVAL WITH NRMAX-TH LARGEST ERROR ESTIMATE.
C
        A1 = ALIST(MAXERR)
        B1 = 5.0E-01*(ALIST(MAXERR)+BLIST(MAXERR))
        B2 = BLIST(MAXERR)
        IF(C.LE.B1.AND.C.GT.A1) B1 = 5.0E-01*(C+B2)
        IF(C.GT.B1.AND.C.LT.B2) B1 = 5.0E-01*(A1+C)
        A2 = B1
        KRULE = 2
        CALL QC25C(F,A1,B1,C,AREA1,ERROR1,KRULE,NEV)
        NEVAL = NEVAL+NEV
        CALL QC25C(F,A2,B2,C,AREA2,ERROR2,KRULE,NEV)
        NEVAL = NEVAL+NEV
C
C           IMPROVE PREVIOUS APPROXIMATIONS TO INTEGRAL AND ERROR
C           AND TEST FOR ACCURACY.
C
        AREA12 = AREA1+AREA2
        ERRO12 = ERROR1+ERROR2
        ERRSUM = ERRSUM+ERRO12-ERRMAX
        AREA = AREA+AREA12-RLIST(MAXERR)
        IF(ABS(RLIST(MAXERR)-AREA12).LT.1.0E-05*ABS(AREA12)
     *    .AND.ERRO12.GE.9.9E-01*ERRMAX.AND.KRULE.EQ.0)
     *    IROFF1 = IROFF1+1
        IF(LAST.GT.10.AND.ERRO12.GT.ERRMAX.AND.KRULE.EQ.0)
     *    IROFF2 = IROFF2+1
        RLIST(MAXERR) = AREA1
        RLIST(LAST) = AREA2
        ERRBND = AMAX1(EPSABS,EPSREL*ABS(AREA))
        IF(ERRSUM.LE.ERRBND) GO TO 15
C
C           TEST FOR ROUNDOFF ERROR AND EVENTUALLY SET ERROR FLAG.
C
        IF(IROFF1.GE.6.AND.IROFF2.GT.20) IER = 2
C
C           SET ERROR FLAG IN THE CASE THAT NUMBER OF INTERVAL
C           BISECTIONS EXCEEDS LIMIT.
C
        IF(LAST.EQ.LIMIT) IER = 1
C
C           SET ERROR FLAG IN THE CASE OF BAD INTEGRAND BEHAVIOUR AT
```

```
C          A POINT OF THE INTEGRATION RANGE.
C
        IF(AMAX1(ABS(A1),ABS(B2)).LE.(1.0E+00+1.0E+03*EPMACH)
     *    *(ABS(A2)+1.0E+03*UFLOW)) IER = 3
C
C          APPEND THE NEWLY-CREATED INTERVALS TO THE LIST.
C
   15   IF(ERROR2.GT.ERROR1) GO TO 20
        ALIST(LAST) = A2
        BLIST(MAXERR) = B1
        BLIST(LAST) = B2
        ELIST(MAXERR) = ERROR1
        ELIST(LAST) = ERROR2
        GO TO 30
   20   ALIST(MAXERR) = A2
        ALIST(LAST) = A1
        BLIST(LAST) = B1
        RLIST(MAXERR) = AREA2
        RLIST(LAST) = AREA1
        ELIST(MAXERR) = ERROR2
        ELIST(LAST) = ERROR1
C
C          CALL SUBROUTINE QSORT TO MAINTAIN THE DESCENDING ORDERING
C          IN THE LIST OF ERROR ESTIMATES AND SELECT THE SUBINTERVAL
C          WITH NRMAX-TH LARGEST ERROR ESTIMATE (TO BE BISECTED NEXT).
C
   30    CALL QSORT(LIMIT,LAST,MAXERR,ERRMAX,ELIST,IORD,NRMAX)
C***JUMP OUT OF DO-LOOP
        IF(IER.NE.0.OR.ERRSUM.LE.ERRBND) GO TO 50
   40 CONTINUE
C
C          COMPUTE FINAL RESULT.
C          ---------------------
C
   50 RESULT = 0.0E+00
      DO 60 K=1,LAST
        RESULT = RESULT+RLIST(K)
   60 CONTINUE
      ABSERR = ERRSUM
   70 IF (AA.EQ.B) RESULT=-RESULT
  999 RETURN
      END
```

```
      SUBROUTINE QFOUR(F,A,B,OMEGA,INTEGR,EPSABS,EPSREL,LIMIT,ICALL,
     *  MAXP1,RESULT,ABSERR,NEVAL,IER,ALIST,BLIST,RLIST,ELIST,IORD,
     *  NNLOG,MOMCOM,CHEBMO)
C
C ..................................................................
C
C 1.     QFOUR
C        COMPUTATION OF OSCILLATORY INTEGRALS
C           STANDARD FORTRAN SUBROUTINE
C
C 2.     PURPOSE
C           THE ROUTINE CALCULATES AN APPROXIMATION  RESULT  TO A GIVEN
C           DEFINITE INTEGRAL
C              I = INTEGRAL OF F(X)*W(X) OVER (A,B)
C                  WHERE W(X) = COS(OMEGA*X)
C                     OR W(X) = SIN(OMEGA*X),
C           HOPEFULLY SATISFYING FOLLOWING CLAIM FOR ACCURACY
C           ABS(I-RESULT).LE.MAX(EPSABS,EPSREL*ABS(I)).
C           QFOUR IS CALLED BY QAWO AND BY QAWF.
C           HOWEVER, IT CAN ALSO BE CALLED DIRECTLY IN A USER-WRITTEN
C           PROGRAM. IN THE LATTER CASE IT IS POSSIBLE FOR THE USER TO
C           DETERMINE THE FIRST DIMENSION OF ARRAY CHEBMO(MAXP1,25).
C           SEE ALSO PARAMETER DESCRIPTION OF MAXP1. ADDITIONALLY SEE
C           PARAMETER DESCRIPTION OF ICALL FOR EVENTUALLY RE-USING
C           CHEBYSHEV MOMENTS COMPUTED DURING FORMER CALL ON SUBINTERVAL
C           OF EQUAL LENGTH ABS(B-A).
C
C 3.     CALLING SEQUENCE
C           CALL QFOUR(F,A,B,OMEGA,INTEGR,EPSABS,EPSREL,LIMIT,ICALL,
C                      MAXP1,RESULT,ABSERR,NEVAL,IER,ALIST,BLIST,RLIST,
C                      ELIST,IORD,NNLOG,MOMCOM,CHEBMO)
C
C        PARAMETERS
C         ON ENTRY
C           F      - REAL
C                    FUNCTION SUBPROGRAM DEFINING THE INTEGRAND
C                    FUNCTION F(X). THE ACTUAL NAME FOR F NEEDS TO BE
C                    DECLARED E X T E R N A L IN THE DRIVER PROGRAM.
C
C           A      - REAL
C                    LOWER LIMIT OF INTEGRATION
C
C           B      - REAL
C                    UPPER LIMIT OF INTEGRATION
C
C           OMEGA  - REAL
C                    PARAMETER IN THE INTEGRAND WEIGHT FUNCTION
C
```

```
C             INTEGR - INTEGER
C                      INDICATES WHICH OF THE WEIGHT FUNCTIONS IS TO BE
C                      USED
C                      INTEGR = 1      W(X) = COS(OMEGA*X)
C                      INTEGR = 2      W(X) = SIN(OMEGA*X)
C                      IF INTEGR.NE.1 AND INTEGR.NE.2, THE ROUTINE
C                      WILL END WITH IER = 6.
C
C             EPSABS - REAL
C                      ABSOLUTE ACCURACY REQUESTED
C             EPSREL - REAL
C                      RELATIVE ACCURACY REQUESTED
C                      IF  EPSABS.LT.0 AND EPSREL.LT.0,
C                      THE ROUTINE WILL END WITH IER = 6.
C
C             LIMIT  - INTEGER
C                      GIVES AN UPPER BOUND ON THE NUMBER OF SUBDIVISIONS
C                      IN THE PARTITION OF (A,B), LIMIT.GE.1.
C
C             ICALL  - INTEGER
C                      IF QFOUR IS TO BE USED ONLY ONCE, ICALL MUST
C                      BE SET TO 1.  ASSUME THAT DURING THIS CALL, THE
C                      CHEBYSHEV MOMENTS (FOR CLENSHAW-CURTIS INTEGRATION
C                      OF DEGREE 24) HAVE BEEN COMPUTED FOR INTERVALS OF
C                      LENGHTS (ABS(B-A))*2**(-L), L=0,1,2,...MOMCOM-1.
C                      THE CHEBYSHEV MOMENTS ALREADY COMPUTED CAN BE
C                      RE-USED IN SUBSEQUENT CALLS, IF QFOUR MUST BE
C                      CALLED TWICE OR MORE TIMES ON INTERVALS OF THE
C                      SAME LENGTH ABS(B-A). FROM THE SECOND CALL ON, ONE
C                      HAS TO PUT THEN ICALL.GT.1.
C                      IF ICALL.LT.1, THE ROUTINE WILL END WITH IER = 6.
C
C             MAXP1  - INTEGER
C                      GIVES AN UPPER BOUND ON THE NUMBER OF
C                      CHEBYSHEV MOMENTS WHICH CAN BE STORED, I.E.
C                      FOR THE INTERVALS OF LENGHTS ABS(B-A)*2**(-L),
C                      L=0,1, ..., MAXP1-2, MAXP1.GE.1.
C                      IF MAXP1.LT.1, THE ROUTINE WILL END WITH IER = 6.
C                      INCREASING (DECREASING) THE VALUE OF MAXP1
C                      DECREASES (INCREASES) THE COMPUTATIONAL TIME BUT
C                      INCREASES (DECREASES) THE REQUIRED MEMORY SPACE.
C
C          ON RETURN
C             RESULT - REAL
C                      APPROXIMATION TO THE INTEGRAL
C
C             ABSERR - REAL
C                      ESTIMATE OF THE MODULUS OF THE ABSOLUTE ERROR,
```

```
C                   WHICH SHOULD EQUAL OR EXCEED ABS(I-RESULT)
C
C          NEVAL  - INTEGER
C                   NUMBER OF INTEGRAND EVALUATIONS
C
C          IER    - INTEGER
C                   IER = 0 NORMAL AND RELIABLE TERMINATION OF THE
C                           ROUTINE. IT IS ASSUMED THAT THE
C                           REQUESTED ACCURACY HAS BEEN ACHIEVED.
C                 - IER.GT.0 ABNORMAL TERMINATION OF THE ROUTINE.
C                           THE ESTIMATES FOR INTEGRAL AND ERROR ARE
C                           LESS RELIABLE. IT IS ASSUMED THAT THE
C                           REQUESTED ACCURACY HAS NOT BEEN ACHIEVED.
C                   IER = 1 MAXIMUM NUMBER OF SUBDIVISIONS ALLOWED
C                           HAS BEEN ACHIEVED. ONE CAN ALLOW MORE
C                           SUBDIVISIONS BY INCREASING THE VALUE OF
C                           LIMIT (AND TAKING ACCORDING DIMENSION
C                           ADJUSTMENTS INTO ACCOUNT). HOWEVER, IF
C                           THIS YIELDS NO IMPROVEMENT IT IS ADVISED
C                           TO ANALYZE THE INTEGRAND, IN ORDER TO
C                           DETERMINE THE INTEGRATION DIFFICULTIES.
C                           IF THE POSITION OF A LOCAL DIFFICULTY CAN
C                           BE DETERMINED (E.G. SINGULARITY,
C                           DISCONTINUITY WITHIN THE INTERVAL) ONE
C                           WILL PROBABLY GAIN FROM SPLITTING UP THE
C                           INTERVAL AT THIS POINT AND CALLING THE
C                           INTEGRATOR ON THE SUBRANGES. IF POSSIBLE,
C                           AN APPROPRIATE SPECIAL-PURPOSE INTEGRATOR
C                           SHOULD BE USED WHICH IS DESIGNED FOR
C                           HANDLING THE TYPE OF DIFFICULTY INVOLVED.
C                       = 2 THE OCCURRENCE OF ROUNDOFF ERROR IS
C                           DETECTED, WHICH PREVENTS THE REQUESTED
C                           TOLERANCE FROM BEING ACHIEVED.
C                           THE ERROR MAY BE UNDER-ESTIMATED.
C                       = 3 EXTREMELY BAD INTEGRAND BEHAVIOUR OCCURS
C                           AT SOME POINTS OF THE INTEGRATION
C                           INTERVAL.
C                       = 4 THE ALGORITHM DOES NOT CONVERGE. ROUNDOFF
C                           ERROR IS DETECTED IN THE EXTRAPOLATION
C                           TABLE. IT IS PRESUMED THAT THE REQUESTED
C                           TOLERANCE CANNOT BE ACHIEVED DUE TO
C                           ROUNDOFF IN THE EXTRAPOLATION TABLE, AND
C                           THAT THE RETURNED RESULT IS THE BEST WHICH
C                           CAN BE OBTAINED.
C                       = 5 THE INTEGRAL IS PROBABLY DIVERGENT, OR
C                           SLOWLY CONVERGENT. IT MUST BE NOTED THAT
C                           DIVERGENCE CAN OCCUR WITH ANY OTHER VALUE
C                           OF IER.GT.0.
```

```
C                         = 6 THE INPUT IS INVALID, BECAUSE
C                             EPSABS.LT.0 AND EPSREL.LT.0,
C                             OR (INTEGR.NE.1 AND INTEGR.NE.2) OR
C                             ICALL.LT.1 OR MAXP1.LT.1.
C                             RESULT, ABSERR, NEVAL, LAST, RLIST(1),
C                             ELIST(1), IORD(1) AND NNLOG(1) ARE SET TO
C                             ZERO. ALIST(1) AND BLIST(1) ARE SET TO A
C                             AND B RESPECTIVELY.
C
C            ALIST  - REAL
C                     VECTOR OF DIMENSION AT LEAST LIMIT, THE FIRST
C                      LAST  ELEMENTS OF WHICH ARE THE LEFT END POINTS
C                     OF THE SUBINTERVALS IN THE PARTITION OF THE GIVEN
C                     INTEGRATION RANGE (A,B)
C
C            BLIST  - REAL
C                     VECTOR OF DIMENSION AT LEAST LIMIT, THE FIRST
C                      LAST  ELEMENTS OF WHICH ARE THE RIGHT END POINTS
C                     OF THE SUBINTERVALS IN THE PARTITION OF THE GIVEN
C                     INTEGRATION RANGE (A,B)
C
C            RLIST  - REAL
C                     VECTOR OF DIMENSION AT LEAST LIMIT, THE FIRST
C                      LAST  ELEMENTS OF WHICH ARE THE INTEGRAL
C                     APPROXIMATIONS ON THE SUBINTERVALS
C
C            ELIST  - REAL
C                     VECTOR OF DIMENSION AT LEAST LIMIT, THE FIRST
C                      LAST  ELEMENTS OF WHICH ARE THE MODULI OF THE
C                     ABSOLUTE ERROR ESTIMATES ON THE SUBINTERVALS
C
C            IORD   - INTEGER
C                     VECTOR OF DIMENSION AT LEAST LIMIT, THE FIRST K
C                     ELEMENTS OF WHICH ARE POINTERS TO THE ERROR
C                     ESTIMATES OVER THE SUBINTERVALS, SUCH THAT
C                     ELIST(IORD(1)), ..., ELIST(IORD(K)), FORM
C                     A DECREASING SEQUENCE, WITH K = LAST
C                     IF LAST.LE.(LIMIT/2+2), AND
C                     K = LIMIT+1-LAST OTHERWISE.
C
C            NNLOG  - INTEGER
C                     VECTOR OF DIMENSION AT LEAST LIMIT, INDICATING THE
C                     SUBDIVISION LEVELS OF THE SUBINTERVALS, I.E.
C                     IWORK(I) = L MEANS THAT THE SUBINTERVAL NUMBERED
C                     I IS OF LENGTH ABS(B-A)*2**(1-L)
C
C         ON ENTRY AND RETURN
C            MOMCOM - INTEGER
```

```
C                    INDICATING THAT THE CHEBYSHEV MOMENTS HAVE BEEN
C                    COMPUTED FOR INTERVALS OF LENGTHS
C                    (ABS(B-A))*2**(-L), L=0,1,2, ..., MOMCOM-1,
C                    MOMCOM.LT.MAXP1
C
C            CHEBMO - REAL
C                    ARRAY OF DIMENSION (MAXP1,25) CONTAINING THE
C                    CHEBYSHEV MOMENTS
C
C 4.      SUBROUTINES OR FUNCTIONS NEEDED
C              - QC25O
C              - QSORT
C              - QEXT
C              - QK15W
C              - QMACO
C              - F(USER PROVIDED FUNCTION)
C              - FORTRAN ABS, AMAX1, AMIN1
C              - QWGTO
C              - QCHEB
C
C ..................................................................
C
      REAL A,ABSEPS,ABSERR,ALIST,AREA,AREA1,AREA12,AREA2,A1,A2,B,BLIST,
     *  B1,B2,CHEBMO,CORREC,DEFAB1,DEFAB2,DEFABS,DOMEGA,DRES,ELIST,
     *  EPMACH,EPSABS,EPSREL,ERLARG,ERLAST,ERRBND,ERRMAX,ERROR1,ERRO12,
     *  ERROR2,ERRSUM,ERTEST,F,OFLOW,OMEGA,RESABS,RESEPS,RESULT,RES3LA,
     *  RLIST,RLIST2,SMALL,UFLOW,WIDTH
      INTEGER ICALL,ID,IER,IERRO,INTEGR,IORD,IROFF1,IROFF2,IROFF3,
     *  JUPBND,K,KSGN,KTMIN,LAST,LIMIT,MAXERR,MAXP1,MOMCOM,NEV,
     *  NEVAL,NNLOG,NRES,NRMAX,NRMOM,NUMRL2
      LOGICAL EXTRAP,NOEXT,EXTALL
C
      DIMENSION ALIST(LIMIT),BLIST(LIMIT),RLIST(LIMIT),ELIST(LIMIT),
     *  IORD(LIMIT),RLIST2(52),RES3LA(3),CHEBMO(MAXP1,25),NNLOG(LIMIT)
C
      EXTERNAL F
C
C           THE DIMENSION OF RLIST2 IS DETERMINED BY  THE VALUE OF
C           LIMEXP IN SUBROUTINE QEXT (RLIST2 SHOULD BE OF DIMENSION
C           (LIMEXP+2) AT LEAST).
C
C           LIST OF MAJOR VARIABLES
C           -----------------------
C
C          ALIST     - LIST OF LEFT END POINTS OF ALL SUBINTERVALS
C                      CONSIDERED UP TO NOW
C          BLIST     - LIST OF RIGHT END POINTS OF ALL SUBINTERVALS
```

```
-----------------------------------------------------------------------
C                         CONSIDERED UP TO NOW
C            RLIST(I)   - APPROXIMATION TO THE INTEGRAL OVER
C                         (ALIST(I),BLIST(I))
C            RLIST2     - ARRAY OF DIMENSION AT LEAST LIMEXP+2 CONTAINING
C                         THE PART OF THE EPSILON TABLE WHICH IS STILL
C                         NEEDED FOR FURTHER COMPUTATIONS
C            ELIST(I)   - ERROR ESTIMATE APPLYING TO RLIST(I)
C            MAXERR     - POINTER TO THE INTERVAL WITH LARGEST ERROR
C                         ESTIMATE
C            ERRMAX     - ELIST(MAXERR)
C            ERLAST     - ERROR ON THE INTERVAL CURRENTLY SUBDIVIDED
C            AREA       - SUM OF THE INTEGRALS OVER THE SUBINTERVALS
C            ERRSUM     - SUM OF THE ERRORS OVER THE SUBINTERVALS
C            ERRBND     - REQUESTED ACCURACY MAX(EPSABS,EPSREL*
C                         ABS(RESULT))
C            *****1     - VARIABLE FOR THE LEFT SUBINTERVAL
C            *****2     - VARIABLE FOR THE RIGHT SUBINTERVAL
C            LAST       - INDEX FOR SUBDIVISION
C            NRES       - NUMBER OF CALLS TO THE EXTRAPOLATION ROUTINE
C            NUMRL2     - NUMBER OF ELEMENTS IN RLIST2. IF AN APPROPRIATE
C                         APPROXIMATION TO THE COMPOUNDED INTEGRAL HAS
C                         BEEN OBTAINED IT IS PUT IN RLIST2(NUMRL2) AFTER
C                         NUMRL2 HAS BEEN INCREASED BY ONE
C            SMALL      - LENGTH OF THE SMALLEST INTERVAL CONSIDERED
C                         UP TO NOW, MULTIPLIED BY 1.5
C            ERLARG     - SUM OF THE ERRORS OVER THE INTERVALS LARGER
C                         THAN THE SMALLEST INTERVAL CONSIDERED UP TO NOW
C            EXTRAP     - LOGICAL VARIABLE DENOTING THAT THE ROUTINE IS
C                         ATTEMPTING TO PERFORM EXTRAPOLATION, I.E. BEFORE
C                         SUBDIVIDING THE SMALLEST INTERVAL WE TRY TO
C                         DECREASE THE VALUE OF ERLARG
C            NOEXT      - LOGICAL VARIABLE DENOTING THAT EXTRAPOLATION
C                         IS NO LONGER ALLOWED (TRUE VALUE)
C
C             MACHINE DEPENDENT CONSTANTS
C             ---------------------------
C
C            EPMACH IS THE LARGEST RELATIVE SPACING.
C            UFLOW IS THE SMALLEST POSITIVE MAGNITUDE.
C            OFLOW IS THE LARGEST POSITIVE MAGNITUDE.
C
C***FIRST EXECUTABLE STATEMENT
      CALL QMACO(EPMACH,UFLOW,OFLOW)
C
C           TEST ON VALIDITY OF PARAMETERS
C           ------------------------------
C
      IER = 0
```

```
      NEVAL = 0
      LAST = 0
      RESULT = 0.0E+00
      ABSERR = 0.0E+00
      ALIST(1) = A
      BLIST(1) = B
      RLIST(1) = 0.0E+00
      ELIST(1) = 0.0E+00
      IORD(1) = 0
      NNLOG(1) = 0
      IF((INTEGR.NE.1.AND.INTEGR.NE.2).OR.(EPSABS.LT.0.0E+00.AND.
     *  EPSREL.LT.0.0E+00).OR.ICALL.LT.1.OR.MAXP1.LT.1) IER = 6
      IF(IER.EQ.6) GO TO 999
C
C           FIRST APPROXIMATION TO THE INTEGRAL
C           -----------------------------------
C
      DOMEGA = ABS(OMEGA)
      NRMOM = 0
      IF (ICALL.GT.1) GO TO 5
      MOMCOM = 0
    5 CALL QC25O(F,A,B,DOMEGA,INTEGR,NRMOM,MAXP1,0,RESULT,ABSERR,NEVAL,
     *  DEFABS,RESABS,MOMCOM,CHEBMO)
C
C           TEST ON ACCURACY.
C
      DRES = ABS(RESULT)
      ERRBND = AMAX1(EPSABS,EPSREL*DRES)
      RLIST(1) = RESULT
      ELIST(1) = ABSERR
      IORD(1) = 1
      IF(ABSERR.LE.1.0E+02*EPMACH*DEFABS.AND.ABSERR.GT.ERRBND) IER = 2
      IF(LIMIT.EQ.1) IER = 1
      IF(IER.NE.0.OR.ABSERR.LE.ERRBND) GO TO 200
C
C           INITIALIZATIONS
C           ---------------
C
      ERRMAX = ABSERR
      MAXERR = 1
      AREA = RESULT
      ERRSUM = ABSERR
      ABSERR = OFLOW
      NRMAX = 1
      EXTRAP = .FALSE.
      NOEXT = .FALSE.
      IERRO = 0
      IROFF1 = 0
```

```
      IROFF2 = 0
      IROFF3 = 0
      KTMIN = 0
      SMALL = ABS(B-A)*7.5E-01
      NRES = 0
      NUMRL2 = 0
      EXTALL = .FALSE.
      IF(5.0E-01*ABS(B-A)*DOMEGA.GT.2.0E+00) GO TO 10
      NUMRL2 = 1
      EXTALL = .TRUE.
      RLIST2(1) = RESULT
   10 IF(2.5E-01*ABS(B-A)*DOMEGA.LE.2.0E+00) EXTALL = .TRUE.
      KSGN = -1
      IF(DRES.GE.(1.0E+00-5.0E+01*EPMACH)*DEFABS) KSGN = 1
C
C           MAIN DO-LOOP
C           ------------
C
      DO 140 LAST = 2,LIMIT
C
C           BISECT THE SUBINTERVAL WITH THE NRMAX-TH LARGEST ERROR
C           ESTIMATE.
C
        NRMOM = NNLOG(MAXERR)+1
        A1 = ALIST(MAXERR)
        B1 = 5.0E-01*(ALIST(MAXERR)+BLIST(MAXERR))
        A2 = B1
        B2 = BLIST(MAXERR)
        ERLAST = ERRMAX
        CALL QC250(F,A1,B1,DOMEGA,INTEGR,NRMOM,MAXP1,0,AREA1,ERROR1,NEV,
     *  RESABS,DEFAB1,MOMCOM,CHEBMO)
        NEVAL = NEVAL+NEV
        CALL QC250(F,A2,B2,DOMEGA,INTEGR,NRMOM,MAXP1,1,AREA2,ERROR2,NEV,
     *  RESABS,DEFAB2,MOMCOM,CHEBMO)
        NEVAL = NEVAL+NEV
C
C           IMPROVE PREVIOUS APPROXIMATIONS TO INTEGRAL AND ERROR AND
C           TEST FOR ACCURACY.
C
        AREA12 = AREA1+AREA2
        ERRO12 = ERROR1+ERROR2
        ERRSUM = ERRSUM+ERRO12-ERRMAX
        AREA = AREA+AREA12-RLIST(MAXERR)
        IF(DEFAB1.EQ.ERROR1.OR.DEFAB2.EQ.ERROR2) GO TO 25
        IF(ABS(RLIST(MAXERR)-AREA12).GT.1.0E-05*ABS(AREA12)
     *  .OR.ERRO12.LT.9.9E-01*ERRMAX) GO TO 20
        IF(EXTRAP) IROFF2 = IROFF2+1
        IF(.NOT.EXTRAP) IROFF1 = IROFF1+1
```

```
   20   IF(LAST.GT.10.AND.ERRO12.GT.ERRMAX) IROFF3 = IROFF3+1
   25   RLIST(MAXERR) = AREA1
        RLIST(LAST) = AREA2
        NNLOG(MAXERR) = NRMOM
        NNLOG(LAST) = NRMOM
        ERRBND = AMAX1(EPSABS,EPSREL*ABS(AREA))
C
C           TEST FOR ROUNDOFF ERROR AND EVENTUALLY SET ERROR FLAG
C
        IF(IROFF1+IROFF2.GE.10.OR.IROFF3.GE.20) IER = 2
        IF(IROFF2.GE.5) IERRO = 3
C
C           SET ERROR FLAG IN THE CASE THAT THE NUMBER OF SUBINTERVALS
C           EQUALS LIMIT.
C
        IF(LAST.EQ.LIMIT) IER = 1
C
C           SET ERROR FLAG IN THE CASE OF BAD INTEGRAND BEHAVIOUR AT
C           A POINT OF THE INTEGRATION RANGE.
C
        IF(AMAX1(ABS(A1),ABS(B2)).LE.(1.0E+00+1.0E+03*EPMACH)
     *  *(ABS(A2)+1.0E+03*UFLOW)) IER = 4
C
C           APPEND THE NEWLY-CREATED INTERVALS TO THE LIST.
C
        IF(ERROR2.GT.ERROR1) GO TO 30
        ALIST(LAST) = A2
        BLIST(MAXERR) = B1
        BLIST(LAST) = B2
        ELIST(MAXERR) = ERROR1
        ELIST(LAST) = ERROR2
        GO TO 40
   30   ALIST(MAXERR) = A2
        ALIST(LAST) = A1
        BLIST(LAST) = B1
        RLIST(MAXERR) = AREA2
        RLIST(LAST) = AREA1
        ELIST(MAXERR) = ERROR2
        ELIST(LAST) = ERROR1
C
C           CALL SUBROUTINE QSORT TO MAINTAIN THE DESCENDING ORDERING
C           IN THE LIST OF ERROR ESTIMATES AND SELECT THE SUBINTERVAL
C           WITH NRMAX-TH LARGEST ERROR ESTIMATE (TO BE BISECTED NEXT).
C
   40   CALL QSORT(LIMIT,LAST,MAXERR,ERRMAX,ELIST,IORD,NRMAX)
C***JUMP OUT OF DO-LOOP
      IF(ERRSUM.LE.ERRBND) GO TO 170
      IF(IER.NE.0) GO TO 150
```

```
      IF(LAST.EQ.2.AND.EXTALL) GO TO 120
      IF(NOEXT) GO TO 140
      IF(.NOT.EXTALL) GO TO 50
      ERLARG = ERLARG-ERLAST
      IF(ABS(B1-A1).GT.SMALL) ERLARG = ERLARG+ERRO12
      IF(EXTRAP) GO TO 70
C
C           TEST WHETHER THE INTERVAL TO BE BISECTED NEXT IS THE
C           SMALLEST INTERVAL.
C
   50 WIDTH = ABS(BLIST(MAXERR)-ALIST(MAXERR))
      IF(WIDTH.GT.SMALL) GO TO 140
      IF(EXTALL) GO TO 60
C
C           TEST WHETHER WE CAN START WITH THE EXTRAPOLATION PROCEDURE
C           (WE DO THIS IF WE INTEGRATE OVER THE NEXT INTERVAL WITH
C           USE OF A GAUSS-KRONROD RULE - SEE SUBROUTINE QC25O).
C
      SMALL = SMALL*5.0E-01
      IF(2.5E-01*WIDTH*DOMEGA.GT.2.0E+00) GO TO 140
      EXTALL = .TRUE.
      GO TO 130
   60 EXTRAP = .TRUE.
      NRMAX = 2
   70 IF(IERRO.EQ.3.OR.ERLARG.LE.ERTEST) GO TO 90
C
C           THE SMALLEST INTERVAL HAS THE LARGEST ERROR.
C           BEFORE BISECTING DECREASE THE SUM OF THE ERRORS OVER THE
C           LARGER INTERVALS (ERLARG) AND PERFORM EXTRAPOLATION.
C
      JUPBND = LAST
      IF (LAST.GT.(LIMIT/2+2)) JUPBND = LIMIT+3-LAST
      ID = NRMAX
      DO 80 K = ID,JUPBND
        MAXERR = IORD(NRMAX)
        ERRMAX = ELIST(MAXERR)
        IF(ABS(BLIST(MAXERR)-ALIST(MAXERR)).GT.SMALL) GO TO 140
        NRMAX = NRMAX+1
   80 CONTINUE
C
C           PERFORM EXTRAPOLATION.
C
   90 NUMRL2 = NUMRL2+1
      RLIST2(NUMRL2) = AREA
      IF(NUMRL2.LT.3) GO TO 110
      CALL QEXT(NUMRL2,RLIST2,RESEPS,ABSEPS,RES3LA,NRES)
      KTMIN = KTMIN+1
      IF(KTMIN.GT.5.AND.ABSERR.LT.1.0E-03*ERRSUM) IER = 5
```

```
        IF(ABSEPS.GE.ABSERR) GO TO 100
        KTMIN = 0
        ABSERR = ABSEPS
        RESULT = RESEPS
        CORREC = ERLARG
        ERTEST = AMAX1(EPSABS,EPSREL*ABS(RESEPS))
C***JUMP OUT OF DO-LOOP
        IF(ABSERR.LE.ERTEST) GO TO 150
C
C           PREPARE BISECTION OF THE SMALLEST INTERVAL.
C
  100   IF(NUMRL2.EQ.1) NOEXT = .TRUE.
        IF(IER.EQ.5) GO TO 150
  110   MAXERR = IORD(1)
        ERRMAX = ELIST(MAXERR)
        NRMAX = 1
        EXTRAP = .FALSE.
        SMALL = SMALL*5.0E-01
        ERLARG = ERRSUM
        GO TO 140
  120   SMALL = SMALL*5.0E-01
        NUMRL2 = NUMRL2+1
        RLIST2(NUMRL2) = AREA
  130   ERTEST = ERRBND
        ERLARG = ERRSUM
  140 CONTINUE
C
C           SET THE FINAL RESULT.
C           ---------------------
C
  150 IF(ABSERR.EQ.OFLOW.OR.NRES.EQ.0) GO TO 170
      IF(IER+IERRO.EQ.0) GO TO 165
      IF(IERRO.EQ.3) ABSERR = ABSERR+CORREC
      IF(IER.EQ.0) IER = 3
      IF(RESULT.NE.0.0E+00.AND.AREA.NE.0.0E+00) GO TO 160
      IF(ABSERR.GT.ERRSUM) GO TO 170
      IF(AREA.EQ.0.0E+00) GO TO 190
      GO TO 165
  160 IF(ABSERR/ABS(RESULT).GT.ERRSUM/ABS(AREA)) GO TO 170
C
C           TEST ON DIVERGENCE.
C
  165 IF(KSGN.EQ.(-1).AND.AMAX1(ABS(RESULT),ABS(AREA)).LE.
     * DEFABS*1.0E-02) GO TO 190
      IF(1.0E-02.GT.(RESULT/AREA).OR.(RESULT/AREA).GT.1.0E+02
     * .OR.ERRSUM.GE.ABS(AREA)) IER = 6
      GO TO 190
C
```

```
C           COMPUTE GLOBAL INTEGRAL SUM.
C
  170 RESULT = 0.0E+00
      DO 180 K=1,LAST
        RESULT = RESULT+RLIST(K)
  180 CONTINUE
      ABSERR = ERRSUM
  190 IF (IER.GT.2) IER=IER-1
  200 IF (INTEGR.EQ.2.AND.OMEGA.LT.0.0E+00) RESULT=-RESULT
  999 RETURN
      END
```

```
      SUBROUTINE QK15(F,A,B,RESULT,ABSERR,RESABS,RESASC)
C
C     ..........................................................
C
C 1.        QK15
C           INTEGRATION RULES
C              STANDARD FORTRAN SUBROUTINE
C
C 2.        PURPOSE
C              TO COMPUTE I = INTEGRAL OF F OVER (A,B), WITH ERROR
C                             ESTIMATE
C                         J = INTEGRAL OF ABS(F) OVER (A,B)
C
C 3.        CALLING SEQUENCE
C              CALL QK15(F,A,B,RESULT,ABSERR,RESABS,RESASC)
C
C           PARAMETERS
C            ON ENTRY
C              F      - REAL
C                       FUNCTION SUBPROGRAM DEFINING THE INTEGRAND
C                       FUNCTION F(X). THE ACTUAL NAME FOR F NEEDS TO BE
C                       DECLARED E X T E R N A L IN THE CALLING PROGRAM.
C
C              A      - REAL
C                       LOWER LIMIT OF INTEGRATION
C
C              B      - REAL
C                       UPPER LIMIT OF INTEGRATION
C
C            ON RETURN
C              RESULT - REAL
C                       APPROXIMATION TO THE INTEGRAL I
C                       RESULT IS COMPUTED BY APPLYING THE 15-POINT
C                       KRONROD RULE (RESK) OBTAINED BY OPTIMAL ADDITION
C                       OF ABSCISSAE TO THE 7-POINT GAUSS RULE (RESG).
C
C              ABSERR - REAL
C                       ESTIMATE OF THE MODULUS OF THE ABSOLUTE ERROR,
C                       WHICH SHOULD NOT EXCEED ABS(I-RESULT)
C
C              RESABS - REAL
C                       APPROXIMATION TO THE INTEGRAL J
C
C              RESASC - REAL
C                       APPROXIMATION TO THE INTEGRAL OF ABS(F-I/(B-A))
C                       OVER (A,B)
C
C 4.        SUBROUTINES OR FUNCTIONS NEEDED
```

```
C                  - F (USER-PROVIDED FUNCTION)
C                  - QMACO
C                  - FORTRAN ABS, AMAX1, AMIN1
C
C     ...........................................................
C
      REAL A,ABSC,ABSERR,B,CENTR,DHLGTH,EPMACH,F,FC,FSUM,FVAL1,FVAL2,
     *  FV1,FV2,HLGTH,OFLOW,RESABS,RESASC,RESG,RESK,RESKH,RESULT,UFLOW,
     *  WG,WGK,XGK
      INTEGER J,JTW,JTWM1
C
      DIMENSION FV1(7),FV2(7),WG(4),WGK(8),XGK(8)
C
C           THE ABSCISSAE AND WEIGHTS ARE GIVEN FOR THE INTERVAL (-1,1).
C           BECAUSE OF SYMMETRY ONLY THE POSITIVE ABSCISSAE AND THEIR
C           CORRESPONDING WEIGHTS ARE GIVEN.
C
C           XGK    - ABSCISSAE OF THE 15-POINT KRONROD RULE
C                    XGK(2), XGK(4), ...  ABSCISSAE OF THE 7-POINT
C                    GAUSS RULE
C                    XGK(1), XGK(3), ...  ABSCISSAE WHICH ARE OPTIMALLY
C                    ADDED TO THE 7-POINT GAUSS RULE
C
C           WGK    - WEIGHTS OF THE 15-POINT KRONROD RULE
C
C           WG     - WEIGHTS OF THE 7-POINT GAUSS RULE
C
      DATA XGK(1),XGK(2),XGK(3),XGK(4),XGK(5),XGK(6),XGK(7),XGK(8)/
     *     9.914553711208126E-01,   9.491079123427585E-01,
     *     8.648644233597691E-01,   7.415311855993944E-01,
     *     5.860872354676911E-01,   4.058451513773972E-01,
     *     2.077849550078985E-01,   0.0E+00                 /
      DATA WGK(1),WGK(2),WGK(3),WGK(4),WGK(5),WGK(6),WGK(7),WGK(8)/
     *     2.293532201052922E-02,   6.309209262997855E-02,
     *     1.047900103222502E-01,   1.406532597155259E-01,
     *     1.690047266392679E-01,   1.903505780647854E-01,
     *     2.044329400752989E-01,   2.094821410847278E-01/
      DATA WG(1),WG(2),WG(3),WG(4)/
     *     1.294849661688697E-01,   2.797053914892767E-01,
     *     3.818300505051189E-01,   4.179591836734694E-01/
C
C
C           LIST OF MAJOR VARIABLES
C           -----------------------
C
C           CENTR  - MID POINT OF THE INTERVAL
C           HLGTH  - HALF-LENGTH OF THE INTERVAL
C           ABSC   - ABSCISSA
```

```
C           FVAL*  - FUNCTION VALUE
C           RESG   - RESULT OF THE 7-POINT GAUSS FORMULA
C           RESK   - RESULT OF THE 15-POINT KRONROD FORMULA
C           RESKH  - APPROXIMATION TO THE MEAN VALUE OF F OVER (A,B),
C                    I.E. TO I/(B-A)
C
C           MACHINE DEPENDENT CONSTANTS
C           ---------------------------
C
C           EPMACH IS THE LARGEST RELATIVE SPACING.
C           UFLOW IS THE SMALLEST POSITIVE MAGNITUDE.
C           OFLOW IS THE LARGEST MAGNITUDE.
C
C***FIRST EXECUTABLE STATEMENT
      CALL QMACO(EPMACH,UFLOW,OFLOW)
C
      CENTR = 5.0E-01*(A+B)
      HLGTH = 5.0E-01*(B-A)
      DHLGTH = ABS(HLGTH)
C
C           COMPUTE THE 15-POINT KRONROD APPROXIMATION TO THE INTEGRAL,
C           AND ESTIMATE THE ABSOLUTE ERROR.
C
      FC = F(CENTR)
      RESG = FC*WG(4)
      RESK = FC*WGK(8)
      RESABS = ABS(RESK)
      DO 10 J=1,3
        JTW = J*2
        ABSC = HLGTH*XGK(JTW)
        FVAL1 = F(CENTR-ABSC)
        FVAL2 = F(CENTR+ABSC)
        FV1(JTW) = FVAL1
        FV2(JTW) = FVAL2
        FSUM = FVAL1+FVAL2
        RESG = RESG+WG(J)*FSUM
        RESK = RESK+WGK(JTW)*FSUM
        RESABS = RESABS+WGK(JTW)*(ABS(FVAL1)+ABS(FVAL2))
   10 CONTINUE
      DO 15 J = 1,4
        JTWM1 = J*2-1
        ABSC = HLGTH*XGK(JTWM1)
        FVAL1 = F(CENTR-ABSC)
        FVAL2 = F(CENTR+ABSC)
        FV1(JTWM1) = FVAL1
        FV2(JTWM1) = FVAL2
        FSUM = FVAL1+FVAL2
        RESK = RESK+WGK(JTWM1)*FSUM
```

```
      RESABS = RESABS+WGK(JTWM1)*(ABS(FVAL1)+ABS(FVAL2))
   15 CONTINUE
      RESKH = RESK*5.0E-01
      RESASC = WGK(8)*ABS(FC-RESKH)
      DO 20 J=1,7
        RESASC = RESASC+WGK(J)*(ABS(FV1(J)-RESKH)+ABS(FV2(J)-RESKH))
   20 CONTINUE
      RESULT = RESK*HLGTH
      RESABS = RESABS*DHLGTH
      RESASC = RESASC*DHLGTH
      ABSERR = ABS((RESK-RESG)*HLGTH)
      IF(RESASC.NE.0.0E+00.AND.ABSERR.NE.0.0E+00)
     *  ABSERR = RESASC*AMIN1(1.0E+00,(2.0E+02*ABSERR/RESASC)**1.5E+00)
      IF(RESABS.GT.UFLOW/(5.0E+01*EPMACH)) ABSERR = AMAX1
     *  ((EPMACH*5.0E+01)*RESABS,ABSERR)
      RETURN
      END
```

```
      SUBROUTINE QK21(F,A,B,RESULT,ABSERR,RESABS,RESASC)
C
C     ..................................................................
C
C 1.        QK21
C           INTEGRATION RULES
C              STANDARD FORTRAN SUBROUTINE
C
C 2.        PURPOSE
C              TO COMPUTE I = INTEGRAL OF F OVER (A,B), WITH ERROR
C                             ESTIMATE
C                         J = INTEGRAL OF ABS(F) OVER (A,B)
C
C 3.        CALLING SEQUENCE
C              CALL QK21(F,A,B,RESULT,ABSERR,RESABS,RESASC)
C
C           PARAMETERS
C            ON ENTRY
C              F      - REAL
C                       FUNCTION SUBPROGRAM DEFINING THE INTEGRAND
C                       FUNCTION F(X). THE ACTUAL NAME FOR F NEEDS TO BE
C                       DECLARED E X T E R N A L IN THE CALLING PROGRAM.
C
C              A      - REAL
C                       LOWER LIMIT OF INTEGRATION
C
C              B      - REAL
C                       UPPER LIMIT OF INTEGRATION
C
C            ON RETURN
C              RESULT - REAL
C                       APPROXIMATION TO THE INTEGRAL I
C                       RESULT IS COMPUTED BY APPLYING THE 21-POINT
C                       KRONROD RULE (RESK) OBTAINED BY OPTIMAL ADDITION
C                       OF ABSCISSAE TO THE 10-POINT GAUSS RULE (RESG).
C
C              ABSERR - REAL
C                       ESTIMATE OF THE MODULUS OF THE ABSOLUTE ERROR,
C                       WHICH SHOULD NOT EXCEED ABS(I-RESULT)
C
C              RESABS - REAL
C                       APPROXIMATION TO THE INTEGRAL J
C
C              RESASC - REAL
C                       APPROXIMATION TO THE INTEGRAL OF ABS(F-I/(B-A))
C                       OVER (A,B)
C
C 4.        SUBROUTINES OR FUNCTIONS NEEDED
```

```
C                    - F (USER-PROVIDED FUNCTION)
C                    - QMACO
C                    - FORTRAN ABS, AMAX1, AMIN1
C
C      ..................................................................
C
      REAL A,ABSC,ABSERR,B,CENTR,DHLGTH,EPMACH,F,FC,FSUM,FVAL1,FVAL2,
     *  FV1,FV2,HLGTH,OFLOW,RESABS,RESASC,RESG,RESK,RESKH,RESULT,UFLOW,
     *  WG,WGK,XGK
      INTEGER J,JTW,JTWM1
C
      DIMENSION FV1(10),FV2(10),WG(5),WGK(11),XGK(11)
C
C            THE ABSCISSAE AND WEIGHTS ARE GIVEN FOR THE INTERVAL (-1,1).
C            BECAUSE OF SYMMETRY ONLY THE POSITIVE ABSCISSAE AND THEIR
C            CORRESPONDING WEIGHTS ARE GIVEN.
C
C            XGK     - ABSCISSAE OF THE 21-POINT KRONROD RULE
C                      XGK(2), XGK(4), ...  ABSCISSAE OF THE 10-POINT
C                      GAUSS RULE
C                      XGK(1), XGK(3), ...  ABSCISSAE WHICH ARE OPTIMALLY
C                      ADDED TO THE 10-POINT GAUSS RULE
C
C            WGK     - WEIGHTS OF THE 21-POINT KRONROD RULE
C
C            WG      - WEIGHTS OF THE 10-POINT GAUSS RULE
C
      DATA XGK(1),XGK(2),XGK(3),XGK(4),XGK(5),XGK(6),XGK(7),XGK(8),
     *  XGK(9),XGK(10),XGK(11)/
     *      9.956571630258081E-01,      9.739065285171717E-01,
     *      9.301574913557082E-01,      8.650633666889845E-01,
     *      7.808177265864169E-01,      6.794095682990244E-01,
     *      5.627571346686047E-01,      4.333953941292472E-01,
     *      2.943928627014602E-01,      1.488743389816312E-01,
     *      0.000000000000000E+00/
C
      DATA WGK(1),WGK(2),WGK(3),WGK(4),WGK(5),WGK(6),WGK(7),WGK(8),
     *  WGK(9),WGK(10),WGK(11)/
     *      1.169463886737187E-02,      3.255816230796473E-02,
     *      5.475589657435200E-02,      7.503967481091995E-02,
     *      9.312545458369761E-02,      1.093871588022976E-01,
     *      1.234919762620659E-01,      1.347092173114733E-01,
     *      1.427759385770601E-01,      1.477391049013385E-01,
     *      1.494455540029169E-01/
C
      DATA WG(1),WG(2),WG(3),WG(4),WG(5)/
     *      6.667134430868814E-02,      1.494513491505806E-01,
     *      2.190863625159820E-01,      2.692667193099964E-01,
```

```
     *      2.955242247147529E-01/
C
C
C           LIST OF MAJOR VARIABLES
C           -----------------------
C
C           CENTR  - MID POINT OF THE INTERVAL
C           HLGTH  - HALF-LENGTH OF THE INTERVAL
C           ABSC   - ABSCISSA
C           FVAL*  - FUNCTION VALUE
C           RESG   - RESULT OF THE 10-POINT GAUSS FORMULA
C           RESK   - RESULT OF THE 21-POINT KRONROD FORMULA
C           RESKH  - APPROXIMATION TO THE MEAN VALUE OF F OVER (A,B),
C                    I.E. TO I/(B-A)
C
C
C           MACHINE DEPENDENT CONSTANTS
C           ---------------------------
C
C           EPMACH IS THE LARGEST RELATIVE SPACING.
C           UFLOW IS THE SMALLEST POSITIVE MAGNITUDE.
C           OFLOW IS THE LARGEST MAGNITUDE.
C
C***FIRST EXECUTABLE STATEMENT
      CALL QMACO(EPMACH,UFLOW,OFLOW)
C
      CENTR = 5.0E-01*(A+B)
      HLGTH = 5.0E-01*(B-A)
      DHLGTH = ABS(HLGTH)
C
C           COMPUTE THE 21-POINT KRONROD APPROXIMATION TO THE
C           INTEGRAL, AND ESTIMATE THE ABSOLUTE ERROR.
C
      RESG = 0.0E+00
      FC = F(CENTR)
      RESK = WGK(11)*FC
      RESABS = ABS(RESK)
      DO 10 J=1,5
        JTW = 2*J
        ABSC = HLGTH*XGK(JTW)
        FVAL1 = F(CENTR-ABSC)
        FVAL2 = F(CENTR+ABSC)
        FV1(JTW) = FVAL1
        FV2(JTW) = FVAL2
        FSUM = FVAL1+FVAL2
        RESG = RESG+WG(J)*FSUM
        RESK = RESK+WGK(JTW)*FSUM
        RESABS = RESABS+WGK(JTW)*(ABS(FVAL1)+ABS(FVAL2))
```

```
   10 CONTINUE
      DO 15 J = 1,5
        JTWM1 = 2*J-1
        ABSC = HLGTH*XGK(JTWM1)
        FVAL1 = F(CENTR-ABSC)
        FVAL2 = F(CENTR+ABSC)
        FV1(JTWM1) = FVAL1
        FV2(JTWM1) = FVAL2
        FSUM = FVAL1+FVAL2
        RESK = RESK+WGK(JTWM1)*FSUM
        RESABS = RESABS+WGK(JTWM1)*(ABS(FVAL1)+ABS(FVAL2))
   15 CONTINUE
      RESKH = RESK*5.0E-01
      RESASC = WGK(11)*ABS(FC-RESKH)
      DO 20 J=1,10
        RESASC = RESASC+WGK(J)*(ABS(FV1(J)-RESKH)+ABS(FV2(J)-RESKH))
   20 CONTINUE
      RESULT = RESK*HLGTH
      RESABS = RESABS*DHLGTH
      RESASC = RESASC*DHLGTH
      ABSERR = ABS((RESK-RESG)*HLGTH)
      IF(RESASC.NE.0.0E+00.AND.ABSERR.NE.0.0E+00)
     *  ABSERR = RESASC*AMIN1(1.0E+00,(2.0E+02*ABSERR/RESASC)**1.5E+00)
      IF(RESABS.GT.UFLOW/(5.0E+01*EPMACH)) ABSERR = AMAX1
     *  ((EPMACH*5.0E+01)*RESABS,ABSERR)
      RETURN
      END
```

```
      SUBROUTINE QK31(F,A,B,RESULT,ABSERR,RESABS,RESASC)
C
C     ............................................................
C
C 1.        QK31
C           INTEGRATION RULES
C              STANDARD FORTRAN SUBROUTINE
C
C 2.        PURPOSE
C              TO COMPUTE I = INTEGRAL OF F OVER (A,B), WITH ERROR
C                             ESTIMATE
C                         J = INTEGRAL OF ABS(F) OVER (A,B)
C
C 3.        CALLING SEQUENCE
C              CALL QK31(F,A,B,RESULT,ABSERR,RESABS,RESASC)
C
C           PARAMETERS
C            ON ENTRY
C              F      - REAL
C                       FUNCTION SUBPROGRAM DEFINING THE INTEGRAND
C                       FUNCTION F(X). THE ACTUAL NAME FOR F NEEDS TO BE
C                       DECLARED E X T E R N A L IN THE CALLING PROGRAM.
C
C              A      - REAL
C                       LOWER LIMIT OF INTEGRATION
C
C              B      - REAL
C                       UPPER LIMIT OF INTEGRATION
C
C            ON RETURN
C              RESULT - REAL
C                       APPROXIMATION TO THE INTEGRAL I
C                       RESULT IS COMPUTED BY APPLYING THE 31-POINT
C                       GAUSS-KRONROD RULE (RESK), OBTAINED BY OPTIMAL
C                       ADDITION OF ABSCISSAE TO THE 15-POINT GAUSS
C                       RULE (RESG).
C
C              ABSERR - REAL
C                       ESTIMATE OF THE MODULUS OF THE ABSOLUTE ERROR,
C                       WHICH SHOULD NOT EXCEED ABS(I-RESULT)
C
C              RESABS - REAL
C                       APPROXIMATION TO THE INTEGRAL J
C
C              RESASC - REAL
C                       APPROXIMATION TO THE INTEGRAL OF ABS(F-I/(B-A))
C                       OVER (A,B)
C
```

```
C 4.        SUBROUTINES OR FUNCTIONS NEEDED
C                 - F (USER-PROVIDED FUNCTION)
C                 - QMACO
C                 - FORTRAN ABS, AMIN1, AMAX1
C
C     ..............................................................
C
      REAL A,ABSC,ABSERR,B,CENTR,DHLGTH,EPMACH,F,FC,FSUM,FVAL1,FVAL2,
     *  FV1,FV2,HLGTH,OFLOW,RESABS,RESASC,RESG,RESK,RESKH,RESULT,UFLOW,
     *  WG,WGK,XGK
      INTEGER J,JTW,JTWM1
C
      DIMENSION FV1(15),FV2(15),WG(8),WGK(16),XGK(16)
C
C           THE ABSCISSAE AND WEIGHTS ARE GIVEN FOR THE INTERVAL (-1,1).
C           BECAUSE OF SYMMETRY ONLY THE POSITIVE ABSCISSAE AND THEIR
C           CORRESPONDING WEIGHTS ARE GIVEN.
C
C           XGK    - ABSCISSAE OF THE 31-POINT KRONROD RULE
C                    XGK(2), XGK(4), ...  ABSCISSAE OF THE 15-POINT
C                    GAUSS RULE
C                    XGK(1), XGK(3), ...  ABSCISSAE WHICH ARE OPTIMALLY
C                    ADDED TO THE 15-POINT GAUSS RULE
C
C           WGK    - WEIGHTS OF THE 31-POINT KRONROD RULE
C
C           WG     - WEIGHTS OF THE 15-POINT GAUSS RULE
C
      DATA XGK(1),XGK(2),XGK(3),XGK(4),XGK(5),XGK(6),XGK(7),XGK(8),
     *  XGK(9),XGK(10),XGK(11),XGK(12),XGK(13),XGK(14),XGK(15),XGK(16)/
     *     9.980022986933971E-01,   9.879925180204854E-01,
     *     9.677390756791391E-01,   9.372733924007059E-01,
     *     8.972645323440819E-01,   8.482065834104272E-01,
     *     7.904185014424659E-01,   7.244177313601700E-01,
     *     6.509967412974170E-01,   5.709721726085388E-01,
     *     4.850818636402397E-01,   3.941513470775634E-01,
     *     2.991800071531688E-01,   2.011940939974345E-01,
     *     1.011420669187175E-01,   0.0E+00                 /
      DATA WGK(1),WGK(2),WGK(3),WGK(4),WGK(5),WGK(6),WGK(7),WGK(8),
     *  WGK(9),WGK(10),WGK(11),WGK(12),WGK(13),WGK(14),WGK(15),WGK(16)/
     *     5.377479872923349E-03,   1.500794732931612E-02,
     *     2.546084732671532E-02,   3.534636079137585E-02,
     *     4.458975132476488E-02,   5.348152469092809E-02,
     *     6.200956780067064E-02,   6.985412131872826E-02,
     *     7.684968075772038E-02,   8.308050282313302E-02,
     *     8.856444305621177E-02,   9.312659817082532E-02,
     *     9.664272698362368E-02,   9.917359872179196E-02,
     *     1.007698455238756E-01,   1.013300070147915E-01/
```

```
      DATA WG(1),WG(2),WG(3),WG(4),WG(5),WG(6),WG(7),WG(8)/
     *     3.075324199611727E-02,   7.036604748810812E-02,
     *     1.071592204671719E-01,   1.395706779261543E-01,
     *     1.662692058169939E-01,   1.861610000155622E-01,
     *     1.984314853271116E-01,   2.025782419255613E-01/
C
C
C           LIST OF MAJOR VARIABLES
C           -----------------------
C           CENTR  - MID POINT OF THE INTERVAL
C           HLGTH  - HALF-LENGTH OF THE INTERVAL
C           ABSC   - ABSCISSA
C           FVAL*  - FUNCTION VALUE
C           RESG   - RESULT OF THE 15-POINT GAUSS FORMULA
C           RESK   - RESULT OF THE 31-POINT KRONROD FORMULA
C           RESKH  - APPROXIMATION TO THE MEAN VALUE OF F OVER (A,B),
C                    I.E. TO I/(B-A)
C
C           MACHINE DEPENDENT CONSTANTS
C           ---------------------------
C           EPMACH IS THE LARGEST RELATIVE SPACING.
C           UFLOW IS THE SMALLEST POSITIVE MAGNITUDE.
C           OFLOW IS THE LARGEST MAGNITUDE.
C
C***FIRST EXECUTABLE STATEMENT
      CALL QMACO(EPMACH,UFLOW,OFLOW)
C
      CENTR = 5.0E-01*(A+B)
      HLGTH = 5.0E-01*(B-A)
      DHLGTH = ABS(HLGTH)
C
C           COMPUTE THE 31-POINT KRONROD APPROXIMATION TO THE INTEGRAL,
C           AND ESTIMATE THE ABSOLUTE ERROR.
C
      FC = F(CENTR)
      RESG = WG(8)*FC
      RESK = WGK(16)*FC
      RESABS = ABS(RESK)
      DO 10 J=1,7
        JTW = J*2
        ABSC = HLGTH*XGK(JTW)
        FVAL1 = F(CENTR-ABSC)
        FVAL2 = F(CENTR+ABSC)
        FV1(JTW) = FVAL1
        FV2(JTW) = FVAL2
        FSUM = FVAL1+FVAL2
        RESG = RESG+WG(J)*FSUM
        RESK = RESK+WGK(JTW)*FSUM
```

```
      RESABS = RESABS+WGK(JTW)*(ABS(FVAL1)+ABS(FVAL2))
   10 CONTINUE
      DO 15 J = 1,8
        JTWM1 = J*2-1
        ABSC = HLGTH*XGK(JTWM1)
        FVAL1 = F(CENTR-ABSC)
        FVAL2 = F(CENTR+ABSC)
        FV1(JTWM1) = FVAL1
        FV2(JTWM1) = FVAL2
        FSUM = FVAL1+FVAL2
        RESK = RESK+WGK(JTWM1)*FSUM
        RESABS = RESABS+WGK(JTWM1)*(ABS(FVAL1)+ABS(FVAL2))
   15 CONTINUE
      RESKH = RESK*5.0E-01
      RESASC = WGK(16)*ABS(FC-RESKH)
      DO 20 J=1,15
        RESASC = RESASC+WGK(J)*(ABS(FV1(J)-RESKH)+ABS(FV2(J)-RESKH))
   20 CONTINUE
      RESULT = RESK*HLGTH
      RESABS = RESABS*DHLGTH
      RESASC = RESASC*DHLGTH
      ABSERR = ABS((RESK-RESG)*HLGTH)
      IF(RESASC.NE.0.0E+00.AND.ABSERR.NE.0.0E+00)
     *  ABSERR = RESASC*AMIN1(1.0E+00,(2.0E+02*ABSERR/RESASC)**1.5E+00)
      IF(RESABS.GT.UFLOW/(5.0E+01*EPMACH)) ABSERR = AMAX1
     *  ((EPMACH*5.0E+01)*RESABS,ABSERR)
      RETURN
      END
```

```
      SUBROUTINE QK41(F,A,B,RESULT,ABSERR,RESABS,RESASC)
C
C     ..............................................................
C
C 1.        QK41
C           INTEGRATION RULES
C              STANDARD FORTRAN SUBROUTINE
C
C 2.        PURPOSE
C              TO COMPUTE I = INTEGRAL OF F OVER (A,B), WITH ERROR
C                             ESTIMATE
C                         J = INTEGRAL OF ABS(F) OVER (A,B)
C
C 3.        CALLING SEQUENCE
C              CALL QK41(F,A,B,RESULT,ABSERR,RESABS,RESASC)
C
C           PARAMETERS
C            ON ENTRY
C              F      - REAL
C                       FUNCTION SUBPROGRAM DEFINING THE INTEGRAND
C                       FUNCTION F(X). THE ACTUAL NAME FOR F NEEDS TO BE
C                       DECLARED E X T E R N A L IN THE CALLING PROGRAM.
C
C              A      - REAL
C                       LOWER LIMIT OF INTEGRATION
C
C              B      - REAL
C                       UPPER LIMIT OF INTEGRATION
C
C            ON RETURN
C              RESULT - REAL
C                       APPROXIMATION TO THE INTEGRAL I
C                       RESULT IS COMPUTED BY APPLYING THE 41-POINT
C                       GAUSS-KRONROD RULE (RESK) OBTAINED BY OPTIMAL
C                       ADDITION OF ABSCISSAE TO THE 20-POINT GAUSS
C                       RULE (RESG).
C
C              ABSERR - REAL
C                       ESTIMATE OF THE MODULUS OF THE ABSOLUTE ERROR,
C                       WHICH SHOULD NOT EXCEED ABS(I-RESULT)
C
C              RESABS - REAL
C                       APPROXIMATION TO THE INTEGRAL J
C
C              RESASC - REAL
C                       APPROXIMATION TO THE INTEGRAL OF ABS(F-I/(B-A))
C                       OVER (A,B)
C
```

```
C 4.        SUBROUTINES OR FUNCTIONS NEEDED
C                 - F (USER-PROVIDED FUNCTION)
C                 - QMACO
C                 - FORTRAN ABS, AMAX1, AMIN1
C
C     ..................................................................
C
      REAL A,ABSC,ABSERR,B,CENTR,DHLGTH,EPMACH,F,FC,FSUM,FVAL1,FVAL2,
     *  FV1,FV2,HLGTH,OFLOW,RESABS,RESASC,RESG,RESK,RESKH,RESULT,UFLOW,
     *  WG,WGK,XGK
      INTEGER J,JTW,JTWM1
C
      DIMENSION FV1(20),FV2(20),WG(10),WGK(21),XGK(21)
C
C           THE ABSCISSAE AND WEIGHTS ARE GIVEN FOR THE INTERVAL (-1,1).
C           BECAUSE OF SYMMETRY ONLY THE POSITIVE ABSCISSAE AND THEIR
C           CORRESPONDING WEIGHTS ARE GIVEN.
C
C           XGK    - ABSCISSAE OF THE 41-POINT GAUSS-KRONROD RULE
C                      XGK(2), XGK(4), ...  ABSCISSAE OF THE 20-POINT
C                      GAUSS RULE
C                      XGK(1), XGK(3), ...  ABSCISSAE WHICH ARE OPTIMALLY
C                      ADDED TO THE 20-POINT GAUSS RULE
C
C           WGK    - WEIGHTS OF THE 41-POINT GAUSS-KRONROD RULE
C
C           WG     - WEIGHTS OF THE 20-POINT GAUSS RULE
C
      DATA XGK(1),XGK(2),XGK(3),XGK(4),XGK(5),XGK(6),XGK(7),XGK(8),
     *  XGK(9),XGK(10),XGK(11),XGK(12),XGK(13),XGK(14),XGK(15),XGK(16),
     *  XGK(17),XGK(18),XGK(19),XGK(20),XGK(21)/
     *      9.988590315882777E-01,    9.931285991850949E-01,
     *      9.815078774502503E-01,    9.639719272779138E-01,
     *      9.408226338317548E-01,    9.122344282513259E-01,
     *      8.782768112522820E-01,    8.391169718222188E-01,
     *      7.950414288375512E-01,    7.463319064601508E-01,
     *      6.932376563347514E-01,    6.360536807265150E-01,
     *      5.751404468197103E-01,    5.108670019508271E-01,
     *      4.435931752387251E-01,    3.737060887154196E-01,
     *      3.016278681149130E-01,    2.277858511416451E-01,
     *      1.526054652409227E-01,    7.652652113349733E-02,
     *      0.0E+00                  /
      DATA WGK(1),WGK(2),WGK(3),WGK(4),WGK(5),WGK(6),WGK(7),WGK(8),
     *  WGK(9),WGK(10),WGK(11),WGK(12),WGK(13),WGK(14),WGK(15),WGK(16),
     *  WGK(17),WGK(18),WGK(19),WGK(20),WGK(21)/
     *      3.073583718520532E-03,    8.600269855642942E-03,
     *      1.462616925697125E-02,    2.038837346126652E-02,
     *      2.588213360495116E-02,    3.128730677703280E-02,
```

```
     *      3.660016975820080E-02,   4.166887332797369E-02,
     *      4.643482186749767E-02,   5.094457392372869E-02,
     *      5.519510534828599E-02,   5.911140088063957E-02,
     *      6.265323755478117E-02,   6.583459713361842E-02,
     *      6.864867292852162E-02,   7.105442355344407E-02,
     *      7.303069033278667E-02,   7.458287540049919E-02,
     *      7.570449768455667E-02,   7.637786767208074E-02,
     *      7.660071191799966E-02/
      DATA WG(1),WG(2),WG(3),WG(4),WG(5),WG(6),WG(7),WG(8),WG(9),WG(10)/
     *      1.761400713915212E-02,   4.060142980038694E-02,
     *      6.267204833410906E-02,   8.327674157670475E-02,
     *      1.019301198172404E-01,   1.181945319615184E-01,
     *      1.316886384491766E-01,   1.420961093183821E-01,
     *      1.491729864726037E-01,   1.527533871307259E-01/
C
C
C           LIST OF MAJOR VARIABLES
C           -----------------------
C
C           CENTR  - MID POINT OF THE INTERVAL
C           HLGTH  - HALF-LENGTH OF THE INTERVAL
C           ABSC   - ABSCISSA
C           FVAL*  - FUNCTION VALUE
C           RESG   - RESULT OF THE 20-POINT GAUSS FORMULA
C           RESK   - RESULT OF THE 41-POINT KRONROD FORMULA
C           RESKH  - APPROXIMATION TO MEAN VALUE OF F OVER (A,B), I.E.
C                    TO I/(B-A)
C
C           MACHINE DEPENDENT CONSTANTS
C           ---------------------------
C
C           EPMACH IS THE LARGEST RELATIVE SPACING.
C           UFLOW IS THE SMALLEST POSITIVE MAGNITUDE.
C           OFLOW IS THE LARGEST MAGNITUDE.
C
C***FIRST EXECUTABLE STATEMENT
      CALL QMACO(EPMACH,UFLOW,OFLOW)
C
      CENTR = 5.0E-01*(A+B)
      HLGTH = 5.0E-01*(B-A)
      DHLGTH = ABS(HLGTH)
C
C           COMPUTE 41-POINT GAUSS-KRONROD APPROXIMATION TO THE
C           THE INTEGRAL, AND ESTIMATE THE ABSOLUTE ERROR.
C
      RESG = 0.0E+00
      FC = F(CENTR)
      RESK = WGK(21)*FC
```

```
      RESABS = ABS(RESK)
      DO 10 J=1,10
        JTW = J*2
        ABSC = HLGTH*XGK(JTW)
        FVAL1 = F(CENTR-ABSC)
        FVAL2 = F(CENTR+ABSC)
        FV1(JTW) = FVAL1
        FV2(JTW) = FVAL2
        FSUM = FVAL1+FVAL2
        RESG = RESG+WG(J)*FSUM
        RESK = RESK+WGK(JTW)*FSUM
        RESABS = RESABS+WGK(JTW)*(ABS(FVAL1)+ABS(FVAL2))
   10 CONTINUE
      DO 15 J = 1,10
        JTWM1 = J*2-1
        ABSC = HLGTH*XGK(JTWM1)
        FVAL1 = F(CENTR-ABSC)
        FVAL2 = F(CENTR+ABSC)
        FV1(JTWM1) = FVAL1
        FV2(JTWM1) = FVAL2
        FSUM = FVAL1+FVAL2
        RESK = RESK+WGK(JTWM1)*FSUM
        RESABS = RESABS+WGK(JTWM1)*(ABS(FVAL1)+ABS(FVAL2))
   15 CONTINUE
      RESKH = RESK*5.0E-01
      RESASC = WGK(21)*ABS(FC-RESKH)
      DO 20 J=1,20
        RESASC = RESASC+WGK(J)*(ABS(FV1(J)-RESKH)+ABS(FV2(J)-RESKH))
   20 CONTINUE
      RESULT = RESK*HLGTH
      RESABS = RESABS*DHLGTH
      RESASC = RESASC*DHLGTH
      ABSERR = ABS((RESK-RESG)*HLGTH)
      IF(RESASC.NE.0.0E+00.AND.ABSERR.NE.0.0E+00)
     *  ABSERR = RESASC*AMIN1(1.0E+00,(2.0E+02*ABSERR/RESASC)**1.5E+00)
      IF(RESABS.GT.UFLOW/(5.0E+01*EPMACH)) ABSERR = AMAX1
     *  ((EPMACH*5.0E+01)*RESABS,ABSERR)
      RETURN
      END
```

```
      SUBROUTINE QK51(F,A,B,RESULT,ABSERR,RESABS,RESASC)
C
C     ...........................................................
C
C 1.        QK51
C           INTEGRATION RULES
C              STANDARD FORTRAN SUBROUTINE
C
C 2.        PURPOSE
C              TO COMPUTE I = INTEGRAL OF F OVER (A,B), WITH ERROR
C                              ESTIMATE
C                         J = INTEGRAL OF ABS(F) OVER (A,B)
C
C 3.        CALLING SEQUENCE
C              CALL QK51(F,A,B,RESULT,ABSERR,RESABS,RESASC)
C
C           PARAMETERS
C            ON ENTRY
C              F      - REAL
C                       FUNCTION SUBROUTINE DEFINING THE INTEGRAND
C                       FUNCTION F(X). THE ACTUAL NAME FOR F NEEDS TO BE
C                       DECLARED E X T E R N A L IN THE CALLING PROGRAM.
C
C              A      - REAL
C                       LOWER LIMIT OF INTEGRATION
C
C              B      - REAL
C                       UPPER LIMIT OF INTEGRATION
C
C            ON RETURN
C              RESULT - REAL
C                       APPROXIMATION TO THE INTEGRAL I
C                       RESULT IS COMPUTED BY APPLYING THE 51-POINT
C                       KRONROD RULE (RESK) OBTAINED BY OPTIMAL ADDITION
C                       OF ABSCISSAE TO THE 25-POINT GAUSS RULE (RESG).
C
C              ABSERR - REAL
C                       ESTIMATE OF THE MODULUS OF THE ABSOLUTE ERROR,
C                       WHICH SHOULD NOT EXCEED ABS(I-RESULT)
C
C              RESABS - REAL
C                       APPROXIMATION TO THE INTEGRAL J
C
C              RESASC - REAL
C                       APPROXIMATION TO THE INTEGRAL OF ABS(F-I/(B-A))
C                       OVER (A,B)
C
C 4.        SUBROUTINES OR FUNCTIONS NEEDED
```

```
C                  - F (USER-PROVIDED FUNCTION)
C                  - QMACO
C                  - FORTRAN ABS, AMAX1, AMIN1
C
C     ..............................................................
C
      REAL A,ABSC,ABSERR,B,CENTR,DHLGTH,EPMACH,F,FC,FSUM,FVAL1,FVAL2,
     *  FV1,FV2,HLGTH,OFLOW,RESABS,RESASC,RESG,RESK,RESKH,RESULT,UFLOW,
     *  WG,WGK,XGK
      INTEGER J,JTW,JTWM1
C
      DIMENSION FV1(25),FV2(25),XGK(26),WGK(26),WG(13)
C
C           THE ABSCISSAE AND WEIGHTS ARE GIVEN FOR THE INTERVAL (-1,1).
C           BECAUSE OF SYMMETRY ONLY THE POSITIVE ABSCISSAE AND THEIR
C           CORRESPONDING WEIGHTS ARE GIVEN.
C
C           XGK    - ABSCISSAE OF THE 51-POINT KRONROD RULE
C                    XGK(2), XGK(4), ...  ABSCISSAE OF THE 25-POINT
C                    GAUSS RULE
C                    XGK(1), XGK(3), ...  ABSCISSAE WHICH ARE OPTIMALLY
C                    ADDED TO THE 25-POINT GAUSS RULE
C
C           WGK    - WEIGHTS OF THE 51-POINT KRONROD RULE
C
C           WG     - WEIGHTS OF THE 25-POINT GAUSS RULE
C
      DATA XGK(1),XGK(2),XGK(3),XGK(4),XGK(5),XGK(6),XGK(7),XGK(8),
     *  XGK(9),XGK(10),XGK(11),XGK(12),XGK(13),XGK(14)/
     *     9.992621049926098E-01,   9.955569697904981E-01,
     *     9.880357945340772E-01,   9.766639214595175E-01,
     *     9.616149864258425E-01,   9.429745712289743E-01,
     *     9.207471152817016E-01,   8.949919978782754E-01,
     *     8.658470652932756E-01,   8.334426287608340E-01,
     *     7.978737979985001E-01,   7.592592630373576E-01,
     *     7.177664068130844E-01,   6.735663684734684E-01/
       DATA XGK(15),XGK(16),XGK(17),XGK(18),XGK(19),XGK(20),XGK(21),
     *  XGK(22),XGK(23),XGK(24),XGK(25),XGK(26)/
     *     6.268100990103174E-01,   5.776629302412230E-01,
     *     5.263252843347192E-01,   4.730027314457150E-01,
     *     4.178853821930377E-01,   3.611723058093878E-01,
     *     3.030895389311078E-01,   2.438668837209884E-01,
     *     1.837189394210489E-01,   1.228646926107104E-01,
     *     6.154448300568508E-02,   0.0E+00                 /
      DATA WGK(1),WGK(2),WGK(3),WGK(4),WGK(5),WGK(6),WGK(7),WGK(8),
     *  WGK(9),WGK(10),WGK(11),WGK(12),WGK(13),WGK(14)/
     *     1.987383892330316E-03,   5.561932135356714E-03,
     *     9.473973386174152E-03,   1.323622919557167E-02,
```

```
     *      1.684781770912830E-02,    2.043537114588284E-02,
     *      2.400994560695322E-02,    2.747531758785174E-02,
     *      3.079230016738749E-02,    3.400213027432934E-02,
     *      3.711627148341554E-02,    4.008382550403238E-02,
     *      4.287284502017005E-02,    4.550291304992179E-02/
       DATA WGK(15),WGK(16),WGK(17),WGK(18),WGK(19),WGK(20),WGK(21),
     *  WGK(22),WGK(23),WGK(24),WGK(25),WGK(26)/
     *      4.798253713883671E-02,    5.027767908071567E-02,
     *      5.236288580640748E-02,    5.425112988854549E-02,
     *      5.595081122041232E-02,    5.743711636156783E-02,
     *      5.868968002239421E-02,    5.972034032417406E-02,
     *      6.053945537604586E-02,    6.112850971705305E-02,
     *      6.147118987142532E-02,    6.158081806783294E-02/
      DATA WG(1),WG(2),WG(3),WG(4),WG(5),WG(6),WG(7),WG(8),WG(9),WG(10),
     *  WG(11),WG(12),WG(13)/
     *      1.139379850102629E-02,    2.635498661503214E-02,
     *      4.093915670130631E-02,    5.490469597583519E-02,
     *      6.803833381235692E-02,    8.014070033500102E-02,
     *      9.102826198296365E-02,    1.005359490670506E-01,
     *      1.085196244742637E-01,    1.148582591457116E-01,
     *      1.194557635357848E-01,    1.222424429903100E-01,
     *      1.231760537267155E-01/
C
C
C           LIST OF MAJOR VARIABLES
C           -----------------------
C
C           CENTR  - MID POINT OF THE INTERVAL
C           HLGTH  - HALF-LENGTH OF THE INTERVAL
C           ABSC   - ABSCISSA
C           FVAL*  - FUNCTION VALUE
C           RESG   - RESULT OF THE 25-POINT GAUSS FORMULA
C           RESK   - RESULT OF THE 51-POINT KRONROD FORMULA
C           RESKH  - APPROXIMATION TO THE MEAN VALUE OF F OVER (A,B),
C                    I.E. TO I/(B-A)
C
C           MACHINE DEPENDENT CONSTANTS
C           ---------------------------
C
C           EPMACH IS THE LARGEST RELATIVE SPACING.
C           UFLOW IS THE SMALLEST POSITIVE MAGNITUDE.
C           OFLOW IS THE LARGEST POSITIVE MAGNITUDE.
C
C***FIRST EXECUTABLE STATEMENT
      CALL QMACO(EPMACH,UFLOW,OFLOW)
C
      CENTR = 5.0E-01*(A+B)
      HLGTH = 5.0E-01*(B-A)
```

```
      DHLGTH = ABS(HLGTH)
C
C           COMPUTE THE 51-POINT KRONROD APPROXIMATION TO THE INTEGRAL,
C           AND ESTIMATE THE ABSOLUTE ERROR.
C
      FC = F(CENTR)
      RESG = WG(13)*FC
      RESK = WGK(26)*FC
      RESABS = ABS(RESK)
      DO 10 J=1,12
        JTW = J*2
        ABSC = HLGTH*XGK(JTW)
        FVAL1 = F(CENTR-ABSC)
        FVAL2 = F(CENTR+ABSC)
        FV1(JTW) = FVAL1
        FV2(JTW) = FVAL2
        FSUM = FVAL1+FVAL2
        RESG = RESG+WG(J)*FSUM
        RESK = RESK+WGK(JTW)*FSUM
        RESABS = RESABS+WGK(JTW)*(ABS(FVAL1)+ABS(FVAL2))
   10 CONTINUE
      DO 15 J = 1,13
        JTWM1 = J*2-1
        ABSC = HLGTH*XGK(JTWM1)
        FVAL1 = F(CENTR-ABSC)
        FVAL2 = F(CENTR+ABSC)
        FV1(JTWM1) = FVAL1
        FV2(JTWM1) = FVAL2
        FSUM = FVAL1+FVAL2
        RESK = RESK+WGK(JTWM1)*FSUM
        RESABS = RESABS+WGK(JTWM1)*(ABS(FVAL1)+ABS(FVAL2))
   15 CONTINUE
      RESKH = RESK*5.0E-01
      RESASC = WGK(26)*ABS(FC-RESKH)
      DO 20 J=1,25
        RESASC = RESASC+WGK(J)*(ABS(FV1(J)-RESKH)+ABS(FV2(J)-RESKH))
   20 CONTINUE
      RESULT = RESK*HLGTH
      RESABS = RESABS*DHLGTH
      RESASC = RESASC*DHLGTH
      ABSERR = ABS((RESK-RESG)*HLGTH)
      IF(RESASC.NE.0.0E+00.AND.ABSERR.NE.0.0E+00)
     *  ABSERR = RESASC*AMIN1(1.0E+00,(2.0E+02*ABSERR/RESASC)**1.5E+00)
      IF(RESABS.GT.UFLOW/(5.0E+01*EPMACH)) ABSERR = AMAX1
     *  ((EPMACH*5.0E+01)*RESABS,ABSERR)
      RETURN
      END
```

```
      SUBROUTINE QK61(F,A,B,RESULT,ABSERR,RESABS,RESASC)
C
C.......................................................................
C
C 1.     QK61
C        INTEGRATION RULE
C           STANDARD FORTRAN SUBROUTINE
C
C 2.     PURPOSE
C           TO COMPUTE I = INTEGRAL OF F OVER (A,B), WITH ERROR
C                          ESTIMATE
C                      J = INTEGRAL OF ABS(F) OVER (A,B)
C
C 3.     CALLING SEQUENCE
C           CALL QK61(F,A,B,RESULT,ABSERR,RESABS,RESASC)
C
C        PARAMETERS
C         ON ENTRY
C           F      - REAL
C                    FUNCTION SUBPROGRAM DEFINING THE INTEGRAND
C                    FUNCTION F(X). THE ACTUAL NAME FOR F NEEDS TO BE
C                    DECLARED E X T E R N A L IN THE CALLING PROGRAM.
C
C           A      - REAL
C                    LOWER LIMIT OF INTEGRATION
C
C           B      - REAL
C                    UPPER LIMIT OF INTEGRATION
C
C         ON RETURN
C           RESULT - REAL
C                    APPROXIMATION TO THE INTEGRAL I
C                    RESULT IS COMPUTED BY APPLYING THE 61-POINT
C                    KRONROD RULE (RESK) OBTAINED BY OPTIMAL ADDITION OF
C                    ABSCISSAE TO THE 30-POINT GAUSS RULE (RESG).
C
C           ABSERR - REAL
C                    ESTIMATE OF THE MODULUS OF THE ABSOLUTE ERROR,
C                    WHICH SHOULD EQUAL OR EXCEED ABS(I-RESULT)
C
C           RESABS - REAL
C                    APPROXIMATION TO THE INTEGRAL J
C
C           RESASC - REAL
C                    APPROXIMATION TO THE INTEGRAL OF ABS(F-I/(B-A))
C
C
C 4.     SUBROUTINES OR FUNCTIONS NEEDED
```

```
C              - F (USER-PROVIDED FUNCTION)
C              - QMACO
C              - FORTRAN ABS, AMAX1, AMIN1
C
C.......................................................................
C
      REAL A,ABSC,ABSERR,B,CENTR,DHLGTH,EPMACH,F,FC,FSUM,FVAL1,FVAL2,
     *  FV1,FV2,HLGTH,OFLOW,RESABS,RESASC,RESG,RESK,RESKH,RESULT,UFLOW,
     *  WG,WGK,XGK
      INTEGER J,JTW,JTWM1
C
      DIMENSION FV1(30),FV2(30),XGK(31),WGK(31),WG(15)
C
C           THE ABSCISSAE AND WEIGHTS ARE GIVEN FOR THE
C           INTERVAL (-1,1). BECAUSE OF SYMMETRY ONLY THE POSITIVE
C           ABSCISSAE AND THEIR CORRESPONDING WEIGHTS ARE GIVEN.
C
C           XGK    - ABSCISSAE OF THE 61-POINT KRONROD RULE
C                    XGK(2), XGK(4)  ... ABSCISSAE OF THE 30-POINT
C                    GAUSS RULE
C                    XGK(1), XGK(3)  ... OPTIMALLY ADDED ABSCISSAE
C                    TO THE 30-POINT GAUSS RULE
C
C           WGK    - WEIGHTS OF THE 61-POINT KRONROD RULE
C
C           WG     - WEIGTHS OF THE 30-POINT GAUSS RULE
C
      DATA XGK(1),XGK(2),XGK(3),XGK(4),XGK(5),XGK(6),XGK(7),XGK(8),
     *   XGK(9),XGK(10)/
     *      9.994844100504906E-01,     9.968934840746495E-01,
     *      9.916309968704046E-01,     9.836681232797472E-01,
     *      9.731163225011263E-01,     9.600218649683075E-01,
     *      9.443744447485600E-01,     9.262000474292743E-01,
     *      9.055733076999078E-01,     8.825605357920527E-01/
      DATA XGK(11),XGK(12),XGK(13),XGK(14),XGK(15),XGK(16),XGK(17),
     *  XGK(18),XGK(19),XGK(20)/
     *      8.572052335460611E-01,     8.295657623827684E-01,
     *      7.997278358218391E-01,     7.677774321048262E-01,
     *      7.337900624532268E-01,     6.978504947933158E-01,
     *      6.600610641266270E-01,     6.205261829892429E-01,
     *      5.793452358263617E-01,     5.366241481420199E-01/
      DATA XGK(21),XGK(22),XGK(23),XGK(24),XGK(25),XGK(26),XGK(27),
     *  XGK(28),XGK(29),XGK(30),XGK(31)/
     *      4.924804678617786E-01,     4.470337695380892E-01,
     *      4.004012548303944E-01,     3.527047255308781E-01,
     *      3.040732022736251E-01,     2.546369261678898E-01,
     *      2.045251166823099E-01,     1.538699136085835E-01,
     *      1.028069379667370E-01,     5.147184255531770E-02,
```

```
     *      0.0E+00                       /
      DATA WGK(1),WGK(2),WGK(3),WGK(4),WGK(5),WGK(6),WGK(7),WGK(8),
     *  WGK(9),WGK(10)/
     *      1.389013698677008E-03,     3.890461127099884E-03,
     *      6.630703915931292E-03,     9.273279659517763E-03,
     *      1.182301525349634E-02,     1.436972950704580E-02,
     *      1.692088918905327E-02,     1.941414119394238E-02,
     *      2.182803582160919E-02,     2.419116207808060E-02/
      DATA WGK(11),WGK(12),WGK(13),WGK(14),WGK(15),WGK(16),WGK(17),
     *  WGK(18),WGK(19),WGK(20)/
     *      2.650995488233310E-02,     2.875404876504129E-02,
     *      3.090725756238776E-02,     3.298144705748373E-02,
     *      3.497933802806002E-02,     3.688236465182123E-02,
     *      3.867894562472759E-02,     4.037453895153596E-02,
     *      4.196981021516425E-02,     4.345253970135607E-02/
      DATA WGK(21),WGK(22),WGK(23),WGK(24),WGK(25),WGK(26),WGK(27),
     *  WGK(28),WGK(29),WGK(30),WGK(31)/
     *      4.481480013316266E-02,     4.605923827100699E-02,
     *      4.718554656929915E-02,     4.818586175708713E-02,
     *      4.905543455502978E-02,     4.979568342707421E-02,
     *      5.040592140278235E-02,     5.088179589874961E-02,
     *      5.122154784925877E-02,     5.142612853745903E-02,
     *      5.149472942945157E-02/
      DATA WG(1),WG(2),WG(3),WG(4),WG(5),WG(6),WG(7),WG(8)/
     *      7.968192496166606E-03,     1.846646831109096E-02,
     *      2.878470788332337E-02,     3.879919256962705E-02,
     *      4.840267283059405E-02,     5.749315621761907E-02,
     *      6.597422988218050E-02,     7.375597473770521E-02/
      DATA WG(9),WG(10),WG(11),WG(12),WG(13),WG(14),WG(15)/
     *      8.075589522942022E-02,     8.689978720108298E-02,
     *      9.212252223778613E-02,     9.636873717464426E-02,
     *      9.959342058679527E-02,     1.017623897484055E-01,
     *      1.028526528935588E-01/
C
C           LIST OF MAJOR VARIABLES
C           -----------------------
C
C           CENTR  - MID POINT OF THE INTERVAL
C           HLGTH  - HALF-LENGTH OF THE INTERVAL
C           ABSC   - ABSCISSA
C           FVAL*  - FUNCTION VALUE
C           RESG   - RESULT OF THE 30-POINT GAUSS RULE
C           RESK   - RESULT OF THE 61-POINT KRONROD RULE
C           RESKH  - APPROXIMATION TO THE MEAN VALUE OF F
C                    OVER (A,B), I.E. TO I/(B-A)
C
C           MACHINE DEPENDENT CONSTANTS
C           ---------------------------
```

```
C
C           EPMACH IS THE LARGEST RELATIVE SPACING.
C           UFLOW IS THE SMALLEST POSITIVE MAGNITUDE.
C           OFLOW IS THE LARGEST MAGNITUDE.
C
C
C***FIRST EXECUTABLE STATEMENT
      CALL QMACO(EPMACH,UFLOW,OFLOW)
      CENTR = 5.0E-01*(B+A)
      HLGTH = 5.0E-01*(B-A)
      DHLGTH = ABS(HLGTH)
C
C           COMPUTE THE 61-POINT KRONROD APPROXIMATION TO THE INTEGRAL,
C           AND ESTIMATE THE ABSOLUTE ERROR.
C
      RESG = 0.0E+00
      FC = F(CENTR)
      RESK = WGK(31)*FC
      RESABS = ABS(RESK)
      DO 10 J=1,15
        JTW = J*2
        ABSC = HLGTH*XGK(JTW)
        FVAL1 = F(CENTR-ABSC)
        FVAL2 = F(CENTR+ABSC)
        FV1(JTW) = FVAL1
        FV2(JTW) = FVAL2
        FSUM = FVAL1+FVAL2
        RESG = RESG+WG(J)*FSUM
        RESK = RESK+WGK(JTW)*FSUM
        RESABS = RESABS+WGK(JTW)*(ABS(FVAL1)+ABS(FVAL2))
   10 CONTINUE
      DO 15 J=1,15
        JTWM1 = J*2-1
        ABSC = HLGTH*XGK(JTWM1)
        FVAL1 = F(CENTR-ABSC)
        FVAL2 = F(CENTR+ABSC)
        FV1(JTWM1) = FVAL1
        FV2(JTWM1) = FVAL2
        FSUM = FVAL1+FVAL2
        RESK = RESK+WGK(JTWM1)*FSUM
        RESABS = RESABS+WGK(JTWM1)*(ABS(FVAL1)+ABS(FVAL2))
  15    CONTINUE
      RESKH = RESK*5.0E-01
      RESASC = WGK(31)*ABS(FC-RESKH)
      DO 20 J=1,30
        RESASC = RESASC+WGK(J)*(ABS(FV1(J)-RESKH)+ABS(FV2(J)-RESKH))
   20 CONTINUE
      RESULT = RESK*HLGTH
```

```
      RESABS = RESABS*DHLGTH
      RESASC = RESASC*DHLGTH
      ABSERR = ABS((RESK-RESG)*HLGTH)
      IF(RESASC.NE.0.0E+00.AND.ABSERR.NE.0.0E+00)
     *  ABSERR = RESASC*AMIN1(1.0E+00,(2.0E+02*ABSERR/RESASC)**1.5E+00)
      IF(RESABS.GT.UFLOW/(5.0E+01*EPMACH)) ABSERR = AMAX1
     *  ((EPMACH*5.0E+01)*RESABS,ABSERR)
      RETURN
      END
```

```
      SUBROUTINE QK15I(F,BOUN,INF,A,B,RESULT,ABSERR,RESABS,RESASC)
C
C     ..................................................................
C
C 1.        QK15I
C           INTEGRATION RULE
C              STANDARD FORTRAN SUBROUTINE
C
C 2.        PURPOSE
C              THE ORIGINAL (INFINITE) INTEGRATION RANGE IS MAPPED
C              ONTO THE INTERVAL (0,1) AND (A,B) IS A PART OF (0,1).
C              IT IS THE PURPOSE TO COMPUTE
C              I = INTEGRAL OF TRANSFORMED INTEGRAND OVER (A,B),
C              J = INTEGRAL OF ABS(TRANSFORMED INTEGRAND) OVER (A,B).
C
C 3.        CALLING SEQUENCE
C              CALL QK15I(F,BOUN,INF,A,B,RESULT,ABSERR,RESABS,RESASC)
C
C           PARAMETERS
C            ON ENTRY
C              F      - REAL
C                       FUNCTION SUBPROGRAM DEFINING THE INTEGRAND
C                       FUNCTION F(X). THE ACTUAL NAME FOR F NEEDS TO BE
C                       DECLARED E X T E R N A L IN THE CALLING PROGRAM.
C
C              BOUN   - REAL
C                       FINITE BOUND OF ORIGINAL INTEGRATION RANGE
C                       (SET TO ZERO IF INF = +2)
C
C              INF    - INTEGER
C                       IF INF = -1, THE ORIGINAL INTERVAL IS
C                                   (-INFINITY,BOUND),
C                       IF INF = +1, THE ORIGINAL INTERVAL IS
C                                   (BOUND,+INFINITY),
C                       IF INF = +2, THE ORIGINAL INTERVAL IS
C                                   (-INFINITY,+INFINITY) AND
C                       THE INTEGRAL IS COMPUTED AS THE SUM OF TWO
C                       INTEGRALS, ONE OVER (-INFINITY,0) AND ONE
C                       OVER (0,+INFINITY).
C
C              A      - REAL
C                       LOWER LIMIT FOR INTEGRATION OVER SUBRANGE
C                       OF (0,1)
C
C              B      - REAL
C                       UPPER LIMIT FOR INTEGRATION OVER SUBRANGE
C                       OF (0,1)
C
```

```
C             ON RETURN
C               RESULT - REAL
C                        APPROXIMATION TO THE INTEGRAL I
C                        RESULT IS COMPUTED BY APPLYING THE 15-POINT
C                        KRONROD RULE(RESK) OBTAINED BY OPTIMAL ADDITION
C                        OF ABSCISSAE TO THE 7-POINT GAUSS RULE (RESG).
C
C               ABSERR - REAL
C                        ESTIMATE OF THE MODULUS OF THE ABSOLUTE ERROR,
C                        WHICH SHOULD EQUAL OR EXCEED ABS(I-RESULT)
C
C               RESABS - REAL
C                        APPROXIMATION TO THE INTEGRAL J
C
C               RESASC - REAL
C                        APPROXIMATION TO THE INTEGRAL OF
C                        ABS((TRANSFORMED INTEGRAND)-I/(B-A)) OVER (A,B)
C
C 4.         SUBROUTINES OR FUNCTIONS NEEDED
C                 - F (USER-PROVIDED FUNCTION)
C                 - QMACO
C                 - FORTRAN ABS, AMAX1, AMIN1, MIN0
C
C     ..............................................................
C
      REAL A,ABSC,ABSC1,ABSC2,ABSERR,B,BOUN,CENTR,DINF,EPMACH,F,FC,
     *  FSUM,FVAL1,FVAL2,FV1,FV2,HLGTH,OFLOW,RESABS,RESASC,RESG,RESK,
     *  RESKH,RESULT,TABSC1,TABSC2,UFLOW,WG,WGK,XGK
      INTEGER INF,J
C
      DIMENSION FV1(7),FV2(7),WGK(8),WG(8),XGK(8)
C
C           THE ABSCISSAE AND WEIGHTS ARE SUPPLIED FOR THE INTERVAL
C           (-1,1).  BECAUSE OF SYMMETRY ONLY THE POSITIVE ABSCISSAE AND
C           THEIR CORRESPONDING WEIGHTS ARE GIVEN.
C
C           XGK    - ABSCISSAE OF THE 15-POINT KRONROD RULE
C                    XGK(2), XGK(4), ... ABSCISSAE OF THE 7-POINT GAUSS
C                    RULE
C                    XGK(1), XGK(3), ...  ABSCISSAE WHICH ARE OPTIMALLY
C                    ADDED TO THE 7-POINT GAUSS RULE
C
C           WGK    - WEIGHTS OF THE 15-POINT KRONROD RULE
C
C           WG     - WEIGHTS OF THE 7-POINT GAUSS RULE, CORRESPONDING
C                    TO THE ABSCISSAE XGK(2), XGK(4), ...
C                    WG(1), WG(3), ... ARE SET TO ZERO.
C
```

```
      DATA XGK(1),XGK(2),XGK(3),XGK(4),XGK(5),XGK(6),XGK(7),XGK(8)/
     *     9.914553711208126E-01,     9.491079123427585E-01,
     *     8.648644233597691E-01,     7.415311855993944E-01,
     *     5.860872354676911E-01,     4.058451513773972E-01,
     *     2.077849550078985E-01,     0.000000000000000E+00/
C
      DATA WGK(1),WGK(2),WGK(3),WGK(4),WGK(5),WGK(6),WGK(7),WGK(8)/
     *     2.293532201052922E-02,     6.309209262997855E-02,
     *     1.047900103222502E-01,     1.406532597155259E-01,
     *     1.690047266392679E-01,     1.903505780647854E-01,
     *     2.044329400752989E-01,     2.094821410847278E-01/
C
      DATA WG(1),WG(2),WG(3),WG(4),WG(5),WG(6),WG(7),WG(8)/
     *     0.000000000000000E+00,     1.294849661688697E-01,
     *     0.000000000000000E+00,     2.797053914892767E-01,
     *     0.000000000000000E+00,     3.818300505051189E-01,
     *     0.000000000000000E+00,     4.179591836734694E-01/
C
C
C           LIST OF MAJOR VARIABLES
C           -----------------------
C
C           CENTR  - MID POINT OF THE INTERVAL
C           HLGTH  - HALF-LENGTH OF THE INTERVAL
C           ABSC*  - ABSCISSA
C           TABSC* - TRANSFORMED ABSCISSA
C           FVAL*  - FUNCTION VALUE
C           RESG   - RESULT OF THE 7-POINT GAUSS FORMULA
C           RESK   - RESULT OF THE 15-POINT KRONROD FORMULA
C           RESKH  - APPROXIMATION TO THE MEAN VALUE OF THE TRANSFORMED
C                    INTEGRAND OVER (A,B), I.E. TO I/(B-A)
C
C           MACHINE DEPENDENT CONSTANTS
C           ---------------------------
C
C           EPMACH IS THE LARGEST RELATIVE SPACING.
C           UFLOW IS THE SMALLEST POSITIVE MAGNITUDE.
C           OFLOW IS THE LARGEST MAGNITUDE.
C
C***FIRST EXECUTABLE STATEMENT
      CALL QMACO(EPMACH,UFLOW,OFLOW)
      DINF = MIN0(1,INF)
C
      CENTR = 5.0E-01*(A+B)
      HLGTH = 5.0E-01*(B-A)
      TABSC1 = BOUN+DINF*(1.0E+00-CENTR)/CENTR
      FVAL1 = F(TABSC1)
      IF(INF.EQ.2) FVAL1 = FVAL1+F(-TABSC1)
```

```
      FC = (FVAL1/CENTR)/CENTR
C
C           COMPUTE THE 15-POINT KRONROD APPROXIMATION TO THE INTEGRAL,
C           AND ESTIMATE THE ERROR.
C
      RESG = WG(8)*FC
      RESK = WGK(8)*FC
      RESABS = ABS(RESK)
      DO 10 J=1,7
        ABSC = HLGTH*XGK(J)
        ABSC1 = CENTR-ABSC
        ABSC2 = CENTR+ABSC
        TABSC1 = BOUN+DINF*(1.0E+00-ABSC1)/ABSC1
        TABSC2 = BOUN+DINF*(1.0E+00-ABSC2)/ABSC2
        FVAL1 = F(TABSC1)
        FVAL2 = F(TABSC2)
        IF(INF.EQ.2) FVAL1 = FVAL1+F(-TABSC1)
        IF(INF.EQ.2) FVAL2 = FVAL2+F(-TABSC2)
        FVAL1 = (FVAL1/ABSC1)/ABSC1
        FVAL2 = (FVAL2/ABSC2)/ABSC2
        FV1(J) = FVAL1
        FV2(J) = FVAL2
        FSUM = FVAL1+FVAL2
        RESG = RESG+WG(J)*FSUM
        RESK = RESK+WGK(J)*FSUM
        RESABS = RESABS+WGK(J)*(ABS(FVAL1)+ABS(FVAL2))
   10 CONTINUE
      RESKH = RESK*5.0E-01
      RESASC = WGK(8)*ABS(FC-RESKH)
      DO 20 J=1,7
        RESASC = RESASC+WGK(J)*(ABS(FV1(J)-RESKH)+ABS(FV2(J)-RESKH))
   20 CONTINUE
      RESULT = RESK*HLGTH
      RESASC = RESASC*HLGTH
      RESABS = RESABS*HLGTH
      ABSERR = ABS((RESK-RESG)*HLGTH)
      IF(RESASC.NE.0.0E+00.AND.ABSERR.NE.0.0E+00) ABSERR = RESASC*
     * AMIN1(1.0E+00,(2.0E+02*ABSERR/RESASC)**1.5E+00)
      IF(RESABS.GT.UFLOW/(5.0E+01*EPMACH)) ABSERR = AMAX1
     * ((EPMACH*5.0E+01)*RESABS,ABSERR)
      RETURN
      END
```

```
      SUBROUTINE QK15W(F,W,P1,P2,P3,P4,KP,A,B,RESULT,ABSERR,RESABS,
     *  RESASC)
C
C     ..................................................................
C
C 1.        QK15W
C           INTEGRATION RULES
C              STANDARD FORTRAN SUBROUTINE
C
C 2.        PURPOSE
C              TO COMPUTE I = INTEGRAL OF F*W OVER (A,B), WITH ERROR
C                             ESTIMATE
C                         J = INTEGRAL OF ABS(F*W) OVER (A,B)
C
C 3.        CALLING SEQUENCE
C              CALL QK15W(F,W,P1,P2,P3,P4,KP,A,B,RESULT,ABSERR,RESABS,
C                         RESASC)
C
C           PARAMETERS
C             ON ENTRY
C              F      - REAL
C                       FUNCTION SUBPROGRAM DEFINING THE INTEGRAND
C                       FUNCTION F(X). THE ACTUAL NAME FOR F NEEDS TO BE
C                       DECLARED E X T E R N A L IN THE CALLING PROGRAM.
C
C              W      - REAL
C                       FUNCTION SUBPROGRAM DEFINING THE INTEGRAND
C                       WEIGHT FUNCTION W(X). THE ACTUAL NAME FOR W
C                       NEEDS TO BE DECLARED E X T E R N A L IN THE
C                       CALLING PROGRAM.
C
C              P1, P2, P3, P4 - REAL
C                       PARAMETERS IN THE WEIGHT FUNCTION
C
C              KP     - INTEGER
C                       KEY FOR INDICATING THE TYPE OF WEIGHT FUNCTION
C
C              A      - REAL
C                       LOWER LIMIT OF INTEGRATION
C
C              B      - REAL
C                       UPPER LIMIT OF INTEGRATION
C
C            ON RETURN
C              RESULT - REAL
C                       APPROXIMATION TO THE INTEGRAL I
C                       RESULT IS COMPUTED BY APPLYING THE 15-POINT
C                       KRONROD RULE (RESK) OBTAINED BY OPTIMAL ADDITION
```

```
C                       OF ABSCISSAE TO THE 7-POINT GAUSS RULE (RESG).
C
C              ABSERR - REAL
C                       ESTIMATE OF THE MODULUS OF THE ABSOLUTE ERROR,
C                       WHICH SHOULD EQUAL OR EXCEED ABS(I-RESULT)
C
C              RESABS - REAL
C                       APPROXIMATION TO THE INTEGRAL OF ABS(F)
C
C              RESASC - REAL
C                       APPROXIMATION TO THE INTEGRAL OF ABS(F-I/(B-A))
C
C 4.        SUBROUTINES OR FUNCTIONS NEEDED
C                  - F (USER-PROVIDED FUNCTION)
C                  - W (QWGTC, QWGTF OR QWGTS, DEPENDING ON THE
C                       CALLING ROUTINE)
C                  - QMACO
C                  - FORTRAN ABS, AMAX1, AMIN1
C
C
C     ............................................................
C
      REAL A,ABSC,ABSC1,ABSC2,ABSERR,B,CENTR,DHLGTH,EPMACH,F,FC,FSUM,
     * FVAL1,FVAL2,FV1,FV2,HLGTH,OFLOW,P1,P2,P3,P4,RESABS,RESASC,RESG,
     * RESK,RESKH,RESULT,UFLOW,W,WG,WGK,XGK
      INTEGER J,JTW,JTWM1,KP
C
      DIMENSION FV1(7),FV2(7),WG(4),WGK(8),XGK(8)
C
C          THE ABSCISSAE AND WEIGHTS ARE GIVEN FOR THE INTERVAL (-1,1).
C          BECAUSE OF SYMMETRY ONLY THE POSITIVE ABSCISSAE AND THEIR
C          CORRESPONDING WEIGHTS ARE GIVEN.
C
C          XGK    - ABSCISSAE OF THE 15-POINT GAUSS-KRONROD RULE
C                     XGK(2), XGK(4), ... ABSCISSAE OF THE 7-POINT GAUSS
C                     RULE
C                     XGK(1), XGK(3), ... ABSCISSAE WHICH ARE OPTIMALLY
C                     ADDED TO THE 7-POINT GAUSS RULE
C
C          WGK    - WEIGHTS OF THE 15-POINT GAUSS-KRONROD RULE
C
C          WG     - WEIGHTS OF THE 7-POINT GAUSS RULE
C
      DATA XGK(1),XGK(2),XGK(3),XGK(4),XGK(5),XGK(6),XGK(7),XGK(8)/
     *     9.914553711208126E-01,     9.491079123427585E-01,
     *     8.648644233597691E-01,     7.415311855993944E-01,
     *     5.860872354676911E-01,     4.058451513773972E-01,
     *     2.077849550789850E-01,     0.000000000000000E+00/
```

```
C
      DATA WGK(1),WGK(2),WGK(3),WGK(4),WGK(5),WGK(6),WGK(7),WGK(8)/
     *     2.293532201052922E-02,     6.309209262997855E-02,
     *     1.047900103222502E-01,     1.406532597155259E-01,
     *     1.690047266392679E-01,     1.903505780647854E-01,
     *     2.044329400752989E-01,     2.094821410847278E-01/
C
      DATA WG(1),WG(2),WG(3),WG(4)/
     *     1.294849661688697E-01,     2.797053914892767E-01,
     *     3.818300505051889E-01,     4.179591836734694E-01/
C
C
C           LIST OF MAJOR VARIABLES
C           -----------------------
C
C           CENTR  - MID POINT OF THE INTERVAL
C           HLGTH  - HALF-LENGTH OF THE INTERVAL
C           ABSC*  - ABSCISSA
C           FVAL*  - FUNCTION VALUE
C           RESG   - RESULT OF THE 7-POINT GAUSS FORMULA
C           RESK   - RESULT OF THE 15-POINT KRONROD FORMULA
C           RESKH  - APPROXIMATION TO THE MEAN VALUE OF F*W OVER (A,B),
C                    I.E. TO I/(B-A)
C
C           MACHINE DEPENDENT CONSTANTS
C           ---------------------------
C
C           EPMACH IS THE LARGEST RELATIVE SPACING.
C           UFLOW IS THE SMALLEST POSITIVE MAGNITUDE.
C           OFLOW IS THE LARGEST MAGNITUDE.
C
C***FIRST EXECUTABLE STATEMENT
      CALL QMACO(EPMACH,UFLOW,OFLOW)
C
      CENTR = 5.0E-01*(A+B)
      HLGTH = 5.0E-01*(B-A)
      DHLGTH = ABS(HLGTH)
C
C           COMPUTE THE 15-POINT KRONROD APPROXIMATION TO THE INTEGRAL,
C           AND ESTIMATE THE ERROR.
C
      FC = F(CENTR)*W(CENTR,P1,P2,P3,P4,KP)
      RESG = WG(4)*FC
      RESK = WGK(8)*FC
      RESABS = ABS(RESK)
      DO 10 J=1,3
        JTW = J*2
        ABSC = HLGTH*XGK(JTW)
```

```
      ABSC1 = CENTR-ABSC
      ABSC2 = CENTR+ABSC
      FVAL1 = F(ABSC1)*W(ABSC1,P1,P2,P3,P4,KP)
      FVAL2 = F(ABSC2)*W(ABSC2,P1,P2,P3,P4,KP)
      FV1(JTW) = FVAL1
      FV2(JTW) = FVAL2
      FSUM = FVAL1+FVAL2
      RESG = RESG+WG(J)*FSUM
      RESK = RESK+WGK(JTW)*FSUM
      RESABS = RESABS+WGK(JTW)*(ABS(FVAL1)+ABS(FVAL2))
   10 CONTINUE
      DO 15 J=1,4
        JTWM1 = J*2-1
        ABSC = HLGTH*XGK(JTWM1)
        ABSC1 = CENTR-ABSC
        ABSC2 = CENTR+ABSC
        FVAL1 = F(ABSC1)*W(ABSC1,P1,P2,P3,P4,KP)
        FVAL2 = F(ABSC2)*W(ABSC2,P1,P2,P3,P4,KP)
        FV1(JTWM1) = FVAL1
        FV2(JTWM1) = FVAL2
        FSUM = FVAL1+FVAL2
        RESK = RESK+WGK(JTWM1)*FSUM
        RESABS = RESABS+WGK(JTWM1)*(ABS(FVAL1)+ABS(FVAL2))
   15 CONTINUE
      RESKH = RESK*5.0E-01
      RESASC = WGK(8)*ABS(FC-RESKH)
      DO 20 J=1,7
        RESASC = RESASC+WGK(J)*(ABS(FV1(J)-RESKH)+ABS(FV2(J)-RESKH))
   20 CONTINUE
      RESULT = RESK*HLGTH
      RESABS = RESABS*DHLGTH
      RESASC = RESASC*DHLGTH
      ABSERR = ABS((RESK-RESG)*HLGTH)
      IF(RESASC.NE.0.0E+00.AND.ABSERR.NE.0.0E+00)
     *  ABSERR = RESASC*AMIN1(1.0E+00,(2.0E+02*ABSERR/RESASC)**1.5E+00)
      IF(RESABS.GT.UFLOW/(5.0E+01*EPMACH)) ABSERR = AMAX1((EPMACH*
     *  5.0E+01)*RESABS,ABSERR)
      RETURN
      END
```

```
      SUBROUTINE QEXT(N,EPSTAB,RESULT,ABSERR,RES3LA,NRES)
C
C     ...............................................................
C
C 1.        QEXT
C           EPSILON ALGORITHM
C              STANDARD FORTRAN SUBROUTINE
C
C 2.        PURPOSE
C              THE ROUTINE DETERMINES THE LIMIT OF A GIVEN SEQUENCE OF
C              APPROXIMATIONS, BY MEANS OF THE EPSILON ALGORITHM
C              OF P. WYNN.
C              AN ESTIMATE OF THE ABSOLUTE ERROR IS ALSO GIVEN.
C              THE CONDENSED EPSILON TABLE IS COMPUTED. ONLY THOSE
C              ELEMENTS NEEDED FOR THE COMPUTATION OF THE NEXT DIAGONAL
C              ARE PRESERVED.
C
C 3.        CALLING SEQUENCE
C              CALL QEXT(N,EPSTAB,RESULT,ABSERR,RES3LA,NRES)
C
C           PARAMETERS
C              N      - INTEGER
C                       EPSTAB(N) CONTAINS THE NEW ELEMENT IN THE
C                       FIRST COLUMN OF THE EPSILON TABLE.
C
C              EPSTAB - REAL
C                       VECTOR OF DIMENSION 52 CONTAINING THE ELEMENTS
C                       OF THE TWO LOWER DIAGONALS OF THE TRIANGULAR
C                       EPSILON TABLE
C                       THE ELEMENTS ARE NUMBERED STARTING AT THE
C                       RIGHT-HAND CORNER OF THE TRIANGLE.
C
C              RESULT - REAL
C                       RESULTING APPROXIMATION TO THE INTEGRAL
C
C              ABSERR - REAL
C                       ESTIMATE OF THE ABSOLUTE ERROR COMPUTED FROM
C                       RESULT AND THE 3 PREVIOUS RESULTS
C
C              RES3LA - REAL
C                       VECTOR OF DIMENSION 3 CONTAINING THE LAST 3
C                       RESULTS
C
C              NRES   - INTEGER
C                       NUMBER OF CALLS TO THE ROUTINE
C                       (SHOULD BE ZERO AT FIRST CALL)
C
C 4.        SUBROUTINES OR FUNCTIONS NEEDED
```

```
C                     - QMACO
C                     - FORTRAN ABS, AMAX1
C
C     ..................................................................
C
      REAL ABSERR,DELTA1,DELTA2,DELTA3,EPMACH,EPSINF,EPSTAB,ERROR,ERR1,
     *  ERR2,ERR3,E0,E1,E1ABS,E2,E3,OFLOW,RES,RESULT,RES3LA,SS,TOL1,
     *  TOL2,TOL3,UFLOW
      INTEGER I,IB,IB2,IE,INDX,K1,K2,K3,LIMEXP,N,NEWELM,NRES,NUM
      DIMENSION EPSTAB(52),RES3LA(3)
C
C            LIST OF MAJOR VARIABLES
C            -----------------------
C
C            E0     - THE 4 ELEMENTS ON WHICH THE
C            E1       COMPUTATION OF A NEW ELEMENT IN
C            E2       THE EPSILON TABLE IS BASED
C            E3                 E0
C                         E3    E1    NEW
C                               E2
C            NEWELM - NUMBER OF ELEMENTS TO BE COMPUTED IN THE NEW
C                     DIAGONAL
C            ERROR  - ERROR = ABS(E1-E0)+ABS(E2-E1)+ABS(NEW-E2)
C            RESULT - THE ELEMENT IN THE NEW DIAGONAL WITH LEAST VALUE
C                     OF ERROR
C
C            MACHINE DEPENDENT CONSTANTS
C            ---------------------------
C
C            EPMACH IS THE LARGEST RELATIVE SPACING.
C            UFLOW IS THE SMALLEST POSITIVE MAGNITUDE.
C            OFLOW IS THE LARGEST POSITIVE MAGNITUDE.
C            LIMEXP IS THE MAXIMUM NUMBER OF ELEMENTS THE EPSILON TABLE
C            CAN CONTAIN. IF THIS NUMBER IS REACHED, THE UPPER DIAGONAL
C            OF THE EPSILON TABLE IS DELETED.
C
C***FIRST EXECUTABLE STATEMENT
      CALL QMACO(EPMACH,UFLOW,OFLOW)
      NRES = NRES+1
      ABSERR = OFLOW
      RESULT = EPSTAB(N)
      IF(N.LT.3) GO TO 100
      LIMEXP = 50
      EPSTAB(N+2) = EPSTAB(N)
      NEWELM = (N-1)/2
      EPSTAB(N) = OFLOW
      NUM = N
      K1 = N
```

```
      DO 40 I = 1,NEWELM
        K2 = K1-1
        K3 = K1-2
        RES = EPSTAB(K1+2)
        E0 = EPSTAB(K3)
        E1 = EPSTAB(K2)
        E2 = RES
        E1ABS = ABS(E1)
        DELTA2 = E2-E1
        ERR2 = ABS(DELTA2)
        TOL2 = AMAX1(ABS(E2),E1ABS)*EPMACH
        DELTA3 = E1-E0
        ERR3 = ABS(DELTA3)
        TOL3 = AMAX1(E1ABS,ABS(E0))*EPMACH
        IF(ERR2.GT.TOL2.OR.ERR3.GT.TOL3) GO TO 10
C
C           IF E0, E1 AND E2 ARE EQUAL TO WITHIN MACHINE ACCURACY,
C           CONVERGENCE IS ASSUMED.
C           RESULT = E2
C           ABSERR = ABS(E1-E0)+ABS(E2-E1)
C
        RESULT = RES
        ABSERR = ERR2+ERR3
C***JUMP OUT OF DO-LOOP
        GO TO 100
   10   E3 = EPSTAB(K1)
        EPSTAB(K1) = E1
        DELTA1 = E1-E3
        ERR1 = ABS(DELTA1)
        TOL1 = AMAX1(E1ABS,ABS(E3))*EPMACH
C
C           IF TWO ELEMENTS ARE VERY CLOSE TO EACH OTHER, OMIT A PART
C           OF THE TABLE BY ADJUSTING THE VALUE OF N
C
        IF(ERR1.LE.TOL1.OR.ERR2.LE.TOL2.OR.ERR3.LE.TOL3) GO TO 20
        SS = 1.0E+00/DELTA1+1.0E+00/DELTA2-1.0E+00/DELTA3
        EPSINF = ABS(SS*E1)
C
C           TEST TO DETECT IRREGULAR BEHAVIOUR IN THE TABLE, AND
C           EVENTUALLY OMIT A PART OF THE TABLE ADJUSTING THE VALUE
C           OF N.
C
        IF(EPSINF.GT.1.0E-04) GO TO 30
   20   N = I+I-1
C***JUMP OUT OF DO-LOOP
        GO TO 50
C
C           COMPUTE A NEW ELEMENT AND EVENTUALLY ADJUST THE VALUE OF
```

```
C           RESULT.
C
   30   RES = E1+1.0E+00/SS
        EPSTAB(K1) = RES
        K1 = K1-2
        ERROR = ERR2+ABS(RES-E2)+ERR3
        IF(ERROR.GT.ABSERR) GO TO 40
        ABSERR = ERROR
        RESULT = RES
   40 CONTINUE
C
C           SHIFT THE TABLE.
C
   50 IF(N.EQ.LIMEXP) N = 2*(LIMEXP/2)-1
      IB = 1
      IF((NUM/2)*2.EQ.NUM) IB = 2
      IE = NEWELM+1
      DO 60 I=1,IE
        IB2 = IB+2
        EPSTAB(IB) = EPSTAB(IB2)
        IB = IB2
   60 CONTINUE
      IF(NUM.EQ.N) GO TO 80
      INDX = NUM-N+1
      DO 70 I = 1,N
        EPSTAB(I)= EPSTAB(INDX)
        INDX = INDX+1
   70 CONTINUE
   80 IF(NRES.GE.4) GO TO 90
      RES3LA(NRES) = RESULT
      ABSERR = OFLOW
      GO TO 100
C
C           COMPUTE ERROR ESTIMATE
C
   90 ABSERR = ABS(RESULT-RES3LA(3))+ABS(RESULT-RES3LA(2))
     *  +ABS(RESULT-RES3LA(1))
      RES3LA(1) = RES3LA(2)
      RES3LA(2) = RES3LA(3)
      RES3LA(3) = RESULT
  100 ABSERR = AMAX1(ABSERR,0.5E+00*EPMACH*ABS(RESULT))
      RETURN
      END
```

```
      SUBROUTINE QSORT(LIMIT,LAST,MAXERR,ERMAX,ELIST,IORD,NRMAX)
C
C     ..................................................................
C
C 1.        QSORT
C           ORDERING ROUTINE
C              STANDARD FORTRAN SUBROUTINE
C
C 2.        PURPOSE
C              THIS ROUTINE MAINTAINS THE DESCENDING ORDERING IN THE
C              LIST OF THE LOCAL ERROR ESTIMATES RESULTING FROM THE
C              INTERVAL SUBDIVISION PROCESS. AT EACH CALL TWO ERROR
C              ESTIMATES ARE INSERTED USING THE SEQUENTIAL SEARCH
C              TOP-DOWN FOR THE LARGEST ERROR ESTIMATE AND BOTTOM-UP
C              FOR THE SMALLEST ERROR ESTIMATE.
C
C 3.        CALLING SEQUENCE
C              CALL QSORT(LIMIT,LAST,MAXERR,ERMAX,ELIST,IORD,NRMAX)
C
C           PARAMETERS (MEANING AT OUTPUT)
C              LIMIT  - INTEGER
C                       MAXIMUM NUMBER OF ERROR ESTIMATES THE LIST CAN
C                       CONTAIN
C
C              LAST   - INTEGER
C                       NUMBER OF ERROR ESTIMATES CURRENTLY IN THE LIST
C
C              MAXERR - INTEGER
C                       MAXERR POINTS TO THE NRMAX-TH LARGEST ERROR
C                       ESTIMATE CURRENTLY IN THE LIST
C
C              ERMAX  - REAL
C                       NRMAX-TH LARGEST ERROR ESTIMATE
C                       ERMAX = ELIST(MAXERR)
C
C              ELIST  - REAL
C                       VECTOR OF DIMENSION LAST CONTAINING THE ERROR
C                       ESTIMATES
C
C              IORD   - INTEGER
C                       VECTOR OF DIMENSION LAST, THE FIRST K ELEMENTS
C                       OF WHICH CONTAIN POINTERS TO THE ERROR ESTIMATES
C                       SUCH THAT ELIST(IORD(1)),... , ELIST(IORD(K))
C                       FORM A DECREASING SEQUENCE, WITH
C                       K = LAST IF LAST.LE.(LIMIT/2+2), AND
C                       K = LIMIT+1-LAST OTHERWISE
C
C              NRMAX  - INTEGER
```

```
C                          MAXERR = IORD(NRMAX)
C
C 4.         NO SUBROUTINES OR FUNCTIONS NEEDED
C
C      ..............................................................
C
      REAL ELIST,ERMAX,ERRMAX,ERRMIN
      INTEGER I,IBEG,IDO,IORD,ISUCC,J,JBND,JUPBN,K,LAST,LIMIT,MAXERR,
     *  NRMAX
      DIMENSION ELIST(LAST),IORD(LAST)
C
C            CHECK WHETHER THE LIST CONTAINS MORE THAN TWO ERROR
C            ESTIMATES.
C
C***FIRST EXECUTABLE STATEMENT
      IF(LAST.GT.2) GO TO 10
      IORD(1) = 1
      IORD(2) = 2
      GO TO 90
C
C            THIS PART OF THE ROUTINE IS ONLY EXECUTED IF, DUE TO A
C            DIFFICULT INTEGRAND, SUBDIVISION INCREASED THE ERROR
C            ESTIMATE. IN THE NORMAL CASE THE INSERT PROCEDURE SHOULD
C            START AFTER THE NRMAX-TH LARGEST ERROR ESTIMATE.
C
   10 ERRMAX = ELIST(MAXERR)
      IF(NRMAX.EQ.1) GO TO 30
      IDO = NRMAX-1
      DO 20 I = 1,IDO
        ISUCC = IORD(NRMAX-1)
C***JUMP OUT OF DO-LOOP
        IF(ERRMAX.LE.ELIST(ISUCC)) GO TO 30
        IORD(NRMAX) = ISUCC
        NRMAX = NRMAX-1
   20    CONTINUE
C
C            COMPUTE THE NUMBER OF ELEMENTS IN THE LIST TO BE MAINTAINED
C            IN DESCENDING ORDER. THIS NUMBER DEPENDS ON THE NUMBER OF
C            SUBDIVISIONS STILL ALLOWED.
C
   30 JUPBN = LAST
      IF(LAST.GT.(LIMIT/2+2)) JUPBN = LIMIT+3-LAST
      ERRMIN = ELIST(LAST)
C
C            INSERT ERRMAX BY TRAVERSING THE LIST TOP-DOWN, STARTING
C            COMPARISON FROM THE ELEMENT ELIST(IORD(NRMAX+1)).
C
      JBND = JUPBN-1
```

```
      IBEG = NRMAX+1
      IF(IBEG.GT.JBND) GO TO 50
      DO 40 I=IBEG,JBND
        ISUCC = IORD(I)
C***JUMP OUT OF DO-LOOP
        IF(ERRMAX.GE.ELIST(ISUCC)) GO TO 60
        IORD(I-1) = ISUCC
   40 CONTINUE
   50 IORD(JBND) = MAXERR
      IORD(JUPBN) = LAST
      GO TO 90
C
C           INSERT ERRMIN BY TRAVERSING THE LIST BOTTOM-UP.
C
   60 IORD(I-1) = MAXERR
      K = JBND
      DO 70 J=I,JBND
        ISUCC = IORD(K)
C***JUMP OUT OF DO-LOOP
        IF(ERRMIN.LT.ELIST(ISUCC)) GO TO 80
        IORD(K+1) = ISUCC
        K = K-1
   70 CONTINUE
      IORD(I) = LAST
      GO TO 90
   80 IORD(K+1) = LAST
C
C           SET MAXERR AND ERMAX.
C
   90 MAXERR = IORD(NRMAX)
      ERMAX = ELIST(MAXERR)
      RETURN
      END
```

```
      SUBROUTINE QC250(F,A,B,OMEGA,INTEGR,NRMOM,MAXP1,KSAVE,RESULT,
     *  ABSERR,NEVAL,RESABS,RESASC,MOMCOM,CHEBMO)
C
C ......................................................................
C
C 1.     QC250
C        INTEGRATION RULES FOR FUNCTIONS WITH COS OR SIN FACTOR
C           STANDARD FORTRAN SUBROUTINE
C
C 2.     PURPOSE
C           TO COMPUTE  THE  INTEGRAL
C              I = INTEGRAL OF F(X)*W(X) OVER (A,B)
C                  WHERE W(X) = COS(OMEGA*X)
C                     OR W(X) = SIN(OMEGA*X),
C           AND TO COMPUTE J = INTEGRAL OF ABS(F) OVER (A,B).
C           FOR SMALL VALUES OF OMEGA OR SMALL INTERVALS (A,B) THE
C           15-POINT GAUSS-KRONROD RULE IS USED. IN ALL OTHER CASES A
C           GENERALIZED CLENSHAW-CURTIS METHOD IS USED, I.E. A
C           TRUNCATED CHEBYSHEV EXPANSION OF THE FUNCTION F IS COMPUTED
C           ON (A,B), SO THAT THE INTEGRAND CAN BE WRITTEN AS A SUM OF
C           TERMS OF THE FORM W(X)T(K,X), WHERE T(K,X) IS THE CHEBYSHEV
C           POLYNOMIAL OF DEGREE K. THE CHEBYSHEV MOMENTS ARE COMPUTED
C           WITH USE OF A LINEAR RECURRENCE RELATION.
C
C 3.     CALLING SEQUENCE
C           CALL QC250(F,A,B,OMEGA,INTEGR,NRMOM,MAXP1,KSAVE,RESULT,
C                      ABSERR,NEVAL,RESABS,RESASC,MOMCOM,CHEBMO)
C
C        PARAMETERS
C         ON ENTRY
C           F      - REAL
C                    FUNCTION SUBPROGRAM DEFINING THE INTEGRAND
C                    FUNCTION F(X). THE ACTUAL NAME FOR F NEEDS TO
C                    BE DECLARED E X T E R N A L IN THE CALLING PROGRAM.
C
C           A      - REAL
C                    LOWER LIMIT OF INTEGRATION
C
C           B      - REAL
C                    UPPER LIMIT OF INTEGRATION
C
C           OMEGA  - REAL
C                    PARAMETER IN THE WEIGHT FUNCTION
C
C           INTEGR - INTEGER
C                    INDICATES WHICH WEIGHT FUNCTION IS TO BE USED
C                       INTEGR = 1   W(X) = COS(OMEGA*X)
C                       INTEGR = 2   W(X) = SIN(OMEGA*X)
```

```
C
C          NRMOM  - INTEGER
C                   THE LENGTH OF INTERVAL (A,B) IS EQUAL TO THE LENGTH
C                   OF THE ORIGINAL INTEGRATION INTERVAL DIVIDED BY
C                   2**NRMOM (WE SUPPOSE THAT THE ROUTINE IS USED IN AN
C                   ADAPTIVE INTEGRATION PROCESS, OTHERWISE SET
C                   NRMOM = 0). NRMOM MUST BE ZERO AT THE FIRST CALL.
C
C          MAXP1  - INTEGER
C                   GIVES AN UPPER BOUND ON THE NUMBER OF CHEBYSHEV
C                   MOMENTS WHICH CAN BE STORED, I.E. FOR THE INTERVALS
C                   OF LENGTHS ABS(BB-AA)*2**(-L), L = 0,1,2, ...,
C                   MAXP1-2.
C
C          KSAVE  - INTEGER
C                   KEY WHICH IS ONE WHEN THE MOMENTS FOR THE
C                   CURRENT INTERVAL HAVE BEEN COMPUTED
C
C        ON RETURN
C          RESULT - REAL
C                   APPROXIMATION TO THE INTEGRAL I
C
C          ABSERR - REAL
C                   ESTIMATE OF THE MODULUS OF THE ABSOLUTE
C                   ERROR, WHICH SHOULD EQUAL OR EXCEED ABS(I-RESULT)
C
C          NEVAL  - INTEGER
C                   NUMBER OF INTEGRAND EVALUATIONS
C
C          RESABS - REAL
C                   APPROXIMATION TO THE INTEGRAL J
C
C          RESASC - REAL
C                   APPROXIMATION TO THE INTEGRAL OF ABS(F-I/(B-A))
C
C        ON ENTRY AND RETURN
C          MOMCOM - INTEGER
C                   FOR EACH INTERVAL LENGTH WE NEED TO COMPUTE
C                   THE CHEBYSHEV MOMENTS. MOMCOM COUNTS THE NUMBER
C                   OF INTERVALS FOR WHICH THESE MOMENTS HAVE ALREADY
C                   BEEN COMPUTED. IF NRMOM.LT.MOMCOM OR KSAVE = 1,
C                   THE CHEBYSHEV MOMENTS FOR THE INTERVAL (A,B)
C                   HAVE ALREADY BEEN COMPUTED AND STORED, OTHERWISE
C                   WE COMPUTE THEM AND WE INCREASE MOMCOM.
C
C          CHEBMO - REAL
C                   ARRAY OF DIMENSION AT LEAST (MAXP1,25) CONTAINING
C                   THE MODIFIED CHEBYSHEV MOMENTS FOR THE FIRST MOMCOM
```

```
C                    INTERVAL LENGTHS
C
C 4.      SUBROUTINES OR FUNCTIONS NEEDED
C              - QK15W
C              - QCHEB
C              - QMACO
C              - F(USER-PROVIDED FUNCTION)
C              - QWGTO
C              - FORTRAN ABS, COS, ALOG, AMAX1, AMIN1, SIN, INT,
C                      MINO
C
C ..........................................................................
C
      REAL A,ABSERR,AC,AN,AN2,AS,ASAP,ASS,B,CENTR,CHEBMO,
     *   CHEB12,CHEB24,CONC,CONS,COSPAR,D,ABS,COS,SIN,QWGTO,
     *   D1,D2,D3,EPMACH,ESTC,ESTS,F,FVAL,HLGTH,OFLOW,OMEGA,PARINT,PAR2,
     *   PAR22,P2,P3,P4,RESABS,RESASC,RESC12,RESC24,RESS12,RESS24,
     *   RESULT,SINPAR,UFLOW,V,X
      INTEGER I,INTEGR,ISYM,J,K,KSAVE,M,MOMCOM,NEVAL,
     *   NOEQU,NOEQ1,NRMOM
C
      DIMENSION CHEBMO(MAXP1,25),CHEB12(13),CHEB24(25),D(43),D1(43),
     *   D2(43),D3(43),FVAL(25),V(43),X(11)
C
      EXTERNAL F,QWGTO
C
C          THE DATA VALUE OF MAXP1 GIVES AN UPPER BOUND
C          ON THE NUMBER OF CHEBYSHEV MOMENTS WHICH CAN BE
C          COMPUTED, I.E. FOR THE INTERVAL (BB-AA), ...,
C          (BB-AA)/2**(MAXP1-2).
C          SHOULD THIS NUMBER BE ALTERED, THE FIRST DIMENSION OF
C          CHEBMO NEEDS TO BE ADAPTED.
C
C
C          THE VECTOR X CONTAINS THE VALUES COS(K*PI/24)
C          K = 1, ...,11, TO BE USED FOR THE CHEBYSHEV EXPANSION OF F
C
      DATA X(1),X(2),X(3),X(4),X(5),X(6),X(7),X(8),X(9),
     *   X(10),X(11)/
     *      9.914448613738104E-01,     9.659258262890683E-01,
     *      9.238795325112868E-01,     8.660254037844386E-01,
     *      7.933533402912352E-01,     7.071067811865475E-01,
     *      6.087614290087206E-01,     5.000000000000000E-01,
     *      3.826834323650898E-01,     2.588190451025208E-01,
     *      1.305261922200516E-01/
C
C
C
```

```
C          LIST OF MAJOR VARIABLES
C          -----------------------
C          CENTR  - MID POINT OF THE INTEGRATION INTERVAL
C          HLGTH  - HALF LENGTH OF THE INTEGRATION INTERVAL
C          FVAL   - VALUE OF THE FUNCTION F AT THE POINTS
C                   (B-A)*0.5*COS(K*PI/12) + (B+A)*0.5
C                   K = 0, ...,24
C          CHEB12 - COEFFICIENTS OF THE CHEBYSHEV SERIES EXPANSION
C                   OF DEGREE 12, FOR THE FUNCTION F, IN THE
C                   INTERVAL (A,B)
C          CHEB24 - COEFFICIENTS OF THE CHEBYSHEV SERIES EXPANSION
C                   OF DEGREE 24, FOR THE FUNCTION F, IN THE
C                   INTERVAL (A,B)
C          RESC12 - APPROXIMATION TO THE INTEGRAL OF
C                   COS(0.5*(B-A)*OMEGA*X)*F(0.5*(B-A)*X+0.5*(B+A))
C                   OVER (-1,+1), USING THE CHEBYSHEV SERIES
C                   EXPANSION OF DEGREE 12
C          RESC24 - APPROXIMATION TO THE SAME INTEGRAL, USING THE
C                   CHEBYSHEV SERIES EXPANSION OF DEGREE 24
C          RESS12 - THE ANALOGUE OF RESC12 FOR THE SINE
C          RESS24 - THE ANALOGUE OF RESC24 FOR THE SINE
C
C
C          MACHINE DEPENDENT CONSTANT
C          --------------------------
C
C          OFLOW IS THE LARGEST POSITIVE MAGNITUDE.
C
      CALL QMACO(EPMACH,UFLOW,OFLOW)
C
      CENTR = 5.0E-01*(B+A)
      HLGTH = 5.0E-01*(B-A)
      PARINT = OMEGA*HLGTH
C
C          COMPUTE THE INTEGRAL USING THE 15-POINT GAUSS-KRONROD
C          FORMULA IF THE VALUE OF THE PARAMETER IN THE INTEGRAND
C          IS SMALL OR IF THE LENGTH OF THE INTEGRATION INTERVAL
C          IS LESS THAN (BB-AA)/2**(MAXP1-2), WHERE (AA,BB) IS THE
C          ORIGINAL INTEGRATION INTERVAL
C
      IF(ABS(PARINT).GT.2.0E+00) GO TO 10
      CALL QK15W(F,QWGTO,OMEGA,P2,P3,P4,INTEGR,A,B,RESULT,
     *   ABSERR,RESABS,RESASC)
      NEVAL = 15
      GO TO 190
C
C          COMPUTE THE INTEGRAL USING THE GENERALIZED CLENSHAW-
C          CURTIS METHOD
```

```
C
   10 CONC = HLGTH*COS(CENTR*OMEGA)
      CONS = HLGTH*SIN(CENTR*OMEGA)
      RESASC = OFLOW
      NEVAL = 25
C
C           CHECK WHETHER THE CHEBYSHEV MOMENTS FOR THIS INTERVAL
C           HAVE ALREADY BEEN COMPUTED
C
      IF(NRMOM.LT.MOMCOM.OR.KSAVE.EQ.1) GO TO 140
C
C           COMPUTE A NEW SET OF CHEBYSHEV MOMENTS
C
      M = MOMCOM+1
      PAR2 = PARINT*PARINT
      PAR22 = PAR2+2.0E+00
      SINPAR = SIN(PARINT)
      COSPAR = COS(PARINT)
C
C           COMPUTE THE CHEBYSHEV MOMENTS WITH RESPECT TO COSINE
C
      V(1) = 2.0E+00*SINPAR/PARINT
      V(2) = (8.0E+00*COSPAR+(PAR2+PAR2-8.0E+00)*SINPAR/
     *   PARINT)/PAR2
      V(3) = (3.2E+01*(PAR2-1.2E+01)*COSPAR+(2.0E+00*
     *   ((PAR2-8.0E+01)*PAR2+1.92E+02)*SINPAR)/
     *   PARINT)/(PAR2*PAR2)
      AC = 8.0E+00*COSPAR
      AS = 2.4E+01*PARINT*SINPAR
      IF(ABS(PARINT).GT.2.4E+01) GO TO 70
C
C           COMPUTE THE CHEBYSHEV MOMENTS AS THE
C           SOLUTIONS OF A BOUNDARY VALUE PROBLEM WITH 1
C           INITIAL VALUE (V(3)) AND 1 END VALUE (COMPUTED
C           USING AN ASYMPTOTIC FORMULA)
C
      NOEQU = MIN0(40,13+INT(-ALOG(EPMACH))/3)
      NOEQ1 = NOEQU-1
      AN = 6.0E+00
      DO 20 K=1,NOEQ1
        AN2 = AN*AN
        D(K) = -2.0E+00*(AN2-4.0E+00)*(PAR22-AN2-AN2)
        D2(K) = (AN-1.0E+00)*(AN-2.0E+00)*PAR2
        D1(K) = (AN+3.0E+00)*(AN+4.0E+00)*PAR2
        V(K+3) = AS-(AN2-4.0E+00)*AC
        AN = AN+2.0E+00
   20 CONTINUE
      AN2 = AN*AN
```

```
      D(NOEQU) = -2.0E+00*(AN2-4.0E+00)*(PAR22-AN2-AN2)
      V(NOEQU+3) = AS-(AN2-4.0E+00)*AC
      V(4) = V(4)-5.6E+01*PAR2*V(3)
      ASS = PARINT*SINPAR
      ASAP = (((((2.10E+02*PAR2-1.0E+00)*COSPAR-(1.05E+02*PAR2
     *   -6.3E+01)*ASS)/AN2-(1.0E+00-1.5E+01*PAR2)*COSPAR
     *   +1.5E+01*ASS)/AN2-COSPAR+3.0E+00*ASS)/AN2-COSPAR)/AN2
      V(NOEQU+3) = V(NOEQU+3)-2.0E+00*ASAP*PAR2*(AN-1.0E+00)*
     *    (AN-2.0E+00)
C
C           SOLVE THE TRIDIAGONAL SYSTEM BY MEANS OF GAUSSIAN
C           ELIMINATION WITH PARTIAL PIVOTING
C
      DO 30 I=1,NOEQU
        D3(I) = 0.0E+00
   30 CONTINUE
      D2(NOEQU) = 0.0E+00
      DO 50 I=1,NOEQ1
        IF (ABS(D1(I)).LE.ABS(D(I))) GO TO 40
        AN = D1(I)
        D1(I) = D(I)
        D(I) = AN
        AN = D2(I)
        D2(I) = D(I+1)
        D(I+1) = AN
        D3(I) = D2(I+1)
        D2(I+1) = 0.0E+00
        AN = V(I+4)
        V(I+4) = V(I+3)
        V(I+3) = AN
  40    D(I+1) = D(I+1)-D2(I)*D1(I)/D(I)
        D2(I+1) = D2(I+1)-D3(I)*D1(I)/D(I)
        V(I+4) = V(I+4)-V(I+3)*D1(I)/D(I)
  50  CONTINUE
      V(NOEQU+3) = V(NOEQU+3)/D(NOEQU)
      V(NOEQU+2) = (V(NOEQU+2)-D2(NOEQ1)*V(NOEQU+3))/D(NOEQ1)
      DO 60 I=2,NOEQ1
        K = NOEQU-I
        V(K+3) = (V(K+3)-D3(K)*V(K+5)-D2(K)*V(K+4))/D(K)
60    CONTINUE
      GO TO 90
C
C           COMPUTE THE CHEBYSHEV MOMENTS BY MEANS OF FORWARD
C           RECURSION
C
   70 AN = 4.0E+00
      DO 80 I=4,13
        AN2 = AN*AN
```

```
        V(I) = ((AN2-4.0E+00)*(2.0E+00*(PAR22-AN2-AN2)*V(I-1)-AC)
     *    +AS-PAR2*(AN+1.0E+00)*(AN+2.0E+00)*V(I-2))/
     *    (PAR2*(AN-1.0E+00)*(AN-2.0E+00))
        AN = AN+2.0E+00
   80 CONTINUE
   90 DO 100 J=1,13
        CHEBMO(M,2*J-1) = V(J)
  100 CONTINUE
C
C           COMPUTE THE CHEBYSHEV MOMENTS WITH RESPECT TO SINE
C
      V(1) = 2.0E+00*(SINPAR-PARINT*COSPAR)/PAR2
      V(2) = (1.8E+01-4.8E+01/PAR2)*SINPAR/PAR2
     *   +(-2.0E+00+4.8E+01/PAR2)*COSPAR/PARINT
      AC = -2.4E+01*PARINT*COSPAR
      AS = -8.0E+00*SINPAR
      CHEBMO(M,2) = V(1)
      CHEBMO(M,4) = V(2)
      IF(ABS(PARINT).GT.2.4E+01) GO TO 120
      DO 110 K=3,12
        AN = K
        CHEBMO(M,2*K) = -SINPAR/(AN*(2.0E+00*AN-2.0E+00))
     *                 -2.5E-01*PARINT*(V(K+1)/AN-V(K)/(AN-1.0E+00))
110   CONTINUE
      GO TO 140
C
C           COMPUTE THE CHEBYSHEV MOMENTS BY MEANS OF
C           FORWARD RECURSION
C
  120 AN = 3.0E+00
      DO 130 I=3,12
        AN2 = AN*AN
        V(I) = ((AN2-4.0E+00)*(2.0E+00*(PAR22-AN2-AN2)*V(I-1)+AS)
     *    +AC-PAR2*(AN+1.0E+00)*(AN+2.0E+00)*V(I-2))
     *    /(PAR2*(AN-1.0E+00)*(AN-2.0E+00))
        AN = AN+2.0E+00
        CHEBMO(M,2*I) = V(I)
  130  CONTINUE
 140  IF(NRMOM.LT.MOMCOM) M = NRMOM+1
      IF(MOMCOM.LT.(MAXP1-1).AND.NRMOM.GE.MOMCOM) MOMCOM=MOMCOM+1
C           COMPUTE THE COEFFICIENTS OF THE CHEBYSHEV EXPANSIONS
C           OF DEGREES 12 AND 24 OF THE FUNCTION F
C
      FVAL(1) = 5.0E-01*F(CENTR+HLGTH)
      FVAL(13) = F(CENTR)
      FVAL(25) = 5.0E-01*F(CENTR-HLGTH)
      DO 150 I=2,12
        ISYM = 26-I
```

```
        FVAL(I) = F(HLGTH*X(I-1)+CENTR)
        FVAL(ISYM) = F(CENTR-HLGTH*X(I-1))
  150 CONTINUE
      CALL QCHEB(X,FVAL,CHEB12,CHEB24)
C
C           COMPUTE THE INTEGRAL AND ERROR ESTIMATES
C
      RESC12 = CHEB12(13)*CHEBMO(M,13)
      RESS12 = 0.0E+00
      ESTC = ABS(CHEB24(25)*CHEBMO(M,25))+ABS((CHEB12(13)-
     * CHEB24(13))*CHEBMO(M,13))
      ESTS = 0.0E+00
      K = 11
      DO 160 J=1,6
        RESC12 = RESC12+CHEB12(K)*CHEBMO(M,K)
        RESS12 = RESS12+CHEB12(K+1)*CHEBMO(M,K+1)
        ESTC = ESTC+ABS((CHEB12(K)-CHEB24(K))*CHEBMO(M,K))
        ESTS = ESTS+ABS((CHEB12(K+1)-CHEB24(K+1))*CHEBMO(M,K+1))
        K = K-2
  160 CONTINUE
      RESC24 = CHEB24(25)*CHEBMO(M,25)
      RESS24 = 0.0E+00
      RESABS = ABS(CHEB24(25))
      K = 23
      DO 170 J=1,12
        RESC24 = RESC24+CHEB24(K)*CHEBMO(M,K)
        RESS24 = RESS24+CHEB24(K+1)*CHEBMO(M,K+1)
        RESABS = RESABS + ABS(CHEB24(K))+ABS(CHEB24(K+1))
        IF(J.LE.5) ESTC = ESTC + ABS(CHEB24(K)*CHEBMO(M,K))
        IF(J.LE.5) ESTS = ESTS + ABS(CHEB24(K+1)*CHEBMO(M,K+1))
        K = K-2
  170 CONTINUE
      RESABS = RESABS*ABS(HLGTH)
      IF(INTEGR.EQ.2) GO TO 180
      RESULT = CONC*RESC24-CONS*RESS24
      ABSERR = ABS(CONC*ESTC)+ABS(CONS*ESTS)
      GO TO 190
  180 RESULT = CONC*RESS24+CONS*RESC24
      ABSERR = ABS(CONC*ESTS)+ABS(CONS*ESTC)
  190 RETURN
      END
```

```
      SUBROUTINE QC25S(F,A,B,BL,BR,ALFA,BETA,RI,RJ,RG,RH,RESULT,ABSERR,
     *  RESASC,INTEGR,NEV)
C
C.......................................................................
C
C 1.     QC25S
C        INTEGRATION RULES FOR INTEGRANDS HAVING ALGEBRAICO-LOGARITHMIC
C        END POINT SINGULARITIES
C           STANDARD FORTRAN SUBROUTINE
C
C 2.     PURPOSE
C           TO COMPUTE I = INTEGRAL OF F*W OVER (BL,BR), WITH ERROR
C           ESTIMATE, WHERE THE WEIGHT FUNCTION W HAS A SINGULAR
C           BEHAVIOUR OF ALGEBRAICO-LOGARITHMIC TYPE AT THE POINTS
C           A AND/OR B. (BL,BR) IS A PART OF (A,B).
C
C 3.     CALLING SEQUENCE
C           CALL QC25S(F,A,B,BL,BR,ALFA,BETA,RI,RJ,RG,RH,RESULT,ABSERR,
C                      RESASC,INTEGR,NEV)
C
C        PARAMETERS
C           F      - REAL
C                    FUNCTION SUBPROGRAM DEFINING THE INTEGRAND
C                    F(X). THE ACTUAL NAME FOR F NEEDS TO BE DECLARED
C                    E X T E R N A L  IN THE DRIVER PROGRAM.
C
C           A      - REAL
C                    LEFT END POINT OF THE ORIGINAL INTERVAL
C
C           B      - REAL
C                    RIGHT END POINT OF THE ORIGINAL INTERVAL, B.GT.A
C
C           BL     - REAL
C                    LOWER LIMIT OF INTEGRATION, BL.GE.A
C
C           BR     - REAL
C                    UPPER LIMIT OF INTEGRATION, BR.LE.B
C
C           ALFA   - REAL
C                    PARAMETER IN THE WEIGHT FUNCTION
C
C           BETA   - REAL
C                    PARAMETER IN THE WEIGHT FUNCTION
C
C           RI,RJ,RG,RH - REAL
C                    MODIFIED CHEBYSHEV MOMENTS FOR THE APPLICATION
C                    OF THE GENERALIZED CLENSHAW-CURTIS METHOD
C                    (COMPUTED IN SUBROUTINE QMOMO)
```

```
C
C            RESULT - REAL
C                     APPROXIMATION TO THE INTEGRAL
C                     RESULT IS COMPUTED BY USING A GENERALIZED
C                     CLENSHAW-CURTIS METHOD IF B1 = A OR BR = B.
C                     IN ALL OTHER CASES THE 15-POINT KRONROD RULE IS
C                     APPLIED, OBTAINED BY OPTIMAL ADDITION OF ABSCISSAE
C                     TO THE 7-POINT GAUSS RULE.
C
C            ABSERR - REAL
C                     ESTIMATE OF THE MODULUS OF THE ABSOLUTE ERROR,
C                     WHICH SHOULD EQUAL OR EXCEED ABS(I-RESULT)
C
C            RESASC - REAL
C                     APPROXIMATION TO THE INTEGRAL OF ABS(F*W-I/(B-A))
C
C            INTEGR - INTEGER
C                     WHICH DETERMINES THE WEIGHT FUNCTION
C                     = 1   W(X) = (X-A)**ALFA*(B-X)**BETA
C                     = 2   W(X) = (X-A)**ALFA*(B-X)**BETA*LOG(X-A)
C                     = 3   W(X) = (X-A)**ALFA*(B-X)**BETA*LOG(B-X)
C                     = 4   W(X) = (X-A)**ALFA*(B-X)**BETA*LOG(X-A)*
C                                  LOG(B-X)
C
C            NEV    - INTEGER
C                     NUMBER OF INTEGRAND EVALUATIONS
C
C 4.      SUBROUTINES OR FUNCTIONS NEEDED
C              - QCHEB
C              - QK15W
C              - F (USER-PROVIDED FUNCTION)
C              - QWGTS
C              - FORTRAN ABS, ALOG, AMAX1
C
C.......................................................................
C
      REAL A,ABSERR,ALFA,B,BETA,BL,BR,CENTR,CHEB12,CHEB24,DC,F,FACTOR,
     *  FIX,FVAL,HLGTH,QWGTS,RESABS,RESASC,RESULT,RES12,RES24,RG,
     *  RH,RI,RJ,U,X
      INTEGER I,INTEGR,ISYM,NEV
C
      DIMENSION CHEB12(13),CHEB24(25),FVAL(25),RG(25),RH(25),RI(25),
     *  RJ(25),X(11)
C
      EXTERNAL F,QWGTS
C
C           THE VECTOR X CONTAINS THE VALUES COS(K*PI/24)
C           K = 1, ..., 11, TO BE USED FOR THE COMPUTATION OF THE
```

```
C           CHEBYSHEV SERIES EXPANSION OF F.
C
      DATA X(1),X(2),X(3),X(4),X(5),X(6),X(7),X(8),X(9),X(10),X(11)/
     *     9.914448613738104E-01,     9.659258262890683E-01,
     *     9.238795325112868E-01,     8.660254037844386E-01,
     *     7.933533402912352E-01,     7.071067811865475E-01,
     *     6.087614290087206E-01,     5.000000000000000E-01,
     *     3.826834323650898E-01,     2.588190451025208E-01,
     *     1.305261922200516E-01/
C
C           LIST OF MAJOR VARIABLES
C           ----------------------
C
C           FVAL   - VALUE OF THE FUNCTION F AT THE POINTS
C                    (BR-BL)*0.5*COS(K*PI/24)+(BR+BL)*0.5
C                    K = 0, ..., 24
C           CHEB12 - COEFFICIENTS OF THE CHEBYSHEV SERIES EXPANSION
C                    OF DEGREE 12, FOR THE FUNCTION F, IN THE INTERVAL
C                    (BL,BR)
C           CHEB24 - COEFFICIENTS OF THE CHEBYSHEV SERIES EXPANSION
C                    OF DEGREE 24, FOR THE FUNCTION F, IN THE INTERVAL
C                    (BL,BR)
C           RES12  - APPROXIMATION TO THE INTEGRAL OBTAINED FROM CHEB12
C           RES24  - APPROXIMATION TO THE INTEGRAL OBTAINED FROM CHEB24
C           QWGTS  - EXTERNAL FUNCTION SUBPROGRAM DEFINING THE FOUR
C                    POSSIBLE WEIGHT FUNCTIONS
C           HLGTH  - HALF-LENGTH OF THE INTERVAL (BL,BR)
C           CENTR  - MID POINT OF THE INTERVAL (BL,BR)
C
C***FIRST EXECUTABLE STATEMENT
      NEV = 25
      IF(BL.EQ.A.AND.(ALFA.NE.0.0E+00.OR.INTEGR.EQ.2.OR.INTEGR.EQ.4))
     * GO TO 10
      IF(BR.EQ.B.AND.(BETA.NE.0.0E+00.OR.INTEGR.EQ.3.OR.INTEGR.EQ.4))
     * GO TO 140
C
C           IF A.GT.BL AND B.LT.BR, APPLY THE 15-POINT GAUSS-KRONROD
C           SCHEME.
C
C
      CALL QK15W(F,QWGTS,A,B,ALFA,BETA,INTEGR,BL,BR,RESULT,ABSERR,
     *  RESABS,RESASC)
      NEV = 15
      GO TO 270
C
C           THIS PART OF THE PROGRAM IS EXECUTED ONLY IF A = BL.
C           ----------------------------------------------------
C
```

```
C           COMPUTE THE CHEBYSHEV SERIES EXPANSION OF THE
C           FOLLOWING FUNCTION
C           F1 = (0.5*(B+B-BR-A)-0.5*(BR-A)*X)**BETA
C                   *F(0.5*(BR-A)*X+0.5*(BR+A))
C
   10 HLGTH = 5.0E-01*(BR-BL)
      CENTR = 5.0E-01*(BR+BL)
      FIX = B-CENTR
      FVAL(1) = 5.0E-01*F(HLGTH+CENTR)*(FIX-HLGTH)**BETA
      FVAL(13) = F(CENTR)*(FIX**BETA)
      FVAL(25) = 5.0E-01*F(CENTR-HLGTH)*(FIX+HLGTH)**BETA
      DO 20 I=2,12
        U = HLGTH*X(I-1)
        ISYM = 26-I
        FVAL(I) = F(U+CENTR)*(FIX-U)**BETA
        FVAL(ISYM) = F(CENTR-U)*(FIX+U)**BETA
   20 CONTINUE
      FACTOR = HLGTH**(ALFA+1.0E+00)
      RESULT = 0.0E+00
      ABSERR = 0.0E+00
      RES12 = 0.0E+00
      RES24 = 0.0E+00
      IF(INTEGR.GT.2) GO TO 70
      CALL QCHEB(X,FVAL,CHEB12,CHEB24)
C
C           INTEGR = 1  (OR 2)
C
      DO 30 I=1,13
        RES12 = RES12+CHEB12(I)*RI(I)
        RES24 = RES24+CHEB24(I)*RI(I)
   30 CONTINUE
      DO 40 I=14,25
        RES24 = RES24+CHEB24(I)*RI(I)
   40 CONTINUE
      IF(INTEGR.EQ.1) GO TO 130
C
C           INTEGR = 2
C
      DC = ALOG(BR-BL)
      RESULT = RES24*DC
      ABSERR = ABS((RES24-RES12)*DC)
      RES12 = 0.0E+00
      RES24 = 0.0E+00
      DO 50 I=1,13
        RES12 = RES12+CHEB12(I)*RG(I)
        RES24 = RES12+CHEB24(I)*RG(I)
   50 CONTINUE
      DO 60 I=14,25
```

```
        RES24 = RES24+CHEB24(I)*RG(I)
   60 CONTINUE
      GO TO 130
C
C           COMPUTE THE CHEBYSHEV SERIES EXPANSION OF THE
C           FOLLOWING FUNCTION
C           F4 = F1*LOG(0.5*(B+B-BR-A)-0.5*(BR-A)*X)
C
   70 FVAL(1) = FVAL(1)*ALOG(FIX-HLGTH)
      FVAL(13) = FVAL(13)*ALOG(FIX)
      FVAL(25) = FVAL(25)*ALOG(FIX+HLGTH)
      DO 80 I=2,12
        U = HLGTH*X(I-1)
        ISYM = 26-I
        FVAL(I) = FVAL(I)*ALOG(FIX-U)
        FVAL(ISYM) = FVAL(ISYM)*ALOG(FIX+U)
   80 CONTINUE
      CALL QCHEB(X,FVAL,CHEB12,CHEB24)
C
C           INTEGR = 3  (OR 4)
C
      DO 90 I=1,13
        RES12 = RES12+CHEB12(I)*RI(I)
        RES24 = RES24+CHEB24(I)*RI(I)
   90 CONTINUE
      DO 100 I=14,25
        RES24 = RES24+CHEB24(I)*RI(I)
  100 CONTINUE
      IF(INTEGR.EQ.3) GO TO 130
C
C           INTEGR = 4
C
      DC = ALOG(BR-BL)
      RESULT = RES24*DC
      ABSERR = ABS((RES24-RES12)*DC)
      RES12 = 0.0E+00
      RES24 = 0.0E+00
      DO 110 I=1,13
        RES12 = RES12+CHEB12(I)*RG(I)
        RES24 = RES24+CHEB24(I)*RG(I)
  110 CONTINUE
      DO 120 I=14,25
        RES24 = RES24+CHEB24(I)*RG(I)
  120 CONTINUE
  130 RESULT = (RESULT+RES24)*FACTOR
      ABSERR = (ABSERR+ABS(RES24-RES12))*FACTOR
      GO TO 270
C
```

```
C           THIS PART OF THE PROGRAM IS EXECUTED ONLY IF B = BR.
C           ------------------------------------------------
C
C           COMPUTE THE CHEBYSHEV SERIES EXPANSION OF THE
C           FOLLOWING FUNCTION
C           F2 = (0.5*(B+BL-A-A)+0.5*(B-BL)*X)**ALFA
C                *F(0.5*(B-BL)*X+0.5*(B+BL))
C
  140 HLGTH = 5.0E-01*(BR-BL)
      CENTR = 5.0E-01*(BR+BL)
      FIX = CENTR-A
      FVAL(1) = 5.0E-01*F(HLGTH+CENTR)*(FIX+HLGTH)**ALFA
      FVAL(13) = F(CENTR)*(FIX**ALFA)
      FVAL(25) = 5.0E-01*F(CENTR-HLGTH)*(FIX-HLGTH)**ALFA
      DO 150 I=2,12
        U = HLGTH*X(I-1)
        ISYM = 26-I
        FVAL(I) = F(U+CENTR)*(FIX+U)**ALFA
        FVAL(ISYM) = F(CENTR-U)*(FIX-U)**ALFA
  150 CONTINUE
      FACTOR = HLGTH**(BETA+1.0E+00)
      RESULT = 0.0E+00
      ABSERR = 0.0E+00
      RES12 = 0.0E+00
      RES24 = 0.0E+00
      IF(INTEGR.EQ.2.OR.INTEGR.EQ.4) GO TO 200
C
C           INTEGR = 1  (OR 3)
C
      CALL QCHEB(X,FVAL,CHEB12,CHEB24)
      DO 160 I=1,13
        RES12 = RES12+CHEB12(I)*RJ(I)
        RES24 = RES24+CHEB24(I)*RJ(I)
  160 CONTINUE
      DO 170 I=14,25
        RES24 = RES24+CHEB24(I)*RJ(I)
  170 CONTINUE
      IF(INTEGR.EQ.1) GO TO 260
C
C           INTEGR = 3
C
      DC = ALOG(BR-BL)
      RESULT = RES24*DC
      ABSERR = ABS((RES24-RES12)*DC)
      RES12 = 0.0E+00
      RES24 = 0.0E+00
      DO 180 I=1,13
        RES12 = RES12+CHEB12(I)*RH(I)
```

```
        RES24 = RES24+CHEB24(I)*RH(I)
  180 CONTINUE
      DO 190 I=14,25
        RES24 = RES24+CHEB24(I)*RH(I)
  190 CONTINUE
      GO TO 260
C
C           COMPUTE THE CHEBYSHEV SERIES EXPANSION OF THE
C           FOLLOWING FUNCTION
C           F3 = F2*LOG(0.5*(B-BL)*X+0.5*(B+BL-A-A))
C
  200 FVAL(1) = FVAL(1)*ALOG(HLGTH+FIX)
      FVAL(13) = FVAL(13)*ALOG(FIX)
      FVAL(25) = FVAL(25)*ALOG(FIX-HLGTH)
      DO 210 I=2,12
        U = HLGTH*X(I-1)
        ISYM = 26-I
        FVAL(I) = FVAL(I)*ALOG(U+FIX)
        FVAL(ISYM) = FVAL(ISYM)*ALOG(FIX-U)
  210 CONTINUE
      CALL QCHEB(X,FVAL,CHEB12,CHEB24)
C
C           INTEGR = 2  (OR 4)
C
      DO 220 I=1,13
        RES12 = RES12+CHEB12(I)*RJ(I)
        RES24 = RES24+CHEB24(I)*RJ(I)
  220 CONTINUE
      DO 230 I=14,25
        RES24 = RES24+CHEB24(I)*RJ(I)
  230 CONTINUE
      IF(INTEGR.EQ.2) GO TO 260
      DC = ALOG(BR-BL)
      RESULT = RES24*DC
      ABSERR = ABS((RES24-RES12)*DC)
      RES12 = 0.0E+00
      RES24 = 0.0E+00
C
C           INTEGR = 4
C
      DO 240 I=1,13
        RES12 = RES12+CHEB12(I)*RH(I)
        RES24 = RES24+CHEB24(I)*RH(I)
  240 CONTINUE
      DO 250 I=14,25
        RES24 = RES24+CHEB24(I)*RH(I)
  250 CONTINUE
  260 RESULT = (RESULT+RES24)*FACTOR
```

```
      ABSERR = (ABSERR+ABS(RES24-RES12))*FACTOR
  270 RETURN
      END
```

```
      SUBROUTINE QC25C(F,A,B,C,RESULT,ABSERR,KRUL,NEVAL)
C
C.......................................................................
C
C 1.     QC25C
C        INTEGRATION RULES FOR THE COMPUTATION OF CAUCHY
C        PRINCIPAL VALUE INTEGRALS
C           STANDARD FORTRAN SUBROUTINE
C
C 2.     PURPOSE
C           TO COMPUTE I = INTEGRAL OF F*W OVER (A,B) WITH ERROR
C           ESTIMATE, WHERE W(X) = 1/(X-C)
C
C 3.     CALLING SEQUENCE
C           CALL QC25C(F,A,B,C,RESULT,ABSERR,KRUL,NEVAL)
C
C        PARAMETERS
C           F      - REAL
C                    FUNCTION SUBPROGRAM DEFINING THE INTEGRAND
C                    F(X). THE ACTUAL NAME FOR F NEEDS TO BE DECLARED
C                    E X T E R N A L  IN THE DRIVER PROGRAM.
C
C           A      - REAL
C                    LEFT END POINT OF THE INTEGRATION INTERVAL
C
C           B      - REAL
C                    RIGHT END POINT OF THE INTEGRATION INTERVAL,
C                    B.GT.A
C
C           C      - REAL
C                    PARAMETER IN THE WEIGHT FUNCTION
C
C           RESULT - REAL
C                    APPROXIMATION TO THE INTEGRAL
C                    RESULT IS COMPUTED BY USING A GENERALIZED
C                    CLENSHAW-CURTIS METHOD IF C LIES WITHIN TEN PERCENT
C                    OF THE INTEGRATION INTERVAL. IN THE OTHER CASE THE
C                    15-POINT KRONROD RULE OBTAINED BY OPTIMAL ADDITION
C                    OF ABSCISSAE TO THE 7-POINT GAUSS RULE, IS APPLIED.
C
C           ABSERR - REAL
C                    ESTIMATE OF THE MODULUS OF THE ABSOLUTE ERROR,
C                    WHICH SHOULD EQUAL OR EXCEED ABS(I-RESULT)
C
C           KRUL   - INTEGER
C                    KEY WHICH IS DECREASED BY 1 IF THE 15-POINT
C                    GAUSS-KRONROD SCHEME HAS BEEN USED
C
```

```
C           NEVAL  - INTEGER
C                    NUMBER OF INTEGRAND EVALUATIONS
C
C 4.     SUBROUTINES OR FUNCTIONS NEEDED
C              - QCHEB
C              - QK15W
C              - F (USER-PROVIDED FUNCTION)
C              - QWGTC
C              - FORTRAN ABS, ALOG, AMAX1, AMIN1
C
C.......................................................................
C
      REAL A,ABSERR,AK22,AMOM0,AMOM1,AMOM2,B,C,CC,CENTR,CHEB12,CHEB24,F,
     *  FVAL,HLGTH,P2,P3,P4,QWGTC,RESABS,RESASC,RESULT,RES12,RES24,U,X
      INTEGER I,ISYM,K,KP,KRUL,NEVAL
C
      DIMENSION CHEB12(13),CHEB24(25),FVAL(25),X(11)
C
      EXTERNAL F,QWGTC
C
C           THE VECTOR X CONTAINS THE VALUES COS(K*PI/24),
C           K = 1, ..., 11, TO BE USED FOR THE CHEBYSHEV SERIES
C           EXPANSION OF F
C
      DATA X(1),X(2),X(3),X(4),X(5),X(6),X(7),X(8),X(9),X(10),X(11)/
     *     9.914448613738104E-01,     9.659258262890683E-01,
     *     9.238795325112868E-01,     8.660254037844386E-01,
     *     7.933533402912352E-01,     7.071067811865475E-01,
     *     6.087614290087206E-01,     5.000000000000000E-01,
     *     3.826834323650898E-01,     2.588190451025208E-01,
     *     1.305261922200516E-01/
C
C           LIST OF MAJOR VARIABLES
C           -----------------------
C           FVAL   - VALUE OF THE FUNCTION F AT THE POINTS
C                    COS(K*PI/24),  K = 0, ..., 24
C           CHEB12 - CHEBYSHEV SERIES EXPANSION COEFFICIENTS, FOR THE
C                    FUNCTION F, OF DEGREE 12
C           CHEB24 - CHEBYSHEV SERIES EXPANSION COEFFICIENTS, FOR THE
C                    FUNCTION F, OF DEGREE 24
C           RES12  - APPROXIMATION TO THE INTEGRAL CORRESPONDING TO THE
C                    USE OF CHEB12
C           RES24  - APPROXIMATION TO THE INTEGRAL CORRESPONDING TO THE
C                    USE OF CHEB24
C           QWGTC  - EXTERNAL FUNCTION SUBPROGRAM DEFINING THE WEIGHT
C                    FUNCTION
C           HLGTH  - HALF-LENGTH OF THE INTERVAL
C           CENTR  - MID POINT OF THE INTERVAL
```

```
C
C
C           CHECK THE POSITION OF C.
C
C***FIRST EXECUTABLE STATEMENT
      CC = (2.0E+00*C-B-A)/(B-A)
      IF(ABS(CC).LT.1.1E+00) GO TO 10
C
C           APPLY THE 15-POINT GAUSS-KRONROD SCHEME.
C
      KRUL = KRUL-1
      CALL QK15W(F,QWGTC,C,P2,P3,P4,KP,A,B,RESULT,ABSERR,RESABS,RESASC)
      NEVAL = 15
      IF (RESASC.EQ.ABSERR) KRUL = KRUL+1
      GO TO 50
C
C           USE THE GENERALIZED CLENSHAW-CURTIS METHOD.
C
   10 HLGTH = 5.0E-01*(B-A)
      CENTR = 5.0E-01*(B+A)
      NEVAL = 25
      FVAL(1) = 5.0E-01*F(HLGTH+CENTR)
      FVAL(13) = F(CENTR)
      FVAL(25) = 5.0E-01*F(CENTR-HLGTH)
      DO 20 I=2,12
        U = HLGTH*X(I-1)
        ISYM = 26-I
        FVAL(I) = F(U+CENTR)
        FVAL(ISYM) = F(CENTR-U)
   20 CONTINUE
C
C           COMPUTE THE CHEBYSHEV SERIES EXPANSION.
C
      CALL QCHEB(X,FVAL,CHEB12,CHEB24)
C
C           THE MODIFIED CHEBYSHEV MOMENTS ARE COMPUTED BY FORWARD
C           RECURSION, USING AMOM0 AND AMOM1 AS STARTING VALUES.
C
      AMOM0 = ALOG(ABS((1.0E+00-CC)/(1.0E+00+CC)))
      AMOM1 = 2.0E+00+CC*AMOM0
      RES12 = CHEB12(1)*AMOM0+CHEB12(2)*AMOM1
      RES24 = CHEB24(1)*AMOM0+CHEB24(2)*AMOM1
      DO 30 K=3,13
        AMOM2 = 2.0E+00*CC*AMOM1-AMOM0
        AK22 = (K-2)*(K-2)
        IF((K/2)*2.EQ.K) AMOM2 = AMOM2-4.0E+00/(AK22-1.0E+00)
        RES12 = RES12+CHEB12(K)*AMOM2
        RES24 = RES24+CHEB24(K)*AMOM2
```

```
        AMOM0 = AMOM1
        AMOM1 = AMOM2
   30 CONTINUE
      DO 40 K=14,25
        AMOM2 = 2.0E+00*CC*AMOM1-AMOM0
        AK22 = (K-2)*(K-2)
        IF((K/2)*2.EQ.K) AMOM2 = AMOM2-4.0E+00/(AK22-1.0E+00)
        RES24 = RES24+CHEB24(K)*AMOM2
        AMOM0 = AMOM1
        AMOM1 = AMOM2
   40 CONTINUE
      RESULT = RES24
      ABSERR = ABS(RES24-RES12)
   50 RETURN
      END
```

```
      SUBROUTINE QMOMO(ALFA,BETA,RI,RJ,RG,RH,INTEGR)
C
C..........................................................................
C
C 1.        QMOMO
C           MODIFIED CHEBYSHEV MOMENTS
C              STANDARD FORTRAN SUBROUTINE
C
C 2.        PURPOSE
C              THIS ROUTINE COMPUTES MODIFIED CHEBYSHEV MOMENTS.
C              THE K-TH MODIFIED CHEBYSHEV MOMENT IS DEFINED AS THE
C              INTEGRAL OVER (-1,1) OF W(X)*T(K,X), WHERE T(K,X) IS THE
C              CHEBYSHEV POLYNOMIAL OF DEGREE K.
C
C 3.        CALLING SEQUENCE
C              CALL QMOMO(ALFA,BETA,RI,RJ,RG,RH,INTEGR)
C
C           PARAMETERS
C              ALFA   - REAL
C                          PARAMETER IN THE WEIGHT FUNCTION W(X), ALFA.GT.(-1)
C
C              BETA   - REAL
C                          PARAMETER IN THE WEIGHT FUNCTION W(X), BETA.GT.(-1)
C
C              RI     - REAL
C                          VECTOR OF DIMENSION 25
C                          RI(K) IS THE INTEGRAL OVER (-1,1) OF
C                          (1+X)**ALFA*T(K-1,X), K = 1, ..., 25.
C
C              RJ     - REAL
C                          VECTOR OF DIMENSION 25
C                          RJ(K) IS THE INTEGRAL OVER (-1,1) OF
C                          (1-X)**BETA*T(K-1,X), K = 1, ..., 25.
C
C              RG     - REAL
C                          VECTOR OF DIMENSION 25
C                          RG(K) IS THE INTEGRAL OVER (-1,1) OF
C                          (1+X)**ALFA*LOG((1+X)/2)*T(K-1,X), K = 1, ...,25.
C
C              RH     - REAL
C                          VECTOR OF DIMENSION 25
C                          RH(K) IS THE INTEGRAL OVER (-1,1) OF
C                          (1-X)**BETA*LOG((1-X)/2)*T(K-1,X), K = 1, ..., 25.
C
C              INTEGR - INTEGER
C                          INPUT PARAMETER INDICATING THE MODIFIED MOMENTS
C                          TO BE COMPUTED
C                          INTEGR = 1 COMPUTE RI, RJ
```

```
C                                  = 2 COMPUTE RI, RJ, RG
C                                  = 3 COMPUTE RI, RJ, RH
C                                  = 4 COMPUTE RI, RJ, RG, RH
C
C 4.      NO SUBROUTINES OR FUNCTIONS NEEDED
C
C.......................................................................
C
      REAL ALFA,ALFP1,ALFP2,AN,ANM1,BETA,BETP1,BETP2,RALF,RBET,RG,RH,RI,
     * RJ
      INTEGER I,IM1,INTEGR
C
      DIMENSION RG(25),RH(25),RI(25),RJ(25)
C
C
C***FIRST EXECUTABLE STATEMENT
      ALFP1 = ALFA+1.0E+00
      BETP1 = BETA+1.0E+00
      ALFP2 = ALFA+2.0E+00
      BETP2 = BETA+2.0E+00
      RALF = 2.0E+00**ALFP1
      RBET = 2.0E+00**BETP1
C
C           COMPUTE RI, RJ USING A FORWARD RECURRENCE RELATION.
C
      RI(1) = RALF/ALFP1
      RJ(1) = RBET/BETP1
      RI(2) = RI(1)*ALFA/ALFP2
      RJ(2) = RJ(1)*BETA/BETP2
      AN = 2.0E+00
      ANM1 = 1.0E+00
      DO 20 I=3,25
        RI(I) = -(RALF+AN*(AN-ALFP2)*RI(I-1))/(ANM1*(AN+ALFP1))
        RJ(I) = -(RBET+AN*(AN-BETP2)*RJ(I-1))/(ANM1*(AN+BETP1))
        ANM1 = AN
        AN = AN+1.0E+00
   20 CONTINUE
      IF(INTEGR.EQ.1) GO TO 70
      IF(INTEGR.EQ.3) GO TO 40
C
C           COMPUTE RG USING A FORWARD RECURRENCE RELATION.
C
      RG(1) = -RI(1)/ALFP1
      RG(2) = -(RALF+RALF)/(ALFP2*ALFP2)-RG(1)
      AN = 2.0E+00
      ANM1 = 1.0E+00
      IM1 = 2
      DO 30 I=3,25
```

```
        RG(I) = -(AN*(AN-ALFP2)*RG(IM1)-AN*RI(IM1)+ANM1*RI(I))/
     *  (ANM1*(AN+ALFP1))
        ANM1 = AN
        AN = AN+1.0E+00
        IM1 = I
   30 CONTINUE
      IF(INTEGR.EQ.2) GO TO 70
C
C           COMPUTE RH USING A FORWARD RECURRENCE RELATION.
C
   40 RH(1) = -RJ(1)/BETP1
      RH(2) = -(RBET+RBET)/(BETP2*BETP2)-RH(1)
      AN = 2.0E+00
      ANM1 = 1.0E+00
      IM1 = 2
      DO 50 I=3,25
        RH(I) = -(AN*(AN-BETP2)*RH(IM1)-AN*RJ(IM1)+
     *  ANM1*RJ(I))/(ANM1*(AN+BETP1))
        ANM1 = AN
        AN = AN+1.0E+00
        IM1 = I
   50 CONTINUE
      DO 60 I=2,25,2
        RH(I) = -RH(I)
   60 CONTINUE
   70 DO 80 I=2,25,2
        RJ(I) = -RJ(I)
   80 CONTINUE
   90 RETURN
      END
```

```
      SUBROUTINE QCHEB(X,FVAL,CHEB12,CHEB24)
C
C.......................................................................
C
C 1.     QCHEB
C        CHEBYSHEV SERIES EXPANSION
C           STANDARD FORTRAN SUBROUTINE
C
C 2.     PURPOSE
C           THIS ROUTINE COMPUTES THE CHEBYSHEV SERIES EXPANSION
C           OF DEGREES 12 AND 24 OF A FUNCTION USING A FAST FOURIER
C           TRANSFORM METHOD
C           F(X) = SUM(K=1, ...,13) (CHEB12(K)*T(K-1,X)),
C           F(X) = SUM(K=1, ...,25) (CHEB24(K)*T(K-1,X)),
C           WHERE T(K,X) IS THE CHEBYSHEV POLYNOMIAL OF DEGREE K.
C
C 3.     CALLING SEQUENCE
C           CALL QCHEB(X,FVAL,CHEB12,CHEB24)
C
C        PARAMETERS
C          ON ENTRY
C           X      - REAL
C                    VECTOR OF DIMENSION 11 CONTAINING THE VALUES
C                    COS(K*PI/24), K = 1, ..., 11
C
C           FVAL   - REAL
C                    VECTOR OF DIMENSION 25 CONTAINING THE FUNCTION
C                    VALUES AT THE POINTS (B+A+(B-A)*COS(K*PI/24))/2,
C                    K = 0, ...,24, WHERE (A,B) IS THE APPROXIMATION
C                    INTERVAL. FVAL(1) AND FVAL(25) ARE DIVIDED BY TWO
C                    (THESE VALUES ARE DESTROYED AT OUTPUT).
C
C          ON RETURN
C           CHEB12 - REAL
C                    VECTOR OF DIMENSION 13 CONTAINING THE CHEBYSHEV
C                    COEFFICIENTS FOR DEGREE 12
C
C           CHEB24 - REAL
C                    VECTOR OF DIMENSION 25 CONTAINING THE CHEBYSHEV
C                    COEFFICIENTS FOR DEGREE 24
C
C 4.     NO SUBROUTINES OR FUNCTIONS NEEDED
C
C.......................................................................
C
      REAL ALAM,ALAM1,ALAM2,CHEB12,CHEB24,FVAL,PART1,PART2,PART3,V,X
      INTEGER I,J
C
      DIMENSION CHEB12(13),CHEB24(25),FVAL(25),V(12),X(11)
```

```
C
C***FIRST EXECUTABLE STATEMENT
      DO 10 I=1,12
        J = 26-I
        V(I) = FVAL(I)-FVAL(J)
        FVAL(I) = FVAL(I)+FVAL(J)
   10 CONTINUE
      ALAM1 = V(1)-V(9)
      ALAM2 = X(6)*(V(3)-V(7)-V(11))
      CHEB12(4) = ALAM1+ALAM2
      CHEB12(10) = ALAM1-ALAM2
      ALAM1 = V(2)-V(8)-V(10)
      ALAM2 = V(4)-V(6)-V(12)
      ALAM = X(3)*ALAM1+X(9)*ALAM2
      CHEB24(4) = CHEB12(4)+ALAM
      CHEB24(22) = CHEB12(4)-ALAM
      ALAM = X(9)*ALAM1-X(3)*ALAM2
      CHEB24(10) = CHEB12(10)+ALAM
      CHEB24(16) = CHEB12(10)-ALAM
      PART1 = X(4)*V(5)
      PART2 = X(8)*V(9)
      PART3 = X(6)*V(7)
      ALAM1 = V(1)+PART1+PART2
      ALAM2 = X(2)*V(3)+PART3+X(10)*V(11)
      CHEB12(2) = ALAM1+ALAM2
      CHEB12(12) = ALAM1-ALAM2
      ALAM = X(1)*V(2)+X(3)*V(4)+X(5)*V(6)+X(7)*V(8)
     *  +X(9)*V(10)+X(11)*V(12)
      CHEB24(2) = CHEB12(2)+ALAM
      CHEB24(24) = CHEB12(2)-ALAM
      ALAM = X(11)*V(2)-X(9)*V(4)+X(7)*V(6)-X(5)*V(8)
     *  +X(3)*V(10)-X(1)*V(12)
      CHEB24(12) = CHEB12(12)+ALAM
      CHEB24(14) = CHEB12(12)-ALAM
      ALAM1 = V(1)-PART1+PART2
      ALAM2 = X(10)*V(3)-PART3+X(2)*V(11)
      CHEB12(6) = ALAM1+ALAM2
      CHEB12(8) = ALAM1-ALAM2
      ALAM = X(5)*V(2)-X(9)*V(4)-X(1)*V(6)
     *  -X(11)*V(8)+X(3)*V(10)+X(7)*V(12)
      CHEB24(6) = CHEB12(6)+ALAM
      CHEB24(20) = CHEB12(6)-ALAM
      ALAM = X(7)*V(2)-X(3)*V(4)-X(11)*V(6)+X(1)*V(8)
     *  -X(9)*V(10)-X(5)*V(12)
      CHEB24(8) = CHEB12(8)+ALAM
      CHEB24(18) = CHEB12(8)-ALAM
      DO 20 I=1,6
        J = 14-I
        V(I) = FVAL(I)-FVAL(J)
```

```
      FVAL(I) = FVAL(I)+FVAL(J)
   20 CONTINUE
      ALAM1 = V(1)+X(8)*V(5)
      ALAM2 = X(4)*V(3)
      CHEB12(3) = ALAM1+ALAM2
      CHEB12(11) = ALAM1-ALAM2
      CHEB12(7) = V(1)-V(5)
      ALAM = X(2)*V(2)+X(6)*V(4)+X(10)*V(6)
      CHEB24(3) = CHEB12(3)+ALAM
      CHEB24(23) = CHEB12(3)-ALAM
      ALAM = X(6)*(V(2)-V(4)-V(6))
      CHEB24(7) = CHEB12(7)+ALAM
      CHEB24(19) = CHEB12(7)-ALAM
      ALAM = X(10)*V(2)-X(6)*V(4)+X(2)*V(6)
      CHEB24(11) = CHEB12(11)+ALAM
      CHEB24(15) = CHEB12(11)-ALAM
      DO 30 I=1,3
        J = 8-I
        V(I) = FVAL(I)-FVAL(J)
        FVAL(I) = FVAL(I)+FVAL(J)
   30 CONTINUE
      CHEB12(5) = V(1)+X(8)*V(3)
      CHEB12(9) = FVAL(1)-X(8)*FVAL(3)
      ALAM = X(4)*V(2)
      CHEB24(5) = CHEB12(5)+ALAM
      CHEB24(21) = CHEB12(5)-ALAM
      ALAM = X(8)*FVAL(2)-FVAL(4)
      CHEB24(9) = CHEB12(9)+ALAM
      CHEB24(17) = CHEB12(9)-ALAM
      CHEB12(1) = FVAL(1)+FVAL(3)
      ALAM = FVAL(2)+FVAL(4)
      CHEB24(1) = CHEB12(1)+ALAM
      CHEB24(25) = CHEB12(1)-ALAM
      CHEB12(13) = V(1)-V(3)
      CHEB24(13) = CHEB12(13)
      ALAM = 1.0E+00/6.0E+00
      DO 40 I=2,12
        CHEB12(I) = CHEB12(I)*ALAM
   40 CONTINUE
      ALAM = 5.0E-01*ALAM
      CHEB12(1) = CHEB12(1)*ALAM
      CHEB12(13) = CHEB12(13)*ALAM
      DO 50 I=2,24
        CHEB24(I) = CHEB24(I)*ALAM
   50 CONTINUE
      CHEB24(1) = 5.0E-01*ALAM*CHEB24(1)
      CHEB24(25) = 5.0E-01*ALAM*CHEB24(25)
      RETURN
      END
```

```
      REAL FUNCTION QWGTO(X,OMEGA,P2,P3,P4,INTEGR)
C
      REAL OMEGA,OMX,P2,P3,P4,X
      INTEGER INTEGR
C***FIRST EXECUTABLE STATEMENT
      OMX = OMEGA*X
      GO TO(10,20),INTEGR
   10 QWGTO = COS(OMX)
      GO TO 30
   20 QWGTO = SIN(OMX)
   30 RETURN
      END
```

```
      REAL FUNCTION QWGTS(X,A,B,ALFA,BETA,INTEGR)
C
      REAL A,ALFA,B,BETA,BMX,X,XMA
      INTEGER INTEGR
C***FIRST EXECUTABLE STATEMENT
      XMA = X-A
      BMX = B-X
      QWGTS = XMA**ALFA*BMX**BETA
      GO TO (40,10,20,30),INTEGR
   10 QWGTS = QWGTS*ALOG(XMA)
      GO TO 40
   20 QWGTS = QWGTS*ALOG(BMX)
      GO TO 40
   30 QWGTS = QWGTS*ALOG(XMA)*ALOG(BMX)
   40 RETURN
      END
```

```
      REAL FUNCTION QWGTC(X,C,P2,P3,P4,KP)
C
      REAL C,P2,P3,P4,X
      INTEGER KP
C***FIRST EXECUTABLE STATEMENT
      QWGTC = 1.0E+00/(X-C)
      RETURN
      END
```

```
      SUBROUTINE QMACO(EPMACH,UFLOW,OFLOW)
C
C.......................................................................
C
C   QMACO IS A MACHINE DEPENDENT SUBROUTINE, PROVIDING THE MACHINE
C   CONSTANTS TO THE QUADPACK ROUTINES
C
C   PARAMETERS
C       ON RETURN
C        EPMACH   - THE LARGEST RELATIVE SPACING
C        UFLOW    - THE SMALLEST POSITIVE MAGNITUDE
C        OFLOW    - THE LARGEST POSITIVE MAGNITUDE
C
C.......................................................................
C
      REAL EPMACH,OFLOW,UFLOW
C
C   MACHINE CONSTANTS FOR PDP-11
C
      EPMACH = 1.0E-7
      UFLOW = 1.0E-37
      OFLOW = 1.0E+38
      RETURN
      END
```

References

Abramowitz, M. (1954)
On the Practical Evaluation of Integrals.
SIAM J. Appl. Math. 2, 20-35.

Abramowitz, M. and Stegun, I.A. (ed) (1964)
Handbook of Mathematical Functions.
Nat. Bur. Stand. Appl. Math. Ser. No. 55,
U.S. Govt. Printing Office, Washington, D.C.

Branders, M. and Piessens, R. (1975)
An Extension of Clenshaw-Curtis Quadrature.
J. Comp. Appl. Math. 1, 55-65.

Clenshaw, C.W. and Curtis, A.R. (1960)
A Method for Numerical Integration on an Automatic Computer.
Num. Math. 2, 197-205.

Davis, P.J. and Rabinowitz, P. (1975)
Methods of Numerical Integration.
Academic Press, New York.

de Boor, C. (1971)
CADRE : An Algorithm for Numerical Quadrature.
in Mathematical Software (Rice, J.R., ed.), 417-449.
Academic Press, New York.

de Doncker, E. (1978)
An Adaptive Extrapolation Algorithm for Automatic Integration.
SIGNUM Newsl. 13, 12-18.

de Doncker, E. and Piessens, R. (1976)
A Bibliography on Automatic Integration.
J. Comp. Appl. Math. 2, 273-280.

Dieudonné, J. (1960)
Foundations of Modern Analysis.
Academic Press, New York and London.

Dixon, V. (1974)
Numerical Quadrature. A Survey of the Available Algorithms.
in Software for Numerical Mathematics (Evans, D.J., ed.), 105-137.
Academic Press, New York.

Engels, H. (1980)
Numerical Quadrature and Cubature.
Academic Press, New York and London.

Fox, L. and Parker, I.B. (1968)
Chebyshev Polynomials in Numerical Analysis.
Oxford Univ. Press, London and New York.

Fritsch, F.N., Kahaner, D.K. and Lyness, J.N. (1979)
Double Integration Using One-Dimensional Adaptive Quadrature Routines : A Software Interface Problem.
ACM Trans. Math. Softw. 7, 46-75.

Gautschi, W. (1967)
Computational Aspects of Three-term Recurrence Relations.
SIAM Review 9, 24-82.

Gautschi, W. (1968)
Construction of Gauss-Christoffel Quadrature Formulas.
Math. Comp. 22, 251-270.

Gautschi, W. (1970)
On the Construction of Gaussian Quadrature Rules from Modified Moments.
Math. Comp. 24, 245-260.

Gentleman, W.M. (1972)
Implementing Clenshaw-Curtis Quadrature.
Comm. Assoc. Comp. Mach. 15, 337-342, 343-346.

Genz, A. (1975)
The Approximate Calculation of Multidimensional Integrals Using Extrapolation Methods.
Ph.D. Thesis, Univ. of Kent.

Golub, G.H. and Welsch, J.H. (1969)
Calculation of Gauss Quadrature Rules.
Math. Comp. 23, 221-230.

Jackson, D. (1930)
The Theory of Approximation.
Amer. Math. Soc., New York.

Iri, M., Moriguti, S. and Takasawa, Y. (1970)
On a Certain Quadrature Formula (in Japanese).
Kokyuroku of the Research Institute for Mathematical Sciences 91, Kyoto University, 82-118.

Kahaner, D.K. (1971)
Comparison of Numerical Quadrature formulas.
in Mathematical Software (Rice, J.R., ed.), 229-259.
Academic Press, New York.

Kahaner, D.K. (1972)
Numerical Quadrature by the ε-Algorithm.
Math. Comp. 26, 689-693.

Kronrod, A.S. (1965)
Nodes and Weights of Quadrature Formulas.
Consultants Bureau, New York.

Krylov, V.I. (1962)
Approximate Calculation of Integrals.
transl. Stroud, A.H., Macmillan, New York.

Lyness, J.N. (1969)
The Effect of Inadequate Convergence Criteria in Automatic Routines.
Computer J. 12, 279-281.

Lyness, J.N. and Kaganove, J.J. (1976)
Comments on the Nature of Automatic Quadrature Routines.
ACM Trans. Math. Softw. 2, 65-81.

Lyness, J.N. and Ninham, B.W. (1967)
Numerical Quadrature and Asymptotic Expansions.
Math. Comp. 21, 162-178.

Lyness, J.N. and Puri, K.K. (1973)
The Euler-Maclaurin Expansion for the Simplex.
Math. Comp. 27, 273-293.

Malcolm, M.A. and Simpson, R.B. (1975)
Local vs. Global Strategies for Adaptive Quadrature.
ACM Trans. Math. Softw. 1, 129-146.

Monegato, G. (1976)
A Note on Extended Gaussian Quadrature Rules.
Math. Comp. 30, 812-817.

Monegato, G. (1978)
Positivity of the Weights of Extended Gauss-Legendre Quadrature Rules.
Math. Comp. 32, 243-245.

Monegato, G. (1982)
Stieltjes Polynomials and Related Quadrature Rules.
SIAM Review 24, 137-158.

Monegato, G. and Lyness, J.N. (1979)
On the Numerical Evaluation of a Particular Singular Two-Dimensional Integral.
Math. Comp. 33, 993-1002.

Olver, F.W.J. (1967)
Numerical Solution of Second-Order Linear Difference Equations.
J. Nat. Bur. Stand. : B. Math. and Math. Phys. 71B, 111-129.

Patterson, T.N.L. (1968)
The Optimum Addition of Points to Quadrature Formulae.
Math. Comp. 22, 847-856.

Patterson, T.N.L. (1980)
Private communication.

Piessens, R. (1973)
An Algorithm for Automatic Integration.
Angewandte Informatik, 399-401.

Piessens, R. and Branders, M. (1973)
The Evaluation and Application of some Modified Moments.
BIT 13, 443-450.

Piessens, R. and Branders, M. (1974)
A Note on the Optimal Addition of Abscissas to Quadrature Formulas of Gauss and Lobatto Type.
Math. Comp. 28, 135-139.

Piessens, R. and Branders, M. (1975)
Computation of Oscillating Integrals.
J. Comp. Appl. Math. 1, 153-164.

Piessens, R. and Criegers, R. (1974)
Estimation Asymptotique des Coefficients du Développement en Série de Polynômes de Chebyshev d'une Fonction ayant Certaines Singularités.
C.R. Acad. Sc. Paris 278 (Série A), 405-407.

Piessens, R., Mertens, I. and Branders, M. (1974)
Automatic Integration of Functions Having Algebraic End Point Singularities.
Angewandte Informatik, 65-68.

Piessens, R., Van Roy-Branders, M. and Mertens, I. (1976)
The Automatic Evaluation of Cauchy Principal Value Integrals.
Angewandte Informatik, 31-35.

Rivlin, T.J. (1969)
An Introduction to the Approximation of Functions.
Blaisdell Publ. Co., Waltham.

Robinson, I. (1979)
A Comparison of Numerical Integration Programs.
J. Comp. Appl. Math. 5, 207-223.

Rutishauser, H. (1967)
Handbook for Automatic Computation, vol. I. Part a : Description of ALGOL 60.
Springer-Verlag, Berlin.

Shanks, D. (1955)
Non-Linear Transformations of Divergent and Slowly Convergent Sequences.
J. Math. and Phys. 34, 1-42.

Smith, D.A. and Ford, W.F. (1979)
Acceleration of Linear and Logarithmic Convergence.
SIAM J. Numer. Anal. 16, 223-240.

Song, C.S. (1969)
Numerical Integration of a Double Integral with a Cauchy-type singularity.
AIAA J. 7, 1389-1390.

Stroud, A.H. and Secrest, D.H. (1966)
Gaussian Quadrature Formulas.
Prentice-Hall, Englewood Cliffs, New Jersey.

Szegö, G. (1934)
Uber Gewisse Orthogonale Polynome, die zu einer Oszillierenden Belegungsfunktion gehoren.
Math. Ann. 110, 501-513.

Szegö, G. (1959)
Orthogonal Polynomials.
Amer. Math. Soc., New York.

Takahasi, H. and Mori, M. (1974)
Double Exponential Formulas for Numerical Integration.
Publ. of the Research Institute for Math. Sciences 9, Kyoto Univ. .

Talbot, A. (1979)
The Accurate Numerical Inversion of Laplace Transforms.
J. Inst. Maths. Applics 23, 97-120.

Tolstov, G.P. (1962)
Fourier Series.
Prentice-Hall, Englewood Cliffs.

Tricomi, F.G. (1955)
Vorlesungen über Orthogonalreihen.
Springer Verlag, Berlin.

Wynn, P. (1956)
On a Device for Computing the $e_m(S_n)$ Transformation.
MTAC 10, 91-96.

Wynn, P. (1966)
On the Convergence and Stability of the Epsilon Algorithm.
SIAM J. Numer. Anal. 3, 91-122.

Springer Series in Computational Mathematics

Forthcoming titles:

E. Allgower, K. Georg
Continuation Methods for Numerically Solving Nonlinear Systems of Equations
ISBN 3-540-12760-7

W. Hackbusch
Multi-Grid Methods
ISBN 3-540-12761-5

A. Iserles
Numerical Analysis of Differential Equations
ISBN 3-540-12762-3

J. Rice, R. F. Boisvert
Soving Elliptic Problems using ELLPACK
ISBN 3-540-90910-9

N. Z. Shor
Methods for Minimization of Non-Differentiable Functions and Their Applications
ISBN 3-540-12763-1

Springer-Verlag
Berlin
Heidelberg
New York
Tokyo

Handbook of Automatic Computation

Editors: **F. L. Bauer, A. S. Householder, F. W. Olver, H. Rutishauser, K. Samelson, E. Stiefel**

Volume 2
J. H. Wilkinson, C. Reinsch
Linear Algebra
Chiefeditor: **F. L. Bauer**
1971. 4 figures. IX, 439 pages. (Grundlehren der mathematischen Wissenschaften, Band 186)
ISBN 3-540-05414-6

Matrix Eigensystem Routines – EISPACK Guide

By **B. T. Smith, J. M. Bovle, J. J. Dongarra, B. S. Garbow, Y. Ikebe, V. C. Klema, C. B. Moler**
2nd edition. 1976. XI, 551 pages (Lecture Notes in Computer Science, Volume 6)
ISBN 3-540-07546-1

Matrix Eigensystem Routines – EISPACK Guide Extension

by **B. S. Garbow, J. M. Boyle, J. J. Dongarra, C. B. Moler**
1977. VIII, 343 pages (Lecture Notes in Computer Science, Volume 51)
ISBN 3-540-08254-9

J. Stoer, R. Bulirsch
Introduction to Numerical Analysis

Translated from the German by R. Bartels, W. Gautschi, C. Witzgall
1980. 30 figures, 8 tables. IX, 609 pages
ISBN 3-540-90420-4

J. Stoer, Ch. Witzgall
Convexity and Optimization in Finite Dimensions I

1970. 19 figures. IX, 293 pages (Grundlehren der mathematischen Wissenschaften, Band 163)
ISBN 3-540-04835-9

Springer-Verlag
Berlin
Heidelberg
New York
Tokyo